Computational Aspects of the Study of Biological Macromolecules by Nuclear Magnetic Resonance Spectroscopy

NATO ASI Series

Advanced Science Institutes Series

A series presenting the results of activities sponsored by the NATO Science Committee, which aims at the dissemination of advanced scientific and technological knowledge, with a view to strengthening links between scientific communities.

The series is published by an international board of publishers in conjunction with the NATO Scientific Affairs Division

A	**Life Sciences**	Plenum Publishing Corporation
B	**Physics**	New York and London
C	**Mathematical and Physical Sciences**	Kluwer Academic Publishers
D	**Behavioral and Social Sciences**	Dordrecht, Boston, and London
E	**Applied Sciences**	
F	**Computer and Systems Sciences**	Springer-Verlag
G	**Ecological Sciences**	Berlin, Heidelberg, New York, London,
H	**Cell Biology**	Paris, Tokyo, Hong Kong, and Barcelona
I	**Global Environmental Change**	

Recent Volumes in this Series

Series A: Life Sciences

Computational Aspects of the Study of Biological Macromolecules by Nuclear Magnetic Resonance Spectroscopy

Edited by

Jeffrey C. Hoch

Rowland Institute for Science
Cambridge, Massachusetts

Flemming M. Poulsen

Carlsberg Laboratory
Copenhagen, Denmark

and

Christina Redfield

University of Oxford
Oxford, England

Plenum Press
New York and London
Published in cooperation with NATO Scientific Affairs Division

Proceedings of a NATO Advanced Research Workshop on
Computational Aspects of the Study of Biological Macromolecules
by Nuclear Magnetic Resonance Spectroscopy,
held June 13–18, 1990,
in Il Ciocco, Italy

QP
519
.9
.N83
C66
1991

Library of Congress Cataloging in Publication Data

Computational aspects of the study of biological macromolecules by nuclear magnetic resonance spectroscopy / edited by Jeffrey C. Hoch, Flemming M. Poulsen, and Christina Redfield.
    p.   cm. — (NATO ASI series. Series A, Life sciences; v. 225)
    "Proceedings of a NATO Advanced Research Workshop on Computational Aspects of the Study
of Biological Macromolecules by Nuclear Magnetic Resonance Spectroscopy, held June 13–18, 1990,
in Il Ciocco, Italy" — T.p. verso.
    "Published in cooperation with NATO Scientific Affairs Division."
    Includes bibliographical references and index.
    ISBN 0-306-44114-4
    1. Nuclear magnetic resonance spectroscopy — Data processing — Congresses. 2. Biomolecules —
Analysis — Data processing — Congresses. I. Hoch, Jeffrey C. II. North Atlantic Treaty Organization.
Scientific Affairs Division. III. NATO Advanced Research Workshop on Computational Aspects of
the Study of Biological Macromolecules by Nuclear Magnetic Resonance Spectroscopy (1990: Il Ciocco, Italy) IV. Series.
QP519.9.N83C66   1992   91-42705
574.19′245 — dc20   CIP

ISBN 0-306-44114-4

© 1991 Plenum Press, New York
A Division of Plenum Publishing Corporation
233 Spring Street, New York, N.Y. 10013

PREFACE

This volume is the scientific chronicle of the NATO Advanced Research Workshop on Computational Aspects of the Study of Biological Macromolecules by Nuclear Magnetic Resonance Spectroscopy, which was held June 3-8, 1990 at Il Ciocco, near Barga, Italy.

The use of computers in the study of biological macromolecules by NMR spectroscopy is ubiquitous. The applications are diverse, including data collection, reduction, and analysis. Furthermore, their use is rapidly evolving, driven by the development of new experimental methods in NMR and molecular biology and by phenomenal increases in computational performance available at reasonable cost. Computers no longer merely facilitate, but are now absolutely essential in the study of biological macromolecules by NMR, due to the size and complexity of the data sets that are obtained from modern experiments. The Workshop, and this proceedings volume, provide a snapshot of the uses of computers in the NMR of biomolecules. While by no means exhaustive, the picture that emerges illustrates both the importance and the diversity of their application.

The Workshop would not have been possible without the generous financial support of the Scientific Affairs Division of NATO and Bruker Instruments. In addition, the efforts of many individuals were indispensable in bringing the Workshop to fruition. The Organizing Committee, which included Dennis Hare, George Levy, Flemming Poulsen, and Christina Redfield, turned a vague concept into a workable plan. Barbara McCaffrey and Lesley Pew of the Rowland Institute set the plan in motion, and Bruno Giannasi and his staff at Il Ciocco helped to execute it.

This proceedings volume benefited from the word processing skills of Lesley Pew. I am grateful to Steven Block and Alan Stern of the Rowland Institute for assistance with other compuational aspects. MaryAnn Nilsson and Jay Scarpetti provided emergency photographic services on more than one occasion. Gregory Safford of Plenum Publishing Corporation provided invaluable guidance along the way. I am also deeply indebted to my co-editors, Flemming Poulsen and Christina Redfield. Christina had the misfortune to be on sabbatical leave at Harvard while much of the work on this volume was being completed, and consequently was particularly vulnerable to being drafted into the pursuit of some tedious task.

Finally, but by no means least of all, much credit must go to Edwin H. Land, who contributed many of the resources of the Rowland Institute for Science to the completion of this volume. Dr. Land died on March 1, 1991, before this volume was completed. Dr. Land believed that the ultimate goal of science is the pursuit of the manifestly important and nearly impossible. The efforts described on these pages will continue to transport the study of biological macromolecules by NMR from the realm of the nearly impossible to the realm of the eminently feasible.

Jeffrey C. Hoch

CONTENTS

WITHOUT COMPUTERS — NO MODERN NMR

R. R. Ernst

Laboratorium für Physikalische Chemie
Eidgenössische Technische Hochschule
CH-8092 Zürich, Switzerland

ABSTRACT

Some very early, mostly unpublished experiments in computer-aided NMR are described.

INTRODUCTORY REMARKS

Computer control and digital data processing have become indispensible in nuclear magnetic resonance to such an extent that the current sophisticated applications in chemistry and molecular biology would be inconceivable without on-line computers. Computers indeed have revolutionized NMR in a more dramatic manner than most other fields. They are indispensible not only as computational tools, they allow radically new approaches, they lead to a new style of life for NMR spectroscopists, and they have changed the philosophy of experimenting. NMR represents one of the most intriguing examples for the successful application of modern concepts of data acquisition and data processing by computer.

It is thus not astonishing that modern NMR facilities often contain more computers than spectrometers. A typical example is the computer network of the author's research group, shown in Fig. 1. A SUN SPARC 490 file server is the heart of the network, linked to both the external (institute) Ethernet and to the private NMR Ethernet. The four NMR spectrometers are connected through Bruker Aspect and X32 computers. More extensive data processing and simulation calculations, using spin dynamics and molecular dynamics programs, are performed on SUN SPARC-1 stations and on a Silicon Graphics IRIS 4D/220 superworkstation. The sizeable data storage facilities, visible in Fig. 1, are needed for the processing of large two-dimensional (2D) and for three-dimensional (3D) data sets. For even more

Computational Aspects of the Study of Biological Macromolecules by Nuclear Magnetic Resonance Spectroscopy, Edited by J.C. Hoch *et al.*, Plenum Press, New York, 1991

extensive jobs, access to external supercomputers, such as a CONVEX 1, a CRAY XMP, and a CRAY II, is easily possible. This network represents a powerful, but indispensible computing environment for a research group of 12-15 scientists.

Obviously, this extent of sophistication was not always available in NMR and has grown step by step in an evolutionary process that lasted so far about 25 years. During this period, the progress of NMR was intimately connected to breakthroughs in computer technology and computer availability.

In the following we afford a few glimpses into the early history of computerized NMR. In the early days it was a rather troublesome love affair among teenagers, with both NMR and the computer being inexperienced and less than twenty years of age. But in the meantime, both have attained maturity, and their companionship has become quite productive.

PREHISTORY: THE COMPUTER OF AVERAGED TRANSIENTS

The first computerlike device, the "computer of average transients" or abbreviated the CAT, entered the NMR laboratory 1962. Its only function was the improvement of sensitivity by coadding signals resulting from repetitive scans through the same spectrum. These time averaging experiments were initially promoted by Jardetzky,[1] Klein,[2] Laszlo,[3] and Allen.[4] They demonstrated that indeed sensitivity could be improved in this manner, although at the expense of performance time T^{tot}.

The proportionality

$$\text{Signal-to-Noise Ratio} \propto \sqrt{T_{tot}} \qquad (1)$$

forms one of the basic laws of experimental science. At that time, there was much discussion about the optimum usage of T_{tot}: Is it of advantage to perform one single slow scan in T_{tot}, or is it better to coadd several rapid scans measured in the same total time? The answer depends on the noise power spectral density $W(\omega)$ produced by the instrument and on the spectrum distortions that can be tolerated. In the presence of excessive low frequency noise with $W(\omega)$ increasing for $\omega \to 0$, rapid scanning helps to render the experiment insensitive to such noise, in full analogy to modulation methods used in other types of spectroscopy (EPR, optical spectroscopy) to suppress low frequency noise.[5]

Figure 1. An example of a modern computer network for NMR spectroscopy. It represents the actual status (August 1990) of the network in the author's laboratory. It serves four NMR spectrometers, MSL 400, CXP 300, a home-made 300 MHz, and a home-made 220 MHz spectrometer.

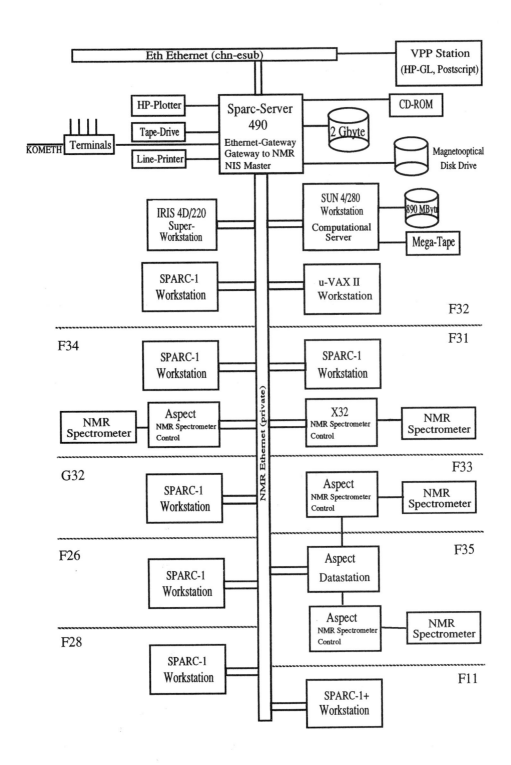

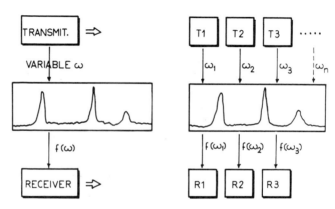

Figure 2. Sweep versus multi-channel spectrometer. The single channel of variable frequency ω is replaced by a set of fixed frequency channels that simultaneously acquire data and reduce the performance time needed for covering the entire spectrum.

Rapid scanning can also enhance the signal power which leads to increased sensitivity, here at the expense of resolution. Roughly speaking, it is of advantage to perform up to one scan per T_1 relaxation time such that the spin system has just barely enough time to recover between scans. Under these conditions, line broadening can be quite severe, leading to a line width b roughly given by

$$b \simeq \Delta\Omega \cdot T_2/T_{tot} \tag{2}$$

where $\Delta\Omega$ is the total sweep width covered in T_{tot} and T_2 is the transverse relaxation time.

It took another 10 years before a solution for the signal distortion to this problem was found, again taking advantage of computers. Dadok and Sprecher[6] realized that the rapid scan signal distortions can be eliminated by correlation with the theoretical response of a single line, leading to correlation spectroscopy.[6,7] Almost the ultimately achievable sensitivity can be reached in this manner. Nevertheless, the experiment did not succeed on a broad scale because it is not particularly suited for extensions, such as multi-dimensional spectroscopy and coherence transfer experiments.

ON THE VERGE OF MODERN HISTORY:
THE CONCEPTION OF FOURIER TRANSFORM SPECTROSCOPY

The pulse Fourier transform experiment[8] that forms the basis of all modern NMR techniques was conceived in 1964 and presented for the first time at the 6th Experimental NMR Conference in February 1965.[9] The motivation for its development had its origin in attempts of W. A. Anderson to

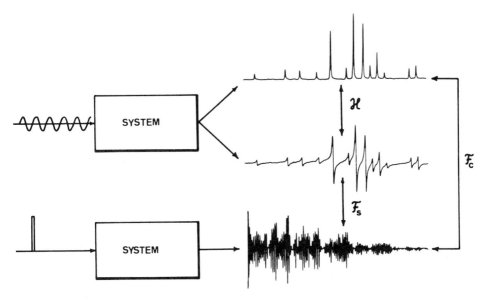

Figure 3. Fourier transform relations for frequency and impulse response of a linear system. Real and imaginary parts of the frequency response are connected by a Hilbert transformation $\mathcal{H}$ while their relation to the impulse response is determined by a cosine Fourier transformation $\mathcal{F}_c$ and by a sine Fourier transformation $\mathcal{F}_s$, respectively.

construct a true multi-channel spectrometer with 64 or more transmitters and receivers that measure simultaneously different points of the spectrum, leading to a corresponding time saving and sensitivity enhancement (Fig. 2). Obviously, such a spectrometer would become exceedingly complicated and expensive, and a simpler physical realization of the multi-channel concept was demanded.

In the course of those studies, it was recognized that a brief radio frequency pulse contains all the necessary transmitter frequencies to simultaneously excite all spins. Instead of using a large number of receivers, the analysis of the simultaneous response of all spins could be performed by a digital computer. The mathematical operation of analysis amounts merely to a Fourier transformation of the multi-frequency response signal (Fig. 3). It was rapidly realized that this touches some of the basic principles of linear response theory, namely that the impulse response $s(t)$, in NMR called free induction decay, and the frequency response $S(\omega)$, called the spectrum, form a Fourier transform pair,

$$S(\omega) = \mathcal{F}\{s(t)\}. \tag{3}$$

Some concern was caused by the nonlinearity of the spin system, manifested in saturation and line broadening effects. But it was soon understood that the nonlinear input-output relations of spin systems under the influence of radio frequency irradiation is important only during the (preparation)

5

Figure 4. "Second generation" set-up used for Fourier transform NMR Spectroscopy in October 1965. The free induction decays produced by a Varian DP 60 spectrometer are coadded in a C1024 time averaging computer. For Fourier transformation, the data are punched into cards by a card punch, operated by the human interface (author), and sent to the computer center.

pulse while during the free induction decay, the system behaves perfectly linearly such that the Fourier transform relations apply exactly. In contrary, traditional slow passage experiments are much more susceptible to nonlinearity effects[10,11] which are fully eliminated by passing to pulse experiments. Thus, in (almost) all respects, pulse experiments turned out to be superior to the old-fashioned sweep experiments.

The expected sensitivity gain can again be estimated based on the multichannel concept. The number of effective "channels" in a pulse Fourier experiment is equal to the number of "spectral elements", i.e. to the number of independent sample values in the spectrum. This number is roughly equal to the total sweep width $\Delta\Omega$ divided by the line width b, and one obtains for the sensitivity gain in a fixed total performance time, in comparison with a slow passage experiment,

$$\text{Sensitivity gain} \simeq \sqrt{\Delta\Omega/b} \qquad (4)$$

and the time saving for obtaining the same sensitivity is

$$\text{Time saving} \simeq \Delta\Omega/b. \qquad (5)$$

Thus, the sensitivity can be improved by one to two orders of magnitude while the time saving amounts to two to four orders of magnitude. Obvi-

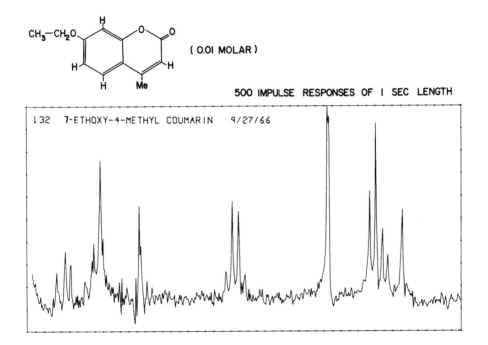

Figure 5. 60 MHz Fourier transform spectrum of 0.01 m 7-ethoxy-4-methyl coumarine obtained by Fourier-transforming 500 free induction decays recorded in 500 s on the equipment of Fig. 4. For comparison, a slow passage spectrum is given that has been recorded on the same spectrometer under optimum conditions also in 500 s.

ously, this can become a quite decisive factor, particularly for rare and low sensitivity nuclei and for dilute solutions of large molecules.

Thus, a real break-through had been achieved, at least on paper. But how to put it into practice? In 1964, Varian Associates in Palo Alto, California, where these concepts were developed, did not possess even one single computer, and all commercial and scientific computations were done externally at Service Bureau Corporation that had an IBM 7090 computer available for customer applications. Nor did Varian have a pulse spectrometer, although pulse experiments were well known at that time. Already in 1946 Bloch[12] had mentioned the possibility to perform pulse experiments. The first ones were then done by Torrey[13] and by Hahn[14] in 1950.

Thus, some pulse equipment had to be built to supplement a DA60 double purpose NMR spectrometer. This was not very difficult for an excellent electronic engineer, such as Bill Siebert. The computing aspects caused somewhat more headache. The only digital device available was a C1024

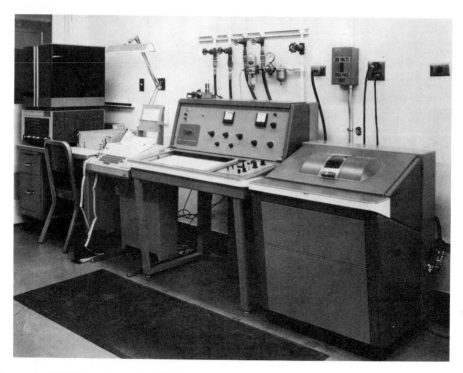

Figure 6. Wedding photo of a PDP 8 computer and a Varian A 60, NMR spectrometer, Palo Alto, California, fall 1966. Between computer and teletype, the interface box is visible.

time averaging computer, the successor of the CAT. It could sample and coadd the free induction decays when triggered synchronously with the radio frequency pulse. Fortunately, we were allowed to rent (!) a Tally paper tape punch that could be connected to the C1024 for the data transfer. Unfortunately, however, Service Bureau Corporation did not have a paper tape reader available. Therefore the tapes had to be sent to IBM in San Jose where they were converted into punched cards (only too often by reading the paper tape in the wrong direction). And with the cards we could finally drive to the computer center where the data were put on magnetic tape for the subsequent Fourier transformation on the computer by a standard implementation of the algorithm (the fast Fourier transform Cooley-Tukey algorithm was not yet available). After plotting the data on a Calcomp plotter a week after the experiment, we often discovered some lethal problems of the experiment such as field instability (we used a home-made fluorine lock while observing protons) or inaccurate triggering, and the entire tedious process had to be started again. Nevertheless, we had the courage (and foresight) to claim significant time saving and sensitivity enhancement.[8] It took almost a year to obtain a card punch to shorten somewhat the connection to the computer. But still a human interface was necessary to close the loop (Fig. 4).

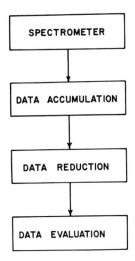

Figure 7. The four stages of NMR spectroscopy, all of which can be computer-controlled or computer-performed.

The possibilities for a commercial success did not look very favorable at that time, although the first spectra, such as the one shown in Fig. 5 had already convincing quality. But the necessary laboratory computers were not yet easily available. Indeed, it took another 5 years before a (competitive) company (Bruker) produced in 1971 a first commercial Fourier transform NMR spectrometer.

THE COMPUTER'S FIRST ENTRY IN THE NMR LABORATORY

In the fall of 1966, finally, we were allowed to purchase a real computer at Varian and to use it for on-line NMR experiments. It was a PDP-8 computer of Digital Equipment Corporation with a memory of just 4096 12 bit words. For programming, a simple assembler was available. As we still did not believe in the success of pulse Fourier transform NMR spectroscopy, we decided to automate processes in conventional continuous wave spectroscopy, and we mated the low performance computer to a low cost spectrometer, an A60, the work-horse at that time; the wedding photo is presented in Fig. 6.

In a short time, in preparation for the 8th Experimental NMR Conference (ENC) in Pittsburgh in March 1967,[15] we programmed numerous computer control and data processing routines that anticipated, in a rudimentary form, much of the NMR computer applications of the following twenty years. However, at that time, this was ridiculed by 'true' scientists as child's play in the same manner as computer scientists were not properly accepted by the 'true' mathematicians. For this reason, we never published this piece of

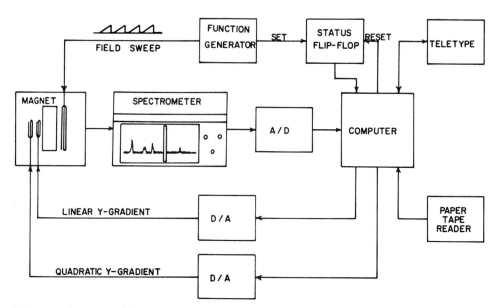

Figure 8. Set-up used for automatic computer-shimming of the A60 magnet. The repetitive field sweep is provided by a sawtooth function generator that triggers the computer for data acquisition through an analog-to-digital converter. The field homogeneity corrections are applied via two digital-to-analog converters to the linear and quadratic y shim coils.

work in the open literature. Perhaps, it is still worthwhile to document some of these early examples of computer applications, if only to put into proper perspective the present mature status of computers in NMR spectroscopy.

The processes that were computerized stem from the four basic stages of NMR spectroscopy, represented in Fig. 7: spectrometer control, data accumulation, data reduction, and data evaluation.

Automatic Homogeneity Adjustment

In the context of computer control of a spectrometer, automatic shimming immediately springs to one's mind as a most desirable goal. Normally, an NMR spectroscopist is spending an appreciable part of his lifetime in front of the spectrometer desperately turning shim knobs, more or less at random, in the hope of improved resolution. A computer could do this job in a much more systematic way.

A simple interface between an A60 and a PDP 11 was constructed that consisted of an analog-to-digital converter for data acquisition and two digital-to-analog converters to drive the two shim coils for the linear and the quadratic y-field gradients, as illustrated in Fig. 8.

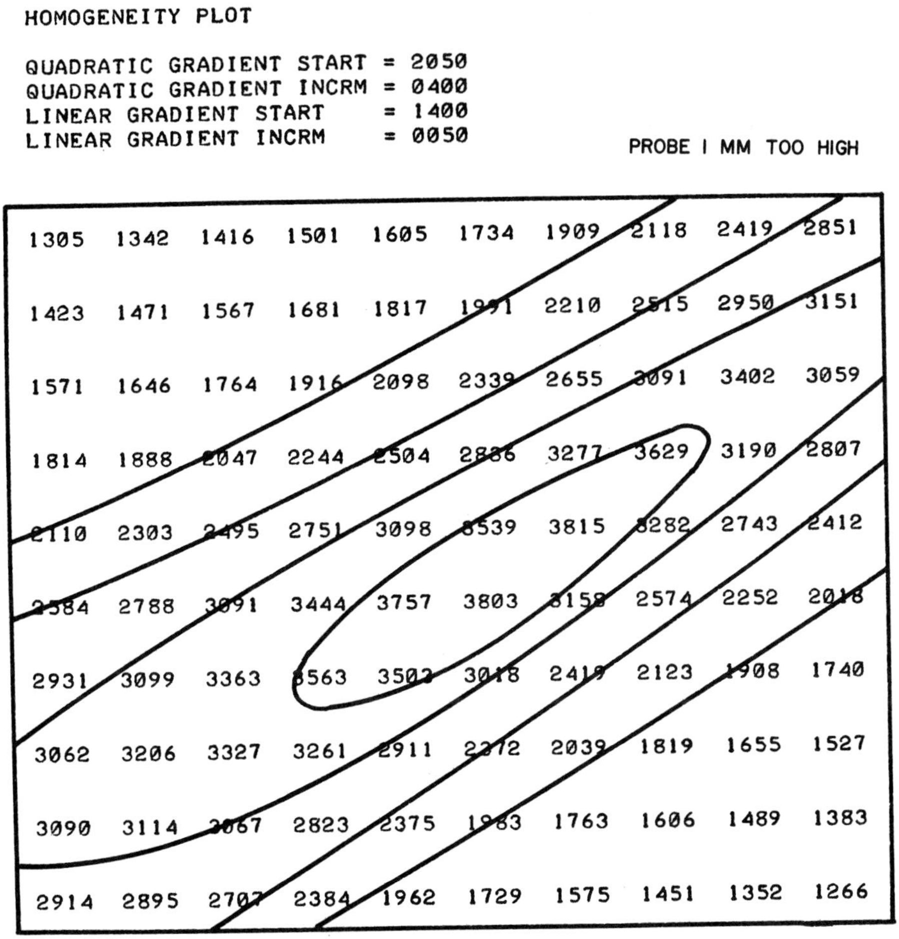

```
QUADRATIC GRADIENT START  = 2050
QUADRATIC GRADIENT INCRM  = 0400
LINEAR GRADIENT START     = 1400
LINEAR GRADIENT INCRM     = 0050                    PROBE 1 MM TOO HIGH
```

Figure 9. Magnetic field homogeneity plot produced by the equipment in Fig. 8. The printed numbers indicate peak height of a doped water reference signal as a function of the linear and quadratic shim currents. The strong dependence of the two shims, indicated by the inclined contours, shows that the probe head has not been properly centered in the magnet.

To obtain a survey on the dependence of the homogeneity on the shim currents, homogeneity plots were produced by a computer routine (EP11). The plot in Fig. 9, for example, shows non-orthogonality of the two shims indicated by the inclined contour lines. This is due to a missetting of the probe head in the magnet. Plots of this type allow the probe head to be properly positioned for optimizing orthogonality of the shim coils.

A simple search algorithm (EP12) was written for maximizing the peak amplitude of a selected line through which the field is swept repetitively by a waveform generator while the shim settings are varied in a systematic and iterative manner. The strategy is visualized in Fig. 10. At the beginning, a sufficiently large rectangle covering 5×5 meshpoints in the two-dimensional

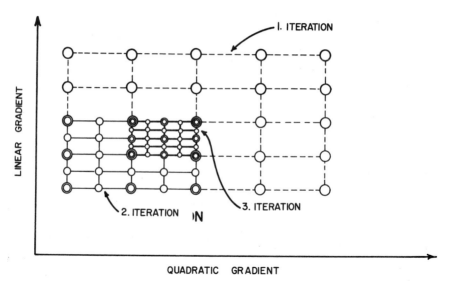

Figure 10. Iterative homogeneity adjustment procedure. For the first iteration, 25 measurement points, arranged in the form of a rectangle in the 2D shim current space, are selected. The new 25 measurement points for the next iteration are centered at the previous point of maximum signal intensity, again arranged in the form of a rectangle but with side lengths reduced by a factor of 2.

shim current space is selected. In a first iteration step, the signal amplitudes are measured for all 25 mesh points and the maximum value is determined. A new rectangle is then defined, again with 5x5 mesh points but with side lengths reduced by a factor two and centered at the shim values for maximum signal amplitude. In the second iteration step, the signal intensities are again measured at all 25 mesh points and their maximum value determined. At this position then, a further reduced rectangle is located, leading to the third iteration step, and so on. The procedure converges in general well provided that the optimum shim settings are within the initially selected rectangle and that the response surface has within this rectangle a single maximum.

The performance is demonstrated in Fig. 11. A sample of acetone with a TMS reference line was used. The TMS line served for shimming. The resolution is successively improved from iteration to iteration as visualized in Fig. 11. After 4 iterations, an optimum is reached that is as good as obtainable by manual adjustment.

When using a narrow line, such as the one of TMS, for shimming, the process is slow, taking about 1 minute per iteration. To speed up the procedure, it is possible to replace the analytical sample by a doped water sample ($\sim$8 Hz line width) for the iterative optimization whereby the time per iteration can be reduced to as little as 5 seconds per iteration. The resolution achievable with this accelerated procedure is not significantly reduced.

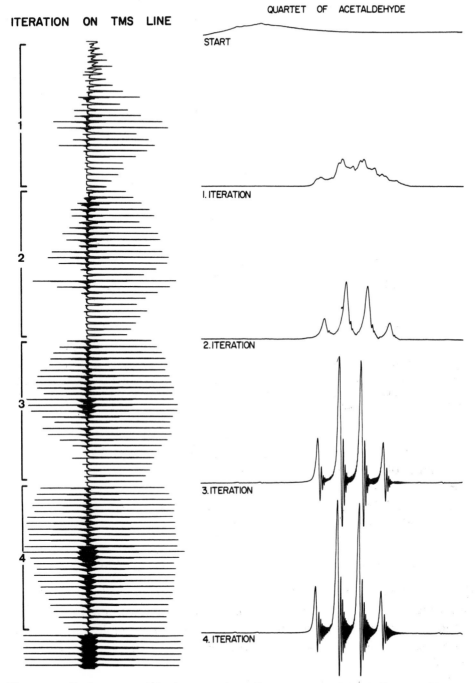

Figure 11. Performance of the homogeneity adjustment procedure of Fig. 10. The resonance of tetramethylsilane (TMS) has been used as a reference line. Before and after each cycle of 25 measurements, the quartet of acetaldehyde is recorded to visualize the homogeneity improvement. After four iterations, no significant further improvement was obtained.

13

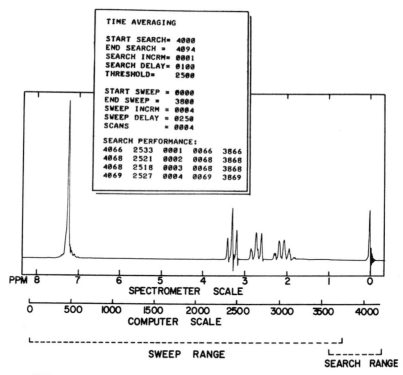

Figure 12. Shift-compensated data acquisition. To allow coaddition of scans despite slow time-dependent shifts in the A60 spectrometer, a reference line is located in a search phase and, depending on its position, the sweep range adjusted. In the insert, input parameters, and the search performance record are given for four scans.

Later, more efficient search algorithms were developed[16] that minimize the number of required measurements which is particularly important when more than two shims have to be iteratively adjusted. The Simplex method that was employed in these more advanced procedures is operative also on more complex response surfaces.

Computer-Controlled Shift-Compensated Data Acquisition

For field stabilization, the A60 spectrometer was equipped with an NMR lock using an external doped water sample. Stability was limited by changes in the field gradient between the location of the lock sample and the analytical sample, and signal averaging was impeded by slight time-dependent line shifts.

A signal averaging routine (EP7) was written that allows synchronous addition of signals irrespective of field shifts. As illustrated in Fig. 12, before each sweep scan through a predefined section of the spectral range (defined by START SWEEP, SWEEP INCRM, and END SWEEP), a search scan was performed for a selected sufficiently strong reference line (defined by

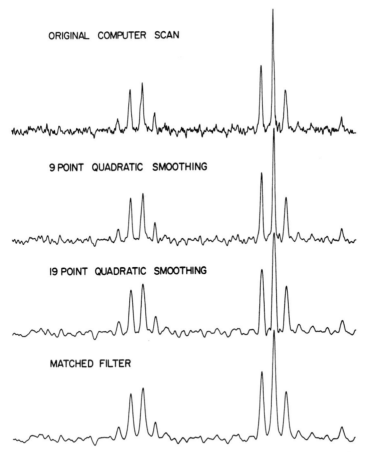

ORIGINAL COMPUTER SCAN

9 POINT QUADRATIC SMOOTHING

19 POINT QUADRATIC SMOOTHING

MATCHED FILTER

Figure 13. Convolution filtering for sensitivity enhancement. The original scan through the spectrum of 1% ethylbenzene, recorded on the A 60, is filtered by 9-point and 19-point quadratic smoothing as well as by a filter matched to the original lineshape.

START SEARCH, SEARCH INCRM, and END SEARCH). The measured position of the reference line allowed then a correcting shift of the sweep range. The data resulting from the search are exemplified in Fig. 12 under SEARCH PERFORMANCE. Each row represents the data for one scan. The first column lists the found reference peak position, the second column the peak intensity, the third column the scan number, the fourth column the shifted start of the sweep, and the last one the shifted end of the sweep. This routine allowed long-time signal averaging unaffected by drifts.

An alternative approach would have been to perform the shift adjustments after storage of a full spectrum in memory. The chosen solution has the advantage that search and sweep conditions concerning range and sweep rate can be selected independently.

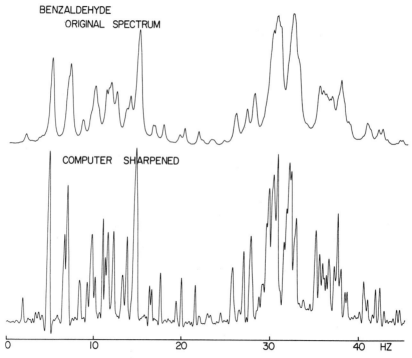

BENZALDEHYDE
ORIGINAL SPECTRUM

COMPUTER SHARPENED

0   10   20   30   40   HZ

Figure 14. Convolution filtering for resolution enhancement. The original slow passage 60 MHz benzaldehyde spectrum is filtered by a Lorentz-Gauss transformation that reduces the linewidth by 40%.

Signal-to-Noise Filtering and Resolution Enhancement

Through several publications,[17,18] it became known at that time that linear filtering processes allow the enhancement of either sensitivity or resolution. Certain compromises are also possible, for example maximizing sensitivity by keeping line shape distortions in bounds.[5] Nowadays linear filtering is performed by multiplication of the free induction decay by a weighting function. Although this possibility was known at that time, initially a direct convolution process of the frequency domain data was programmed (EP8). To avoid special treatment of the fringe points of the spectrum, a cyclic convolution process was used where the missing points at the beginning of a spectrum were supplemented by the last ones and vice versa. This, by the way, is an inherent feature when convolution is replaced by sequential Fourier transformation — multiplication — reverse Fourier transformation.

Fig. 13 presents some filtering attempts executed on a 1% ethylbenzene spectrum recorded on the A60. As predicted, the matched filter produces maximum signal-to-noise ratio while 9-point and 19-point quadratic smoothing functions[17] lead to high-frequency filtering with less line broadening. On the other hand, resolution enhancement could be reached by line shape

16

transformations, such as a Lorentz-Gauss transformation.[5] Fig. 14 shows the result of such an effort applied to a proton-resonance A60 spectrum of benzaldehyde. The improvement of apparent resolution is convincing although the line width has been reduced by only 38%. Many more lines become clearly visible. Already this early example shows that with the resolution also noise and artifacts are enhanced, leading to a rather uneven baseline.

The experts in the field were well aware that linear filtering processing merely amounts to cosmetics improving the appearance and facilitating the visual analysis of spectra. The same is true also for more modern approaches, such as maximum entropy methods and linear prediction. The information content can not be increased. However, sometimes, it is possible to exploit supplementary information to guide the eye in the analysis procedure. An intelligent data processing procedure can use this information to enhance 'likely' and suppress 'unlikely' features. This may involve knowledge concerning line shape, number of (possibly overlapping) lines, and regular patterns in a spectrum, such as multiplets. Fortunately, some of the wilder claims of overoptimistic signal improvement magicians have, in the meantime, faded out and been disproved, and the field has returned again to reality.

Data Reduction

The goal of data reduction is here the determination of the positions and intensities of the peaks in a spectrum. A simple routine (EP4) has been written for this purpose. Its performance is visualized in Fig. 15. It uses two reference peaks, one at the left and one at the right end of the spectrum, of which the chemical shifts are input parameters (FIRST VALUE and LAST VALUE). All intensities are normalized to the first peak. The program progresses point by point from left to right through the spectrum and continuously stores the highest amplitude value found so far. A peak is recognized when (i) its intensity is higher than the THRESHOLD and (ii) when the intensity drops on the right side of the peak by at least the value SENSITIVITY. After a peak has been accepted, at first the minimum signal amplitude on its right is searched for. When the amplitude rises again by at least the amount of THRESHOLD above this minimum, it is assumed that a shoulder of the next peak is encountered and a new search for the next maximum is started. Obviously, overlapping peaks can not be identified in this manner.

Multiplet Identification

Nowadays it is very clear that spin coupling multiplets are best identified and analyzed by 2D spectroscopy. In earlier times, double resonance experiments

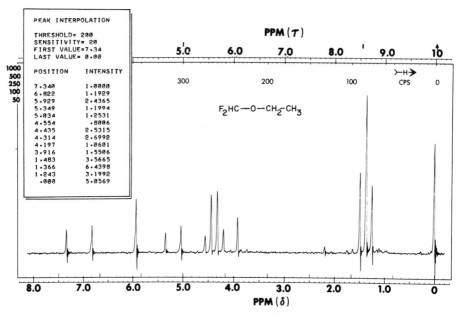

Figure 15. Data reduction of a slow passage 60 MHz spectrum of difluoromethylethyl-ether. The peak positions and intensities are determined by interpolation between the assumed chemical shifts of chloroform (7.34 ppm) and TMS (0 ppm). A threshold and a sensitivity level are used to discriminate peaks from noise as described in the text.

were used for the same purpose. It is known that one-dimensional spectra contain a priori insufficient information to distinguish a spin coupling multiplet from a set of independent chemical shifts. Nevertheless, an attempt was made to identify multiplets exclusively based on a one-dimensional spectrum, as shown in Fig. 16.

The program EP14 processes further the output of the routine EP4 (preceeding subsection) that contains peak positions and intensities. The program searches for groups of equal line separation by scanning systematically through all pairs of identified peaks. The allowed deviation in line separation can be specified by the quantity ERROR. The groups of line pairs belonging to the same splitting family are then printed out in the format given in Fig. 16. The line numbers are connected by the corresponding spin-spin coupling values, given in Hz. At first, the two aromatic doublets with a coupling constant of 9.0 Hz are found, and also the quartet and triplet of the ethyl group with 7.0 Hz splitting are identified. In addition, a pair of doublets with a coupling of 14 Hz is located which however is a subset of the ethyl group lines.

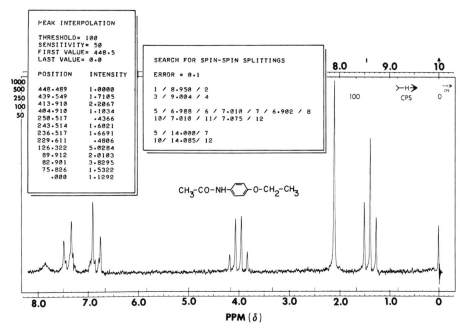

Figure 16. Multiplet analysis by computer. The different multiplets in the 60 MHz proton resonance spectrum of N-acetyl-p-phenetidine are identified based on regular splitting patterns and printed out in the form of a table shown in the insert, numbering the lines from left to right.

Spectrum Coding and Library Search

The final step in the analysis of NMR spectra is the identification of the compound under investigation. In cases where a full analysis in terms of chemical shifts and multiplet splittings and in terms of a resulting chemical structure is difficult, it is still possible to employ the spectrum as a fingerprint in a search through a library of prerecorded and identified spectra. As it is difficult and tedious to directly compare full spectra, it is necessary to determine at first a code that characterizes a spectrum sufficiently well. The code can then be compared with the codes stored in the library.

The special code implemented in the routine EP17 consists of eight numbers between 0 and 100 that represent the partial integrals of the (base-line-corrected) spectrum from 8 ppm to 0 ppm, each integral extending over a 1 ppm range. The highest value is normalized to 100. An example for 1,2-propanediol is shown in Fig. 17.

A Fortran program (E76) has been written for establishing the library and for the library search based on this eight-number code. A sample print-out is given in Fig. 18. It shows that the library compounds with the four best fits to the input code are identified and listed together with their code and the fit error indicating the absolute deviation between the code words. The test library was quite small with a total of 83 entries.

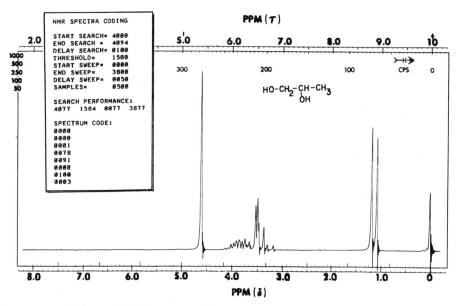

Figure 17. Coding of NMR spectra. The 60 MHz proton resonance spectrum of 1,2 propanediol is coded by a set of eight numbers that represent the partial integrals in one-ppm intervals from 8 ppm to 0 ppm.

Automatic Phase Adjustment by Computer

In 1967 attempts towards automatic phase correction of NMR spectra were started.[19] The crucial observation which allowed such a procedure was that even without quadrature detection, a single cw spectrum, irrespective of the sweep rate, can be phase-corrected by computational means. It is possible to compute from the recorded signal, possibly with misadjusted phase, by a numerical Hilbert transformation a quadrature phase signal that can be combined with the original signal to correct its phase at will.

Two approaches were investigated (i) phase correction by convolution in the frequency domain and (ii) phase correction through Fourier transformation. For convolution in the frequency domain, a discrete Hilbert transformation has been found in analogy to the well-known continuous Hilbert transformation. It allows the computation of the quadrature and the phase-corrected signals by a convolution sum with an array of simple kernel coefficients. An example using the Fortran routine E87 is shown in Fig. 19. The originally recorded trace (a) has been phase-corrected by discrete convolution to obtain trace (b).

```
CODE OF SPECTRUM TO BE IDENTIFIED

  1      10-30-66              -0   -0   19   100   -0   79   -0   -0   -0

SPECTRA WITH 1. BEST FIT
ERROR=    0

 24      ALLYLBROMIDE          -0   -0   19   100   -0   79   -0   -0   -0

SPECTRA WITH 2. BEST FIT
ERROR=   32

 56      ALLYL CYANIDE         -0   -0   -0   100   -0   66   -0   -0   -0

SPECTRA WITH 3. BEST FIT
ERROR=   52

 26      ALLYL IODIDE          -0   -0   25   100   25   100  -0   -0   -0

SPECTRA WITH 4. BEST FIT
ERROR=  107

  2      1,1,2-TRICHLOROETHANE  -0   -0   -0   66    33   100  -0   -0   -0
```

Figure 18. Library search for NMR spectra. Based on an eight-number code exemplified in Fig. 17, a search is performed in a spectra library. The four best matches, based on a comparison of the absolute value deviations of the codes, are printed out. Allylbromide was the compound to be identified. It leads to a perfect match with the corresponding library spectrum.

The phase correction through Fourier transformation requires the Fourier transformation of the original signal. To invoke the causality principle that is the basis for the Hilbert transform relation between real and imaginary part of a frequency response function, it is necessary to set the obtained time domain signal equal to zero for $t < 0$ and perform a reverse Fourier transformation. This delivers again a quadrature phase signal that can be used for the phase correction. A phase-adjusted signal obtained in this (computationally more economical) manner is represented in Fig. 19c.

For automatic phase adjustment by computer, a suitable criterion to detect phase deviations is needed. It was found that the ratio of the areas above and below the signal trace is a good measure.[19] For accurate absorption mode phase, the area below the trace should become minimum while the area above reaches its maximum.

Prehistoric Attempts towards Two-Dimensional Spectroscopy

Many years before Jeener proposed for the first time a time-domain two-dimensional (2D) NMR experiment,[20] ideas were floating around in the NMR community to represent spin-coupling connectivity information in the form of a 2D correlation diagram. In some way, it is possible to regard the systematic spin tickling plots by Anderson and Freeman[21] as predecessors of 2D spectra.

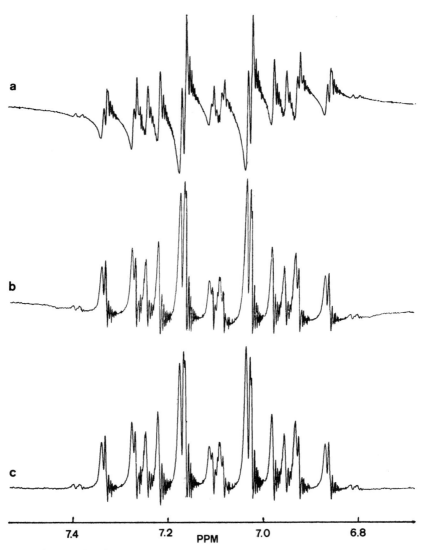

Figure 19. Automatic phase correction of a 60 MHz slow passage spectrum of acryloni-
trile. (a) Original, incorrectly phased spectrum. (b) Phase-corrected spectrum obtained
with the convolution algorithm. (c) Phase-corrected spectrum obtained with the Fourier
transformation algorithm.

In early 1968 computer-controlled double resonance experiments were started and programs written for the PDP8 computer (EP30). By means of double field modulation on a DP60 spectrometer, irradiation at two spots within one spectrum was achieved. The field modulation frequencies were computer-controlled by means of a voltage-controlled oscillator (Hewlett-Packard HP 3300A). The expected outcome would have resembled an 'Andersen-Freeman plot'.[21] However, due to the imminent departure of the author from California, it was impossible to finish these measurements.

The Anderson-Freeman plots as practical spectroscopic tools, in comparison to standard 2D spectra recorded in time domain, have the disadvantage of producing extended signal ridges in two dimensions instead of narrow cross peaks. Tickling effects require relatively strong radio frequency fields and do not permit reaching the high resolution customary in modern 2D spectroscopy.

An earlier attempt in March 1967 intended to produce 2D spectra displaying population transfers for the identification of connected transitions. A routine (EP13) was written to perform such an experiment with the A60-PDP 11 combination. The resonance conditions were adjusted to a particular peak in the spectrum under partially saturating conditions. For a short time, the magnetic field was then displaced to a different position in the spectrum where it was left for a time just sufficient to cause a 180° rotation under the influence of the strong rf field. In a repetitive sequence of experiments, the field displacement was successively increased. Whenever a resonance line with an energy level in common with the saturated one was reached by the field jump, a polarization transfer by the 180° rotation and a corresponding desaturation of the saturated line was expected. This experiment has much in common with ENDOR spectroscopy. After completion of the field jump sequence, the next line in the spectrum was saturated and the experiment repeated. Unfortunately, due to spurious undesired responses of the A60 spectrometer, no satisfactory 2D spectra were obtained and the experiment was abandoned. Apparently, the time was not yet ripe for 2D spectroscopy.

CONCLUDING REMARKS

In 1971, finally, the first commercial Fourier transform NMR spectrometer was introduced by Bruker. It lead to an unforeseen breakthrough and to an explosion of new NMR techniques and applications. The computer became for the first time an indispensible and integral part of the spectrometer. Initially it was primarily used for basic operations, such as Fourier transformation, signal filtering by application of weighting functions to free induction decays, and phase adjustment. More sophisticated applications developed slowly but steadily.

A further milestone in computer applications was undoubtedly the introduction of two-dimensional NMR techniques[20,22,23] which started, in some fields of application, to replace almost completely the traditional one-dimensional methods. The demand for extensive computer power and data storage was rising sharply. Also the need for automatic processing of the very large data sets with a vast information content became more and more obvious, and much research was devoted to the development of suitable computer algorithms.[24-26]

Even more acute became the demand for high performance computing equipment in the context of three-dimensional spectroscopy.[27-29] Here it is nearly impossible to adequately represent the data on two-dimensional paper, and highly developed computer graphical tools are indispensible for working in an efficient way with these enormous amounts of data.

It is hardly needed to exemplify in this account all these more recent developments. They are authoritatively covered in the following contributions. It is certain that the general trend towards computerized NMR spectroscopy will continue in parallel with the further development of computer hardware. However, major efforts are needed in the software domain where still much has to be done to match the interpretative power of an experienced spectroscopist who still has a considerable edge over even the most sophisticated software packages existing today.

ACKNOWLEDGMENTS

The author is grateful to NATO for exerting brute force onto him to write up this account by temporarily withholding his travel and living expenses. A more honest acknowledgment refers to the inspiration and support provided by Dr. W. A. Anderson during these early days of computer-aided NMR at Varian Associates in Palo Alto, California. Numerous discussions with colleagues, in particular with Dr. Ray Freeman, were helpful, and the technical support by Mr. Bill Siebert was indispensible.

REFERENCES

1. O. Jardetzky, N. G. Wade, and J. J. Fisher, *Nature* **197**, 183 (1963).
2. M. P. Klein and G. W. Barton, *Rev. Sci. Instr.* **34**, 754 (1963).
3. P. Laszlo and P. R. Schleyer, *J. Am. Chem. Soc.* **85**, 2017 (1963).
4. L. C. Allen and L. F. Johnson, *J. Am. Chem. Soc.* **85**, 2668 (1963).
5. R. R. Ernst, *Adv. Magn. Reson.* **2**, 1 (1966).
6. J. Dadok and R. F. Sprecher, *J. Magn. Reson.* **13**, 243 (1974).
7. R. K. Gupta, J. A. Ferretti, and E. D. Becker, *J. Magn. Reson.* **13**, 275 (1974).
8. R. R. Ernst and W. A. Anderson, *Rev. Sci. Instr.* **37**, 93 (1966).

9. R. R. Ernst, "Sensitivity Improvement by Fourier Transform Techniques", 6th Experimental NMR Conference, 25-27 February 1965 Mellon Institute, Pittsburgh.

10. R. R. Ernst, *Rev. Sci. Instr.* **36**, 1689 (1965).

11. R. R. Ernst and W. A. Anderson, *Rev. Sci. Instr.* **36**, 1696 (1965).

12. F. Bloch, *Phys. Rev.* **70**, 460 (1946).

13. H. C. Torrey, *Phys. Rev.* **76**, 1059 (1949).

14. E. L. Hahn, *Phys. Rev.* **77**, 297 (1950).

15. R. R. Ernst, "Online NMR Data Processing by Means of Small-Scale Computers", 8th Experimental NMR Conference, 2-4 March 1967, Mellon Institute, Pittsburgh.

16. R. R. Ernst, *Rev. Sci. Instr.* **39**, 998 (1968).

17. A. Savitzky and M. J. E. Golay, *Anal. Chem.* **36**, 1627 (1964).

18. L. C. Allen, M. M. Gladney, and S. M. Glarum, *J. Chem. Phys.* **40**, 3134 (1964).

19. R. R. Ernst, *J. Magn. Reson.* **1**, 7 (1969).

20. J. Jeener, Ampere International Summer School II, Basko Polje, Yugoslavia (1971).

21. W. A. Anderson and R. Freeman, *J. Chem. Phys.* **37**, 85 (1962).

22. L. Müller, A. Kumar, and R. R. Ernst, *J. Chem. Phys.* **63**, 5490 (1975).

23. W. P. Aue, E. Bartholdi, and R. R. Ernst, *J. Chem. Phys.* **64**, 2229 (1976).

24. P. Pfändler, G. Bodenhausen, and R. R. Ernst, *Anal. Chem.* **57**, 2510 (1985).

25. K. P. Neidig, H. Bodenmüller, and H. R. Kalbitzer, *Biochem. Biophys. Res. Commun.* **125**, 1143 (1984).

26. J. C. Hoch, S. Hengyi, M. Kjær, S. Ludvigsen, and F. M. Poulsen, *Carlsberg Res. Commun.* **52**, 111 (1987).

27. C. Griesinger, O. W. Sørensen, and R. R. Ernst, *J. Am. Chem. Soc.* **109**, 7227 (1987).

28. H. Oschkinat, C. Griesinger, P. J. Kraulis, O. W. Sørensen, R. R. Ernst, A. M. Gronenborn, and G. M. Clore, *Nature* **332**, 374 (1988).

29. C. Griesinger, O. W. Sørensen, and R. R. Ernst, *J. Magn. Reson.* **84**, 14 (1989).

PARAMETRIC ESTIMATION IN 1-D, 2-D, AND 3-D NMR

David Cowburn, John Glushka, Frank DiGennaro, and
Carlos B. Rios

The Rockefeller University
New York, NY, 10021, U.S.A.

ABSTRACT

The increased power of nmr as an analytical and structure determining method arises from the development of two-dimensional methods in which spectral maps reflect through-bond and through-space connectivities between atoms. The interpretation of these maps is made complex by many factors, including the large number of such connectivities in many applications, the multiplicity of the spectral elements, and the limited signal-to-noise ratio. As one element of solving these problems, we have developed methods to simplify the cataloging of positions and intensities of peaks, and for examining and selecting among them. Linear Predictive Singular Value Decomposition is one numerical method used for the extraction of peak characteristics. Simple tools have been developed to select among spectral characteristics, to reduce multiplets to single peaks, and to incorporate pre-existing chemical information into the analysis. Characteristics of this approach of parametric estimation permit somewhat different approaches to experimental design. This may be particularly valuable in the determination of multiple vicinal coupling constants about single bonds.

INTRODUCTION

The development of two dimensional NMR[1,2] has permitted remarkable developments in the determination of biopolymer structure[3,4] and many other chemical areas[5]. Applications to even larger biopolymers[6], the development of 3-D methods[7] and spatially resolved spectroscopy[8], promise to produce volumes of data which cannot be effectively analyzed by current methods. Several aspects of the interpretation of data are areas of active research interest, including pattern recognition[9] and heuristic refinement of derived structures[10] and others[11]. For biological macromolecules, improved data processing methods may be expected to achieve increased speed, accuracy, and sensitivity of determination. Increased speed is desirable to match the time required for production of new proteins by protein engineering, (only a

Computational Aspects of the Study of Biological Macromolecules by Nuclear Magnetic Resonance Spectroscopy, Edited by J.C. Hoch *et al.,* Plenum Press, New York, 1991

27

few days in favorable cases). Increased sensitivity is always welcome, given the intrinsic low sensitivity of nmr, and the desirability of working at lower concentrations, or with smaller quantities of materials.

A common feature of these nmr experiments is the extensive processing of data after acquisition. Typically, free induction decays (fids) are corrected for DC offset, subject to digital filtering to optimize resolution and signal-to-noise ratio (snr), zero filled to extend the processed vector length, Fourier transformed, and phase corrected[12]. The resulting interferogram is then subject to a similar process in the remaining dimensions, along with recombination and elimination of data to achieve phase sensitive detection in multiple dimensions, with useful line shapes[2]. Peaks of the spectra are usually identified by hand, and positions and intensities recorded. These processes are repeated multiple times for different experimental types and variables. Usually there is considerable involvement of the operating spectroscopist, both in selecting variables used in the processing, and in the identification of peaks, including the rejection of artifact and "uninteresting" signals. These actions of the operator are guided by a mixture of specific spectroscopic experience (e.g., identifying artifacts arising from large solvent peaks), and more general chemical knowledge (e.g., the methyl group of interest will have a particular set of shift, coupling, and intensity characteristics). One approach to improving the harvesting of the spectral characteristics from the original data is to split the task into two parts — (i) the optimal extraction of signal characteristics, using standard methods of signal analysis, combined with knowledge of the general characteristics of nmr signals, and (ii) the applications of filters to select (and possibly recombine and reanalyze) these signal data, using preexisting spectroscopic and chemical knowledge. A preliminary system demonstrates the basic feasibility of the approach, using Linear Predictive Singular Value Decomposition as the means of obtaining signal characteristics.

Nmr signals from high resolution pulsed experiments can be adequately represented by sums of sinusoids[13]. After Fourier transformation, the spectral characteristics can sometimes be obtained by non-linear least squares techniques in the frequency domain. This deconvolution in the frequency domain is particularly complicated by the need to set initial parameters for the non-linear least squares analysis[11,14]. Alternative methods of parametric estimation include simulated annealing[15,16], and direct methods, especially Linear Prediction. In general, simulated annealing is excessively computationally intensive[17], although properly conducted, it avoids the local minimum problem associated with other minimization methods including least squares. The technique of linear predictive singular value decomposition (LPSVD)[13] applied previously in nmr[19], permits the extraction of frequency information from a time-domain signal. While a Fourier transform of this

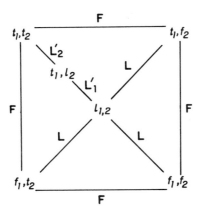

Figure 1. Topograph of the interconversions of two-dimensional spectral information between the time domain (t_1, t_2), the interferograms $\{(t_1, f_2)$ and $(f_1, t_2)\}$ the frequency domain (f_1, f_2) and the list representation $(l_{1,2})$ of roots. The operators are **F** for forward/inverse Fourier transform, **L** for decomposition/expression between a list of characteristic roots and the respective domain. In the case described here, **L** is the LPSVD algorithm for the interferogram/frequency domain transform. The $(t_1, t_2) - (l_{1,2})$ path represents the direct use of LPSVD in 2-D spectra[19,23] The path for analysis of 2-D spectra in this article is $(t_1, t_2) - (t_1, f_2) - (l_{1,2})$. The analysis of 3-D spectra is comparably done by applying **L** to f_1, t_2, t_2 planes.

signal produces another related signal, equivalent to the previous one but viewed in the frequency domain, this LPSVD technique presents the same information in a different and sometimes more useful manner. The algorithm assumes that the signal is made up of a certain number of exponentially damped sinusoids of the form

$$z(t) = \sum_{j=1}^{l} a_j e^{t/\beta_j} e^{i(\omega_j t + \phi_j)}. \tag{1}$$

The details of the algorithm and of several related methods have been discussed elsewhere.[19,20,21] The actual roots of the signal, (i.e., the back solution of Eq. 1) then contain the frequency (ω), damping factor (β), amplitude (a), and phase (ϕ) of each. In contrast to the extraction of $n/2$ real and $n/2$ imaginary intensities by Fourier transformation of a vector of n real (in quadrature) data points, LPSVD returns $4l$ values for l roots, ω_j, β_j, a_j, and ϕ_j. In cases of adequate signal to noise ratio, the restriction of l to the number of known contributing sinusoids leads to roots whose expression in a simulated spectrum is noiseless. The precision of the coefficients of these roots is, naturally, limited by the snr.

In our approach[22], we have chosen to use a single vector LPSVD, employing a Fourier transform performed as a digital filter on each row in the t_1, t_2 space of a 2-D spectrum. (See Fig. 1.) While the LP method may be generalized to n-dimensions[19,23,24], it is not yet evident that its numerical

Table 1. The Structure of the Derived Table of LPSVD roots

<frequency>, <damping>, <amplitude> and <phase> are the values derived from the LPSVD operation for that vector.

<flag-word>:==	set of logical indicators for
	(1) out of frequency range of filter,
	(2) out of damping range,
	(3) out of amplitude range,
	(4) out of phase range,
	(5) out of range of calculated intensity
	(integrated amplitude and damping
	for an estimated time,
	(6) root marked invalid by graphical
	or other user input,
	(7) root invalidated by peak identifier,
	(8) root invalidated by distance range
	from diagonal
<index>:==	identifies the position in the 2 or 3D
	structure from which the root is
	obtained.
<root>:==	<flag-word> <frequency> <damping>
	<amplitude> <phase> <index>
<roots>:==	<root> [<roots>]
<cluster>:==	<roots> of same index value
<clusters>:==	<cluster> [<clusters>]
<empties>:==	<roots> with negative index values.
<block>:==	<clusters> [<empties>]
<blocks>:==	<block> [<blocks>]
<header-record>:==	48 byte record with appropriate initial
	information on sweep widths, maximum
	number of roots, etc.
<starting-block>:==	<header-record> <clusters>
	[<empties>]
<root-data-structure>:==	<starting-block> [<blocks>]

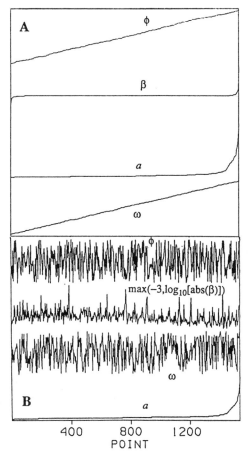

Figure 2. Distributions of values of coefficients for roots obtained by processing the $^{1}H\{^{15}N\}$ multiquantum coherence detected 2-D spectrum of 50 mM pyrollidinone in $95\%H_2O/5\%D_2O^{34}$. The spectra resulting from various stages of filtering are shown in Fig. 5 of (22), (A) Each coefficient, amplitude (a), damping (β), frequency (ω), and phase (ϕ), is separately sorted and graphed between their extreme values. (B) The amplitudes are sorted, and the root-associated other coefficients are displayed above the sorted vector. The most significant roots are at the right hand side, and most of the small amplitude roots elsewhere are noise.

stability is adequate for lower snr 2-D spectra[18]. In addition, the combined FT-LP approach provides a general purpose tool for different kinds of multi-dimensional spectra, in which the additional dimensions are represented by frequency spaced indexing of the resultant roots of vectors. The results of the FT-LP are stored in files containing the amplitude, frequency, damping, and phase of the roots extracted for each column as well as a declaration field. (See Table 1). A set of flags in the declaration field is used to store the results of the various filters and selections which are subsequently performed on the roots. A root that fails any of these imposed constraints is marked as "inactive" and subsequently ignored when the roots are used to generate a simulated column of the matrix. In this way, a flexible structure is developed which may employ an unlimited number of routines to process the roots and remove spurious results. The methods of accessing these files has been improved from the original implementation[22] and any size constraints remaining are unlikely to be limiting for applications foreseen in the near future.

The first process usually acted on the roots is a simple linear filter that passes through the entire matrix "deactivating" roots not meeting the prescribed constraints. All components of a root, its frequency, amplitude, phase, and damping factor, are subject to constraint. Typical filters may select roots with negative damping factors and with amplitudes and intensities above some predefined limit. A procedure for arriving at the proper filters uses the graphing of a profile of the roots. Similarly, frequency selection can be used to limit the analysis to regions of interest. In determining the proper limits for an effective amplitude filter, for example, the sorted amplitude values would be graphed. The proper lower-limit may then be inferred from viewing such a graph for, in practice, there is a gradually sloping section of low-amplitude roots followed by a dramatic break in the graph separating the significant roots. This is complementary to the graph of equivalent singular values[18,19] which may not clearly show such transitions. The sorted values for one coefficient can be used to display the other coef-

---

Figure 3. Contour plots comparing conventional FT, FT-LP, and FT-LP filtering incorporating simple matching using chemical information. Original data are the ^{1}H{^{15}N} FIDs of a 22 mM sample of bovine pancreatic trypsin inhibitor, pH 4.6, 68° C. Full experimental details are given in (25). The dataset used in that reference was an improved set from that used here, with lower systematic artifacts, believed to arise from improved temperature stability in the collection of the better set. (A) Shows the original FT processed 2-D spectra. Absolute value mode is shown. (B) Result of applying FT-LP process with filtering roots of amplitude greater than 150,000 (arbitrary units), and damping of less than 1 msec. A peak filter was used, scanning two adjacent rows for other portions of a peak, with a tolerance of 25 Hz for matching. These filters resulted in 1,377 active roots, and the inactivation of 18,714 roots. (C) Effect of applying a pairing method, in which active roots were scanned for similar frequencies in f_1, separated by 91 Hz (the expected $^1J_{NH}$) in f_2, with a tolerance of 4 Hz.

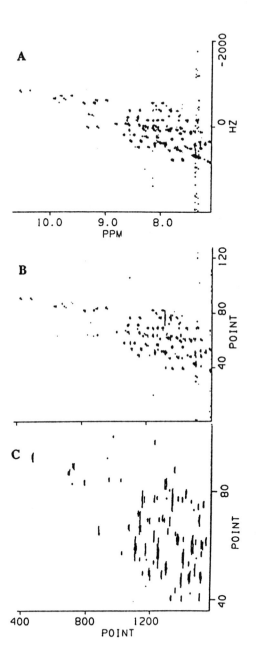

33

ficients of the roots in the same sorted order, permitting evaluation of the relationships between phase, frequency, damping and amplitude of "true" signals, compared to noise coefficients. As an example, Fig. 2A illustrates the sorted coefficients for a simple analysis of an experiment in which only two dominant signals are seen, distributed over several f_2 indices. Most of the roots are of small amplitude with randomly distributed phase, and frequency. Their damping distribution is more complex, with most values around zero, illustrating that the algorithmic analysis treats noise predominantly as sinusoids of zero damping. In Fig. 2B, these coefficients are sorted only on amplitude (lowest trace), and the corresponding coefficients are displayed above. (The β's are displayed on a log scale, for ease of examination in 2B). We note that the largest amplitude signals have similar frequency and damping, but that phase is more random. This has been seen in several such analyzes, and probably reflects the lower precision associated with the derivation of phase by the LP method, compared to the other coefficients[13].

A method is also available to select peaks complying with basic notions of peak shape while removing spikes, likely to be artifacts. A basic constraint is that in addition to having a threshold level of amplitude, a peak should have width in the second, f_2, dimension. Each root is examined to see if it has neighboring roots in f_2 with equivalent frequencies and with intensities above a certain level. (n-dimensional LP methods use another subset of frequency and damping variables of the root to represent this information directly from the signal analysis.)

These methods appear to have several advantages in addition to those identified earlier over conventional Fourier processing and peak identification. These include frequency resolution limited only by snr, avoidance of truncation artifact as a result of zero-filling, and reduction of the need for phase sensitive detection. These characteristics are especially important for their potential to reduce the number of incremented steps in two or higher dimensional experiments. The basic concept of using root coefficients as input to filtering methods, have been previously demonstrated for simple cases[22].

Figure 4. 3D NOESY-HMQC spectrum of 2.5 mM ^{15}N-labeled adenylate kinase in 10% D_2O/H_2O at pH 7.5, 25° C. The experiment was acquired on a GE GN-500 NMR spectrometer. The data were collected as a series of HMQC planes. A 1 s presaturation time and a mixing time of 80 ms was used. The t_1, t_2, and t_3 dimensions consisted of 64×64×2048 data points. A linear baseline correction and gausian multiplication with a line broadening of -2 Hz and gaussian broadening of 0.05 in the f_3 dimension keeping the amide region and a 60° shifted sine bell and zero filling to 128 points was applied to the f_1 and f_2 dimensions yielding a final absorptive spectrum of the amide region consisting of 512×128×128 data points in the f_3, f_2, and f_1 dimensions. (A) 3D-NOESY-^{15}N-HMQC spectra with 256×128×128 data points in f_2, f_2, and f_1 displayed, (B) f_2, f_3 plane at 122.4 ppm (^{15}N), see arrow in panel (A), processed with conventional FT methods. (C) f_2, f_3 plane as in (B) formed by reconstruction of the LP-roots analyzed from the t_2, f_3 interferogram.

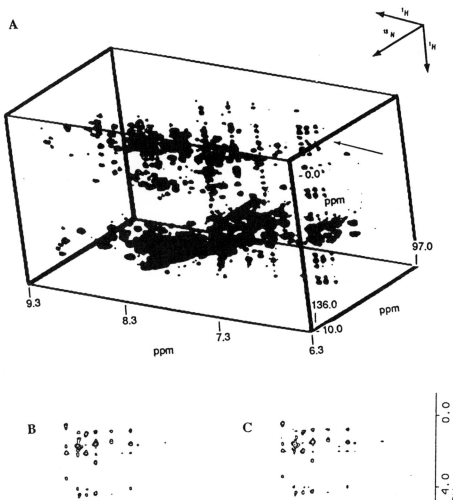

A

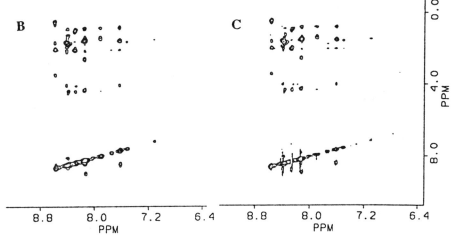

B

C

35

As a more complex case, we have attempted to fully analyze the ^{1}H{^{15}N}-HMQ-COSY spectrum of bovine pancreatic trypsin inhibitor, which we have studied extensively by conventional processing methods[25]. In Fig. 3, the conventional FT contour spectrum from a dataset with known artifacts is shown. The artifacts appear as a vertical (t_1-noise) band around 7.4 ppm, and as a slanted diagonal at about -20 degrees to the horizontal through the (0 Hz, 0 Hz) position (proton carrier at 8.77 ppm). As described in the figure legend, the FT-LP roots were filtered, and re-expressed in Fig. 3B. It may be noted that the two artifacts are much reduced, since they do not share the amplitude, damping, and peak clustering characteristics of the real signals. As noted in simpler cases[22], the imprecision of the damping coefficients makes the apparent linewidths in f_1 much more variable than apparent in conventional FT plots. Lastly, those peaks separated by the one-bond N-H coupling constant in that data can be detected and collapsed to singlet positions (Fig. 3C). While this spectrum contains many features common to an exhaustively characterized spectrum using conventional FT methods[25], it has not yet proved practical to fully automate the extraction of every feature. The major cause of this appears to be the limited capability of the LP algorithm to analyze fids of relatively low snr (as was this set), a feature pointed out in detail by its originators[18]. A small number of fids (typically about 0.3%) are not correctly analyzed. The current algorithm does seem to be entirely robust for spectra whose snr is greater than 5. Some of these failures arise from strictly computational considerations[26], but many arise from the intrinsic limitations of the algorithm, and may be detected by the large values of agreement factors between observed and calculated given by

$$R = \sum (I_{obs} - I_{calc})^2 / \sum (I_{calc})^2. \tag{2}$$

In these failure cases, simulated annealing may be substituted. Nonetheless, the general approach of using an LP-like method appears sound, and further attempts to modify this algorithm, or to use related non-linear least squares methods are worthy of pursuit.

As a last example, planes of 3D spectra may be analyzed in the same way. Fig. 4 shows a 3D spectrum and views of planes processed conventionally, and via FT-LP. An area of significant application of these methods lies in the extraction of vicinal coupling constants. The usefulness of this approach for investigating the backbone dihedral angle, ϕ, is well known. Approaches to measuring ψ are now practical in ^{15}N labelled proteins[27]. The use of other heteronuclear vicinal couplings for the angle χ^1 has been widely suggested[28] and shown to be feasible experimentally in small peptides for the unequivocal determination of rotamer states or determination of fixed angle, in conjunction with derivation of the stereospecific assignment of the two β protons, where present[29,30]. The extrapolation of these methods to proteins may be expected to improve significantly the resolution of structures derived from the combination of coupling and nOe data. The measurement of such couplings in larger uniformly enriched materials is technically difficult, although several approaches have been described. For homonuclear couplings, frequency separation of multiplets in 2-D experiments[31], and 3-D equivalents have been described[32]. Indeed, the homo- and heteronuclear J-resolved 3D and 4D spectra[33] would be readily obtained were it not for the relaxation problem associated with the long time sampling needed for the low modulation frequency associated with the small coupling constant. As pointed out[32], the decay curve in the J-resolved dimension can be numerically analyzed, and LPSVD is especially promising for this purpose since it can be used with such a low number of data incrementation steps.

CONCLUSION

The use of very simple filter techniques on machine-extracted peak positions characterized as sinusoids promises to be an important tool in speeding the analysis of spectra, and in combining preexisting spectral information into the analysis in a parametric fashion.

ACKNOWLEDGEMENTS

Supported by NIH AM-20357, and NSF PCM 8413457. Nmr and computer equipment was purchased with grants from NSF, NIH, and the Keck Foundation. We are grateful to Dr. Dennis Hare, and to Dr. R. De Beer for provision of source codes for their programs, which served as the bases for this work. The software described is available though the authors, but is not documented or supported.

REFERENCES

1. W. P. Aue, E. Bartholdi, and R. R. Ernst, *J. Chem. Phys.* **64**, 2229-46 (1976).
2. R. R. Ernst, G. Bodenhausen, and A. Wokaun, *in* "Principles of Nuclear Magnetic Resonance in One and Two Dimensions", Oxford University Press, Oxford (1987).
3. K. Wüthrich, "NMR of Proteins and Nucleic Acids", Wiley-Interscience, New York (1986).
4. A. Bax, *Ann. Rev. Biochem.*, **58**, 223-256 (1989).
5. R. R. Ernst, *Chimia* **41**, 323-340 (1987).
6. L. P. McIntosh, F. W. Dahlquist, and A. G. Redfield, *J. Biomolec. Struct Dyn.* **5**, 21-34 (1987).
7. H. Oschkinat, C. Griesinger, P. J. Kraulis, O. W. Sørensen, R. R. Ernst, A. M. Gronenborn, and G. M. Clore, *Nature* **332**, 374-376 (1988).
8. P. A. Bottomley, H. C. Charles, P. B. Roemer, D. Flamig, H. Engeseth, W. A. Edelstein, and O. M. Mueller, *Mag. Res. Med.* **7**, 319-336 (1988).

9. P. Pfändler, G. Bodenhausen, B. U. Meier, and R. R. Ernst, *Anal. Chem.*, **57**, 2510, (1985).

10. O. Lichtarge, C. W. Cornelius, B. G. Buchanan, and O. Jardetzky, *Proteins* **2**, 340-358 (1987).

11. T. J. Harner, G. C. Levy, E. J. Dudewicz, F. Delaglio, and A. Kumar, Artificial Intelligence, Logic Programming, and Statistics in Magnetic Resonance Imaging and Spectrsocopic Analysis, *in* "Artificial Intelligence Applications in Chemistry," T. H. Pierce and B. A. Hohne, ed., pp. 337-349, American Chemical Society, Washington, D. C. (1986).

12. R. R. Ernst, *Adv. Magn. Res.*, 2, (1968).

13. R. R. Ernst and W. A. Anderson, *The Review of Scientific Instruments* **37**, 93-102, (1966).

14. W. Dietrich, B. Frohlich, U. Gunther, and M. Wiecken, *Fresnius Z. Anal. Chem.*, **316**, 227-230 (1983).

15. R. E. Hoffman and G. C. Levy, *J. Mag. Res.* **83**, 411-417 (1989).

16. P. J. M. van Laarhoven and E. H. L. Aarts, "Simulated Annealing: The Theory and Applications," D. Reidel, Dordrecht (1987).

17. F. DiGennaro and D. Cowburn, *J. Magn. Res.*, in press (1991).

18. R. Kumaresan and D. W. Tufts, *IEEE Trans. Acoust., Speech, Signal Processing*, **ASSP-30**, 833-840 (1982).

19. H. Barkuijsen, R. de Beer, W. M. M. J. Bovee, and D. van Ormondt, *J. Magn. Res.* **61**, 465-481 (1985).

20. J. Tang, C. W. Lin, M. K. Bowman, and J. R. Norris, *J. Magn. Res.* **62**, 167-171 (1985).

21. J. Tang and J. R. Norris, *J. Magn. Res.* **78**(1), 23-30 (1988).

22. A. E. Schussheim and D. Cowburn, *J. Magn. Res.* **71**, 371-378 (1987).

23. H. Gesmar and J. J. Led, *J. Mag. Res.* **83**, 53-64 (1989).

24. J. Gorcester and J. H. Freed, *J. Magn. Res.*, **78**(2), 292-301 (1988).

25. J. Glushka and D. Cowburn, *J. Am. Chem. Soc.* **109**, 7879-7888 (1987).

26. D. Cowburn and J. Chaudhary, *Quantum Chemistry Program Exchange Notes*, **9**, 97 (1989).

27. G. M. Clore, A. Bax, P. Wingfield, and A. M. Gronenborn, *FEBS Letters* **238**, 17-21, (1988).

28. V. F. Bystrov, "Progress in NMR Spectroscopy" **10**, 41-81, Pergamon Press (1976).

29. A. J. Fischman, D. H. Live, H. R. Wyssbrod, W. C. Agosta, and D. Cowburn, *J. Am. Chem. Soc.* **102**, 2533-2599 (1980).

30. D. Cowburn, D. H. Live, A. J. Fischman, and W. C. Agosta, *J. Am. Chem. Soc.*, **105**, 7435-7442, (1983).

31. G. T. Montelione and G. Wagner, *J.A.C.S.* **111**, 5474-5475, (1989).

32. D. Neri, G. Otting, and K. Wüthrich, *J. Am. Chem. Soc.* **112**, 3663-3665 (1990).

33. G. W. Vuister and R. Boelens, *J. Mag. Res.*, **73**, 382 (1987).

34. D. H. Live, D. G. Davis, W. C. Agosta, and D. Cowburn, *J. Am. Chem. Soc.* **106**, 6104-6105, (1984).

COMPUTATIONAL ASPECTS OF MULTINUCLEAR NMR SPECTROSCOPY OF PROTEINS AT NMRFAM[¶]

John L. Markley, Prashanth Darba, Jasna Fejzo[†],
Andrzej M. Krezel, Slobodan Macura[‡],
Charles W. McNemar, Ed S. Mooberry, Beverly R. Seavey,
William M. Westler, and Zsolt Zolnai*

National Magnetic Resonance Facility at Madison
and Biochemistry Department
College of Agricultural and Life Sciences
University of Wisconsin
420 Henry Mall, Madison, WI 53706
[†]Institute of Physical Chemistry, University of Belgrade
11000 Beograd, POB 550, Yugoslavia
[‡]Department of Biochemistry and Molecular Biology
Mayo Foundation
200 First St. SW, Rochester, MN 55905
*Mathematical Institute, Knez Mihajlova 35
11000 Beograd, Yugoslavia

ABSTRACT

In recent studies, we have explored various strategies for using isotope labeling in conjunction with 1D, 2D, and 3D NMR spectroscopy to obtain information about how proteins work. With larger proteins, a concerted approach to the assignment of protein spectra that makes use of all available information about through-bond and through-space connectivities appears to provide the most reliable results. We summarize here various ways in which we have been using computers to process and analyze data and to derive information about protein structure and dynamics. We also describe in brief our protein NMR database project.

[¶]This work was supported by grants RR02301, GM35976, and LM04958 from the U.S. National Institutes of Health, grant 88-37262-3406 from the U.S. Department of Agriculture, and a grant from NEDO, Japan.

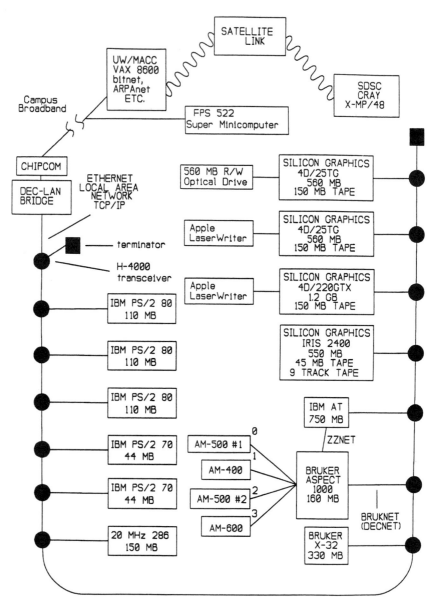

Figure 1. Diagram of the local area network in NMRFAM, its connections to the campus broadband network, and satellite link to a remote supercomputer (San Diego Supercomputer Facility). Shown are the NMR spectrometers (Bruker AM-600, AM-500 #1, AM-500 #2, and AM-400), computer workstations, personal computers, and peripherals.

HARDWARE AND SOFTWARE CONSIDERATIONS

Our laboratory, the National Magnetic Resonance Facility at Madison (NM-RFAM), contains a heterogeneous mixture of NMR spectrometers, workstations, and computers purchased over the past few years (Fig. 1). We have had to face the serious problem of rapid obsolescence of computer workstations. The useful lifetime of workstations currently is under five years without costly upgrades. Our first design goal has been that computers associated with spectrometers should be dedicated to data collection and that data processing and analysis should take place offline on appropriate workstations. Our NMR spectrometers have Bruker AM consoles each with an Aspect 3000 computer. Data are routed from these by fiber-optic cables ("Lightnet") to a central Aspect 1000 computer. Data can be off-loaded directly from the Aspect 1000 onto 1/2" magnetic tape for storage. The Aspect 1000 cannot be set up conveniently as a node for transferring data to other computers. Our solution to this problem has been to develop a parallel data transfer port with associated software (ZZNET)[1] between the Aspect 1000 and a personal computer (IBM AT clone) that has an Ethernet interface, PC NFS software, a 760 Mb Winchester disk, and a 1/4" cartridge streaming tape (Maynstream) for archival storage. A wiring diagram has been presented for the adapter we use to link the Aspect 1000 to a commercially available parallel port for Intel 80×86 computers[1]. The ZZNET software allows for the transfer of data from the Aspect 1000 through the personal computer (PC) to any other node on our local area network. An annoying problem with this arrangement is that the Winchester disk fills up frequently with intermediate data from users. The proposed solution will be a Sony 560 Mb optical read/write disk for the PC which will enable users to store their data on optical disks; the 760 Mb Winchester will then become a scratch disk.

A second general design goal was to have multiple workstations of a given kind so as to minimize the number of different operating systems and computer programs that need to be learned by a proficient user and to minimize maintenance problems. Our initial workstation configuration consisted of two Aspect 1000 workstations. Since each Aspect 1000 was equivalent to the computer in each spectrometer console, users did not have to learn additional commands for processing data. A serious disadvantage with the Aspect 1000 is that it is most efficiently programmed in an arcane assembly language. Unfortunately, the Aspect 1000 workstations became obsolete without any upgrade possibility. Our largest problem with the concept of all equivalent workstations has been that computers needed for molecular graphics and structure calculations were so expensive that we never had sufficient funds to purchase more than two or three of them.

On the other hand, user demand was for several times as many workstations.

As a compromise, we adopted a two-tier strategy in which we use personal computers for as many purposes as possible and then purchase high-end workstations to carry out more sophisticated operations. IBM PC-AT clones operating under *MS-DOS* can be used for word processing, for simple computations, and for display and analysis of 1D, 2D, and 3D (slices) data sets. These machines are cost effective and offer an upgrade path (80286, to 80386, to 80486). Their peripherals also are inexpensive. For NMR data analysis and display on PC's we use the PC version of *FELIX* (Hare Research) and *PIXI*[1] (described below). The PC's have shown themselves to be adequate for most routine applications. Each has an Ethernet interface and PC NFS (Sun Microsystems) for file transfers. Each can serve as a terminal for communication with the *UNIX* workstations in our laboratory or with any of the campus computers on the broad-band network (or off-campus supercomputer facilities). Array processor boards (model PL 800, Eighteen Eight Laboratories) have been installed on three of the PC computers. These enable Fourier transforms to be carried out with processing times that are competitive with those on the more expensive workstations.

High-end workstations with good graphics are essential for molecular graphics and for structural calculations. They also are excellent, if available, for processing and displaying NMR data. Our choice for high-end workstations has been Silicon Graphics (SGI). (We still have one of their earlier Turbo workstations, which has been superseded by newer models with RISC architecture and with completely different software requirements). SGI workstations installed in the laboratory include two Personal IRIS workstations (IRIS 4D/25TG with a 550 Mb disk, 1/4" 150 Mb cartridge tape drive, and 24 Mb cpu memory) and a two-processor Power Series IRIS workstation (IRIS 4D/220GTX). A Bruker X32 data station, which runs under the AT&T version of *UNIX*, is used for processing 2D NMR data and NMR microimages. A Sony optical read/write disk drive (280 Mb on each side) is connected to one of the personal IRIS workstations. Two Apple Laserwriters, which are physically connected to IRIS workstations but set up as nodes on the network, handle remote printing from any computer on the LAN.

We run a variety of programs on the general-purpose *UNIX* workstations: distance geometry (*DSPACE, DISMAN*[2], *DIANA*[3]), or restrained dynamics (*AMBER*[4], *DISCOVER* [Biosym Technologies, Inc.], molecular modeling (*INSIGHT* [Biosym Technologies, Inc.], *MIDAS*[5]), NMR processing and analysis (*FELIX, MADNMR*).

In addition, we have access to a Floating Point Systems model 522 mini-supercomputer, which is shared with several other biophysicists on campus.

We use this computer mainly for molecular mechanics and dynamics calculations.

NMR DATA PROCESSING AND POSTPROCESSING

We recently demonstrated that use of Hilbert transforms in phasing of 2D and 3D NMR data can reduce computer memory requirements and save considerable processing time[6]. Various routines have been developed for postprocessing of NMR data to remove noise[7], flatten the baseplane by a spline method[8], and compress the data without losing information[9].

The *ZOOM* routine[10] in which one cuts out a piece of a 2D spectrum, inverse Fourier transforms, zero fills, and Fourier transforms, has proved useful for achieving the level of digital resolution needed for accurate measurements of NOE cross peak intensities and coupling constants in proteins. We have used this approach to measure three-bond 1H-13C coupling constants in the 19 kDa protein flavodoxin by a 2D ^{1}H$\{^{13}$C$\}$ NOE experiment[11] and in a 10 kDa cytochrome by a 3D ^{1}H$\{^{13}$C$\}$ ^{1}H-^{1}H-TOCSY experiment.[12]

Heteronuclear 2D and 3D NMR spectroscopy are rapidly becoming the methods of choice for NMR studies of larger proteins that can be labeled uniformly with ^{13}C and ^{15}N.[13] We have implemented these pulse sequences on standard Bruker hardware with the addition of an external amplifier on the X-nucleus channel for decoupling. Bruker EXE files, created from a PASCAL program, are used to automatically collect 2D planes with incrementation of the dwell time in the "vertical" dimension. Typical data sizes are from 24 to 96 Mb. Data collection rates on the AM consoles are slowed by wasteful overhead time used for disk I/O; this problem has been alleviated by incorporating an "AM timer box" from Tschudin Associates. This hardware eliminates the disk overhead and reduces the collection time significantly. Other hardware modifications include a third channel based on the Bruker X-nucleus decoupler and hardware GARP decoupling boxes from Tschudin Associates. Modification of the console for fast RF switching has been developed in house.

One of the major stumbling blocks in the early days of 3D NMR was the lack of commercial software for the three dimensional transformation of the data. Our first approach was to write a PASCAL program for the Aspect 1000 to carry out an FFT of the third dimension[14]. Even though the FFT code was optimized by using a radix 4 transform, this approach proved to be too slow since the calculation was done without the aid of the Bruker array processor. The program tied up too much valuable workstation time (10 h or more for each data set). To get around this, we wrote a processing program for the 8 Mflop array processor on the PC computer[15]. This proved to be efficient and useful; the FFT that required 10 h on the Aspect 1000 could be done in about 10 min on a PC AT clone with the array processor. In the meantime, 3D (and higher dimension) processing routines (in *FELIX* and *MADNMR*[16] have become available for the *UNIX* workstations.

SPECTRAL ANALYSIS AND ASSIGNMENT

Personal computers. We developed a fast, multi-dimensional NMR spectral analysis program (*PIXI*) that runs on personal computers built around the Intel 80×86 family of microprocessors with EGA compatible video graphics controllers[1]. This program enables us to display, pick peaks, and analyze 2D or slices of 3D spectra on the various personal computers in the lab. The program operates on compressed data sets that can be extracted from spectra processed by any NMR processing software package. Up to eight different spectra can be loaded simultaneously, and each spectrum can be expanded into nine regions. All of the eight full spectra and the 72 expanded regions are accessible for rapid display with three or fewer keystrokes. Features of *PIXI* include contour plots with up to 16 levels for all expansions, overlaying of up to four spectra or expansions, peak picking with storage to a disk file, and the simultaneous display of symmetric regions of spectra.

Silicon Graphics workstations. An interactive NMR processing and analysis package (*MADNMR*) [16] has been developed that operates on the Silicon Graphics 4D IRIS workstation family. The package includes FFT processing of 1D, 2D, or 3D data or linear prediction in one dimension. Spectra can be viewed in a two-dimensional window while interactive phase corrections are applied along a third dimension. The program takes approximately 75 minutes to process a 3D-NMR data set of the dimensions 256×64×256 to a final size of 512×64×512. This is equivalent to a 1024×128×1024 phase sensitive transform along the three time domains.

Processed data can be displayed in various modes; multiple spectra can be overlaid or plotted side-by-side in different orientations so as to assist in the concerted assignment of NMR data[13]. Manual and automatic peak-picking and volume computation facilities are included. Routines have been written that sort the picked peak files and display trial spin system networks or sequential connectivities against contour plots of the original data.

Visualization and analysis are major bottle-necks in working with nD-NMR data. *MADNMR* contains a graphics program called *3dview* which has a suite of graphics tools to visualize processed nD spectra. Some of these include: rapid display of all or a sub set of 2D-planes from a nD-data in mono or stereo views, selective expansion of spectrum from a desired volume of 3D-data, and display of intensity levels across different 2D-planes with a wide choice of colors. These tools enable quick inspection of processed data for determining the quality of phasing without the need for generating plots.

The first step in automated analysis procedure is peak picking. *MAD-NMR* has automated peak picking routines for 2D (*search2D*) and 3D (*search-3D*) data sets. In the *search2D* routine, a grid search procedure locates points whose intensity is greater than the intensities of the eight neighboring points surrounding it. In the *search3d* routine, each point in the 3D spectrum is checked to determine if it is the largest of all the points adjacent to it. Imagine a body centered and face centered cube of 3×3×3 points. If the intensity of the point at the body center is greater than the intensities of all the other 26 points surrounding it, then it will be considered as a peak. The algorithm is efficient and uses points from only three adjacent 2D-planes for every search. By taking data as groups of three 2D-planes, the entire 3D spectrum is searched to identify all the peak tops. Intensity-weighted averaging is used to assign the chemical shift values. The output of this program consists of a table containing the location and intensities of the picked peaks. The x, y, z coordinates for every picked peak are reported in terms of data position (data point numbers) and chemical shift (ppm). The user can (a) choose the threshold above which peaks are picked, (b) search specific regions of the 3D spectrum, or (c) ignore diagonal peaks or specified (water) ridges.

The second step in automated analysis procedures is identification of spin system networks for individual amino acids. An interactive search program (*cohosrch*) in *MADNMR* is designed to assist the picking and assignment of COSY- and HOHAHA-type spin systems. An idealized spin system (chemical shifts and connectivities) from a library can be displayed over the region whose peaks are being picked. The idealized spin systems can be derived from standard peptide chemical shifts or from the assignments of a related protein from a protein NMR database. Every peak in the manually picked spin system is compared with peaks in the spin system library, and a tentative assignment table is produced based on this comparison. A user-defined search tolerance (in ppm) is used for deciding the match between picked peaks and peaks in the library spin system. The most likely assignment can be decided on the basis of the number of matched peaks and the figure of merit of the match (computed as the sum of squares of the differences between matched peaks weighted by the total number of matched peaks). As a test of this approach, we have compared the experimentally assigned spin systems of proteins (oxidized *Anabaena* 7120 ferredoxin and *Anabaena* 7120 cytochrome c_{553}) against synthetic spin systems created from a table of amino acid chemical shifts. Most of the spin systems matched reasonably well with the standard values. This approach is being used to develop tentative assignments in the spectrum of the reduced ferredoxin from comparisons of the experimental spin systems of the oxidized protein. The tentative assignments can be tested against additional information from heteronuclear 2D and 3D experiments.

A similar program (*noesysrch*) has been developed for the identification of NOE connectivities. This routine makes use of the total (or partial) chemical shift assignment table. The program is run interactively with the display of the NOESY spectrum. A cross hair cursor is used to select a peak in the spectrum, and the program searches for matched peaks along the ω_1 and ω_2 directions and reports them as possible NOE's to that peak. If the coordinates of a preliminary NMR structure or X-ray structure of the protein (or that of a homologous protein) are available in Protein Data Bank format, then the program will compute the distances for the full list of potential NOE cross peaks.

In a commonly used method for measuring cross peak volumes one has to define the line widths along ω_1 and ω_2 and the footprint of the baseplane of the cross peak. In *MADNMR*, an automated procedure directly computes the volume of cross peaks by numerically integrating the intensities of the points enclosed within the bounds of a contour drawn in accordance with a user-defined threshold above the baseplane. This algorithm works automatically for well resolved cross peaks. However, in the case of partially overlapped peaks, the user has to specify the bounds of the footprint. With this algorithm, it takes less than 10 min to compute the volumes of all cross peaks in a typical 2048×2048 data matrix.

The major stumbling blocks to automated spectral analysis are missing peaks and chemical shift redundancies along one dimension. The latter necessitate the analysis of tree structures that can so complicated as to be insoluble by currently available computational methods. The obvious remedies for these problems are to use higher dimension spectra to minimize the redundancies and to collect redundant data sets so that peaks in a particular data set can be inferred from others.

STRUCTURE CALCULATIONS AND REFINEMENT

A semi-automated routine for the Aspect 1000 computer has been developed for use in determining interproton distances from NOESY and ROESY data obtained at different mixing times.[17] The program has been used to determine accurate distances in turkey ovomucoid third domain.[18]

One of the noesysrch utilities of *MADNMR*[16] prepares an input file of NOE constraints to be used with different NMR structure calculation program. In setting up data for distance calculations, it is useful to be able to handle various input data options: actual distance constraints, numbers of contour levels, or intensities of cross peaks. The interconversion routine can handle any of these cases. Correspondence between contour level numbers and distance constraints can be defined by the user. Intensities can be converted to distances using the proportionality between NOE cross peak intensity and the inverse of the sixth power of interproton distances. The program can sort constraints and eliminate duplications. Distances can be grouped into a number of ranges, with the range boundaries defined by the user. Parameters unique to a given program, such as the force constants used in the molecular dynamics program *DISCOVER* or the correction for pseudo-atoms in the variable target routine *DISMAN*, can also be set up with this routine.

Structure calculations typically involve several steps. By using an initial set of distances a family of NMR structures is computed with one type of distance geometry algorithm (e.g., *DISMAN*). These preliminary structures are used with *noesysrch* or other routines to check assignments and assist in the identification and assignment of additional NOE cross peaks that yield more distance constraints for the next round of refinement. Data from other kinds of experiments (e.g., coupling measurements for computing dihedral constraints) provide additional distance constraints. Calculations are repeated with the expanded set of constraints as the refinement progresses.

Most of the NMR structure calculation programs have very specific requirements for atom naming conventions as well as input data format. If one wishes to repeat a structure computation with a different algorithm (e.g., *DSPACE*), it is ordinarily a tedious task to recreate all the input files with different formats. We have written a general program that can be used to translate one format to another[19] This program interconverts data from *DSPACE*, EMBED/VEMBED, *DISMAN*, *DIANA*, and *DISCOVER* and can be easily modified to include additional formats. The program has an interface to *MADNMR* that facilitates the construction of distance matrices.

We have explored methods for separating out various components of exchange spectroscopy so that they can be interpreted in terms of pure cross relaxation (for distance measurements) or pure chemical exchange (for studies of dynamics). We developed and evaluated a method that involves the measurement of NOE build-up in the laboratory and rotating frames and fitting of results to a quadratic approximation to the full relaxation matrix.[17] It is our impression that this approach, given current data quality, produces distances that are about as good as those from the full relaxation matrix approach. The approach was used to measure about 90 distances in OMTKY3 (turkey ovomucoid third domain, M_r 6,000); these were compared with those available from an X-ray structure.[18]

Another concern of ours has been the separation of cross-relaxation and chemical-exchange information in exchange spectra. One simple solution is to use ^{13}C labeling and to obtain exchange information from that nucleus: either by observing ^{13}C directly or by indirect detection of a ^1H nucleus coupled to the ^{13}C. Several pulse sequences that can be used in such an approach were tested and compared[20]. Another approach is to make use of the fact that cross relaxation in the rotating and laboratory frames have opposite sign. A pulse sequence was developed by computer simulation of the system according to average Hamiltonian theory. The method has been applied to two proteins to obtain "pure exchange spectra" that show signals only from residues that undergo slow internal motions.[11,21].

The converse problem is to remove chemical exchange effects from cross-relaxation spectra. In situations where there is only pure chemical exchange and pure cross relaxation, cross peaks for chemical exchange can be removed by decoupling. A method for doing this has been developed and tested on OMTKY3[22,23]

Recent papers in the literature have dealt the effects of internal motions on NMR cross relaxation and structure calculation. The issue of hybrid chemical exchange and cross relaxation (chemical exchange mediated cross relaxation) has been raised, but few experimental data are available to evaluate the problem. We have used OMTKY3 to acquire such data; the molecule is convenient for this purpose since the internal rotation rate of Tyr^{31} can be controlled by changing the pH and/or temperature.[23] Our results show clearly that unrecognized internal motions can lead to significant distortions in structures obtained by standard distance-geometry methods; the distortions can be both local (near the site of internal motion) or global (transmitted away from mobile groups).[24] The time scale of motions that produce such distortions is intermediate between the faster motions studied by molecular dynamics and slower motions that give rise to structural heterogeneity. Thus in refining NMR solution structures, it will be advisable to test for the presence of such motions and to correct the cross relaxation data to remove their adverse effects.

By forming linear combinations of rotating-frame and laboratory-frame magnetization exchange data, one can obtain "direct cross-relaxation spectra" (D.NOESY) in which, to a first approximation, spin-diffusion effects have been removed.[24] Such an approach may be useful for obtaining initial solution structures that can later be refined by a more sophisticated analysis of the magnetization exchange data.

PROTEIN NMR DATABASE PROJECT

We have adapted earlier ideas on a protein NMR database[26] to a relational database format and have begun its implementation.[27] The database (Bio-MagResBank) is an attempt to catalog sequence related NMR data on peptides (12 or more residues) and proteins for which one or more atom-specific assignment have been made. Data files will include primary NMR parameters (e.g., chemical shifts, coupling constants, relaxation rates, and cross relaxation rates) and derived quantities (e.g., pK_a values, hydrogen exchange rates, binding constants, and molecular structures). Attention is being paid to associating relevant experimental information with the results (e.g., origin of the peptide or protein, solution conditions, referencing standards, protocol for the NMR experiment). The database will provide cross references to entries for the same molecule in other databases such as the Protein Data Bank, Protein Identification Resource, and Chemical Abstracts Service. A useful feature of a relational database is that new entities can be added without rearranging preexisting work. This will enable us to modify the database to keep up with new developments in the protein NMR field. Although we currently are storing cross relaxation data and lists of distance constraints, we have decided to wait for the development of more of a consensus on protocols before entering 3D coordinates derived from NMR data.

In recent months, the database progressed from a small prototype created to test the design into a more comprehensive, organized collection. The current database contains information from over 800 papers. Information can be retrieved indexed by authors, protein, organism, sequence, amino acid, or any other characteristic described in the database.

The prototype database was started under the *INGRES* relational database management package on the campus VAX computer. Because *INGRES* is not supported on Silicon Graphics or PC computers, we have converted the database to another relational database package (*ORACLE*). Our goal is to begin releasing the database in late 1991.

SOFTWARE REQUESTS

Requests for software developed at NMRFAM should be addressed to: Operations Assistant, NMRFAM, Biochemistry Department, University of Wisconsin, 420 Henry Mall, Madison, WI 53706.

ACKNOWLEDGMENTS

The authors thank colleagues who have contributed to the work reviewed here, in particular, Arthur S. Edison, Elizabeth A. Farr, Byung-Ha Oh, John B. Olson, Michael D. Reily, Brian J. Stockman, Eldon L. Ulrich, and Micheal H. Zehfus. We also thank the developers of other NMR software and hardware packages that our group uses: Ad Bax, Werner Braun, N. Go, Peter Güntert, Dennis Hare, Timothy Havel, I. D. Kuntz, John Thomasson, Rolf Tschudin, and Kurt Wüthrich.

REFERENCES

1. Z. Zolnai, W. M. Westler, E. L. Ulrich, and J. L. Markley, *J. Magn. Reson.* **88**, 511 (1990).
2. W. Braun and N. Go, *J. Mol. Biol.* **186**, 611 (1985).
3. P. Güntert, W. Braun, and K. Wüthrich, *J. Mol. Biol.*, in press.
4. U. C. Singh, P. K. Weiner, J. W. Caldwell, and P. A. Kollman, *AMBER* (UCSF), Version 3.0, Department of Pharmaceutical Chemistry, University of California, San Francisco, 1986.
5. T. E. Ferrin, C. C. Huang, L. E. Jarvis, and R. Langridge, *J. Mol. Graphics* **6**, 1 (1988); T. E. Ferrin, C. C. Huang, L. E. Jarvis, and R. Langridge, *J. Mol. Graphics* **6**, 13 (1988).
6. Z. Zolnai, S. Macura, and J. L. Markley, *J. Magn. Reson.* **89**, 94 (1990).
7. Z. Zolnai, S. Macura, and J. L. Markley, *Computer Enhanced Spectroscopy* **3**, 141 (1986).
8. Z. Zolnai, S. Macura, and J. L. Markley, *J. Magn. Reson.* **82**, 496 (1989).
9. Z. Zolnai, S. Macura, and J. L. Markley, *J. Magn. Reson.* **80**, 60 (1988).
10. Z. Zolnai, S. Macura, and J. L. Markley, unpublished.
11. B. J. Stockman, A. M. Krezel, J. L. Markley, K. G. Leonhardt, and N. A. Straus, *Biochemistry* **29**, 9600 (1990).
12. A. S. Edison, W. M. Westler, and J. L. Markley, *J. Magn. Reson.*, in press.
13. B. J. Stockman and J. L. Markley, "Stable-Isotope Assisted Protein NMR Spectroscopy in Solution," Advances in Biophysical Chemistry, Vol. 1, pp. 1-46 (Bush, C. A., Ed.), JAI Press Inc., Greenwich, Connecticut (1990); L. P. McIntosh and F. W. Dahlquist, *Quart. Rev. Biophys.* **23**, 1 (1990); S. W. Fesik and E. R. P. Zuiderweg, *Quart. Rev. Biophys* **23**, 97 (1990); L. E. Kay, M. Ikura, and A. Bax, *J. Magn. Reson.* **91**, 84 (1991); and references therein.
14. Z. Zolani, S. Macura, and J. L. Markley, unpublished.
15. W. M. Westler, unpublished.
16. P. Darba, unpublished.
17. J. Fejzo, Z. Zolnai, S. Macura, and J. L. Markley, *J. Magn. Reson.* **82**, 518 (1989).
18. J. Fejzo, Z. Zolnai, S. Macura, and J. L. Markley, *J. Magn. Reson.* **88**, 93 (1990).
19. A. M. Krezel and P. Darba, unpublished.
20. A. T. Alexandrescu, S. N. Loh, and J. L. Markley, *J. Magn. Reson.* **87**, 523 (1990).
21. J. Fejzo, W. M. Westler, S. Macura, and J. L. Markley, *J. Am. Chem. Soc.* **112**, 2574 (1990).
22. J. Fejzo, W. M. Westler, S. Macura, and J. L. Markley, *J. Magn. Reson.*, in press.
23. J. Fejzo, W. M. Westler, S. Macura, and J. L. Markley, *J. Magn. Reson.*, in press.
24. J. Fejzo, A. M. Krezel, W. M. Westler, S. Macura, and J. L. Markley, "Refinement of the NMR Solution Structure of a Protein to Remove Distortions Arising from Neglect of Internal Motion," submitted.
25. J. Fejzo, A. M. Krezel, W. M. Westler, S. Macura, and J. L. Markley, *J. Magn. Reson.*, in press.
26. E. L. Ulrich, J. L. Markley, and Y. Kyogoku, *Prot. Seq. Data Anal.* **2**, 23 (1989); J. L. Markley, B. R. Seavey, A. T. Alexandrescu, P. Darba, A. P. Hinck, S. N. Loh, C. W. McNemar, E. S. Mooberry, B.-H. Oh, B. J. Stockman, J. Wang, W. M. Westler, M. H. Zehfus, and Z. Zolnai, in "Protein Engineering. Protein Design in Basic Research, Medicine, and Industry," M. Ikehara, T. Oshima, and K. Titani, eds., pp. 285-290, Springer verlag, New York (1990).
27. B. R. Seavey, E. A. Farr, W. M. Westler, and J. L. Markley, "A Relational Database for Sequence-Specific Protein NMR Data," submitted.

PRINCIPLES OF MULTIDIMENSIONAL NMR TECHNIQUES FOR MEASUREMENT OF J COUPLING CONSTANTS

Ole W. Sørensen

Laboratorium für Physikalische Chemie
Eidgenössische Technische Hochschule
8092 Zürich, Switzerland

ABSTRACT

The common features of multidimensional NMR experiments suitable for measurement of J coupling constants are described. In the limit of weak or no perturbation of passive spins cross-peak multiplets are constructed in the same way independent of the experiment and only differ in the so-called basic pattern that typically is the response of a two-spin system.

INTRODUCTION

Since shortly after their discovery it has been known that J coupling constants contain most valuable structural information. Unfortunately, the extraction of this information in larger proteins has until recently been difficult and in some cases even impossible because of a devastating ratio between J and linewidth. However, progress in experimental techniques and in particular in the possibility of incorporating ^{15}N and ^{13}C isotopes in the molecules have opened new routes to measurement of J coupling constants.

Even with the new techniques and isotopically enriched molecules the focus seems to be exclusively on homo- and hetero-nuclear three-bond couplings due to their well-known information about dihedral angles through so-called Karplus relations. One- and two-bond coupling constants exhibit small relative conformation-dependent fluctuations. In contrast, the relative fluctuation of four-bond couplings are huge and may change sign. However, until a few years ago these J's were almost impossible to measure because of their small magnitudes. Future efforts in the direction of determining ^{4}J's using the new techniques could turn out to be useful in structure analysis.

In the light of the fact that lectures at this NATO workshop and the recent literature have shown applications of several of the new techniques we

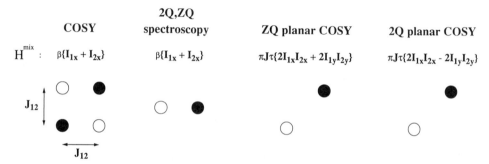

Figure 1. 2D basic patterns and mixing Hamiltonians for various experiments. All vertical and horizontal peak separations are J_{12}. The phases of peaks represented by open and filled circles differ by π.

want in this paper to describe only common features and principles of these methods. The focus will be on homonuclear techniques but the principles also apply for heteronuclear experiments.

THE BASIC PATTERN

We define the basic pattern of a multidimensional spectrum as the cross-peak multiplet of the smallest and simplest spin system that can be observed in the particular experiment. For the sake of simplicity we restrict the description to two-dimensional (2D) experiments. Extension to higher dimensional spectroscopy is easily possible since basic patterns are determined by mixing sequences of which higher dimensional experiments contain several of the same type as in 2D experiments.

Fig. 1 shows basic patterns, all arising from two-spin systems, and required mixing Hamiltonians for different experiments. The ω_1 and ω_2 axis of the 2D spectra are vertical and horizontal, respectively. The coherence transfers underlying the cross-peak multiplet components are illustrated in Fig. 2.

The mixing Hamiltonian $\beta\{I_{1x} + I_{2x}\}$ is realized by a nonselective pulse of flip angle β. In the COSY experiment each of the two magnetizations of spin I_1 is transferred equally well to the two transitions of spin I_2 such that the basic pattern consists of four components. The phases of the resonances are due to the following transformation formulas:

$$I_{1y}I_2^\alpha \xrightarrow{\beta(I_{1x}+I_{2x})} \frac{1}{2}\sin^2\beta\left\{-I_1^\alpha I_{2y} + I_1^\beta I_{2y}\right\} \tag{1a}$$

$$I_{1y}I_2^\beta \xrightarrow{\beta(I_{1x}+I_{2x})} \frac{1}{2}\sin^2\beta\left\{I_1^\alpha 1 I_{2y} - I_1^\beta I_{2y}\right\} \tag{1b}$$

where only the cross-peak relevant terms are included.

In two-spin systems there is only one double-quantum (2Q) and one zero-quantum (ZQ) transition such that only one resonance frequency occurs in

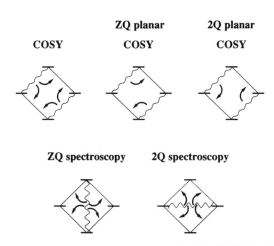

Figure 2. Coherence transfers leading to cross peaks. In COSY and ZQ and 2Q spectroscopy there are no restrictions on the coherence transfer. The 2Q and ZQ planar COSY experiments do only allow transfer between progressively and regressively connected transitions, respectively. Furthermore, the 2Q planar COSY transfers invert the coherence order whereas it is conserved in the ZQ planar COSY experiment.

the ω_1 dimension of 2Q and ZQ 2D spectra. By the mixing pulse the 2Q or ZQ coherence is transferred equally well to both transitions of each spin, and the antiphase character in ω_2 follows from the formulas

$$2I_{1x}I_{2y} \pm 2I_{1y}I_{2x} \xrightarrow{\beta(I_{1x}+I_{2x})} \sin\beta\{I_{1x}I_2^\alpha - I_{1x}I_2^\beta \pm \left(I_1^\alpha I_{2x} - I_1^\beta I_{2x}\right)\} \qquad (2a)$$

$$2I_{1x}I_{2x} \mp 2I_{1y}I_{2y} \xrightarrow{\beta(I_{1x}+I_{2x})} \mp\frac{1}{2}\sin 2\beta\left\{I_{1y}I_2^\alpha - I_{1y}I_2^\beta + I_1^\alpha I_{2y} - I_1^\beta I_{2y}\right\} \qquad (2b)$$

where the upper and lower sign applies for 2Q and ZQ coherences, respectively.

The ZQ planar and 2Q planar COSY experiments (introduced by Schulte-Herbrüggen et al.[1]) restrict coherence transfer as outlined in Fig. 2. The mixing Hamiltonian of the ZQ planar COSY experiment is effectively a zero-quantum pulse exchanging populations of the two levels with equal magnetic quantum number. Therefore any of the two I_1 coherences can be transferred to only one of the I_2 coherences giving rise to a two-component basic pattern. In analogy the 2Q planar COSY experiment involves population exchange between the highest and lowest energy level.

An interesting feature of the planar COSY experiments relates to the sign of the coherence orders. The upper and lower energy level of any of the coherences in the ZQ planar COSY part of Fig. 2 retain their orientation with respect to the vertical energy axis after the coherence transfer. More formally this means that the coherence order is unchanged and peaks are observed only in the antiecho part of the 2D spectrum. In contrast the upper level (lower level) of a coherence in the 2Q planar COSY experiment becomes the lower level (upper level) after the coherence transfer process. In other words cross peaks are associated with inversion of coherence order and appear thereby only in the echo part of the 2D spectrum. Diagonal peaks, on the other hand, experience no inversion of coherence order and show up only in the antiecho part. Hence application of the 2Q planar COSY sequence generates a 2D echo spectrum without diagonal peaks.

Experimentally the planar COSY experiments can be realized by the following mixing sequences with the I_1 and I_2 pulses applied simultaneously.

ZQ planar COSY:

$$\begin{cases} I_1: \left(\frac{\pi}{2}\right)_y - \frac{\tau}{2} - (\pi)_y - \frac{\tau}{2} - \left(\frac{\pi}{2}\right)_y \left(\frac{\pi}{2}\right)_x - \frac{\tau}{2} - (\pi)_x - \frac{\tau}{2} - \left(\frac{\pi}{2}\right)_x \\ I_2: \left(\frac{\pi}{2}\right)_y - \frac{\tau}{2} - (\pi)_y - \frac{\tau}{2} - \left(\frac{\pi}{2}\right)_y \left(\frac{\pi}{2}\right)_x - \frac{\tau}{2} - (\pi)_x - \frac{\tau}{2} - \left(\frac{\pi}{2}\right)_x \end{cases}$$

2Q planar COSY:

$$\begin{cases} I_1: \left(\frac{\pi}{2}\right)_y - \frac{\tau}{2} - (\pi)_y - \frac{\tau}{2} - \left(\frac{\pi}{2}\right)_y \left(\frac{\pi}{2}\right)_x - \frac{\tau}{2} - (\pi)_x - \frac{\tau}{2} - \left(\frac{\pi}{2}\right)_x \\ I_2: \left(\frac{\pi}{2}\right)_y - \frac{\tau}{2} - (\pi)_y - \frac{\tau}{2} - \left(\frac{\pi}{2}\right)_y \left(\frac{\pi}{2}\right)_{-x} - \frac{\tau}{2} - (\pi)_{-x} - \frac{\tau}{2} - \left(\frac{\pi}{2}\right)_{-x} \end{cases}$$

It should be noticed that the second part of the 2Q planar COSY sequence involves separate selective irradiation of the two spins I_1 and I_2.

Finally, the socalled bilinear COSY experiment has a mixing Hamiltonian equal to the average of the ZQ and 2Q planar Hamiltonians (i.e., $\pi J \tau 2 I_{1x} I_{2x}$) and can be understood as simultaneous ZQ and 2Q pulses. This allows diagonal peaks to occur in the echo part and the echo or antiecho selectivity of the planar COSY experiments is lost, but the basic pattern is unchanged.

THE PASSIVE SPINS

In this section we address the problem of what happens when the pulse sequences of the previous section are applied to larger spin systems. The spins I_3, I_4, ... are called passive for the $I_1 \rightarrow I_2$ cross peaks. First, we consider the case where the mixing sequences are selective to the spins I_1 and I_2. Then the magnetic quantum numbers of the passive spins are good quantum numbers and the I_1 and I_2 transitions subdivide into independent two-spin systems according to the spin states of the passive spins.

The consequences of this are illustrated in Fig. 3 for three- and four-spin systems. For a three-spin system the two possible spin states (α, β) of I_3 lead to two identical basic patterns separated by a displacement vector $\mathbf{J}_3$. In analogy the four possible spin states $(\alpha\alpha, \alpha\beta, \beta\alpha, \beta\beta)$ for the two passive spins I_3 and I_4 in a four-spin system result in four occurrences of the $I_1 - I_2$ basic pattern according to the displacement vectors $\mathbf{J}_3$ and $\mathbf{J}_4$.

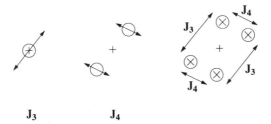

J₃   J₄

Figure 3. The structure of cross-peak multiplets in experiments where the passive spins are unperturbed during the mixing sequence. The actual multiplet pattern is obtained by replacing all crossed circles by the basic pattern of the experiment in question. Such multiplets are commonly referred to as E.COSY-type multiplets. The centers of the multiplets are indicated by a + and the displacement vectors $\mathbf{J}_i$ are explained in the text.

The coordinates of the two-dimensional displacement vector $\mathbf{J}_i$ are determined by the J couplings between spin I_i and the spins active in ω_1 and ω_2. For COSY and the planar COSY experiments with single-quantum coherence we have for the $I_1 \rightarrow I_2$ cross peak:

$$\mathbf{J}_i = (J_{1i}, J_{2i}) \qquad (3)$$

On the other hand for the cross peaks $\{ZQ\}_{12} \rightarrow I_2$ and $\{2Q_{12}\} \rightarrow I_2$ in ZQ and $2Q$ $2D$ spectra, respectively, the coordinates are

$$ZQ: \mathbf{J}_i = (J_{1i} - J_{2i}, J_{2i})$$
$$2Q: \mathbf{J}_i = (J_{1i} + J_{2i}, J_{2i}) \qquad (4)$$

and correspondingly for the transfers $\{ZQ\}_{12} \rightarrow I$ and $\{2Q\}_{12} \rightarrow I_1$.

The cross-peak multiplets in Fig. 3 can be thought of as being constructed in the following way. Starting from the center of the multiplet (e.g., $(\Omega_{I_1}, \Omega_{I_2})$ in COSY-type spectra) basic pattern centers are placed at positions $\frac{1}{2}\mathbf{J}_3$ and $-\frac{1}{2}\mathbf{J}_3$. In the next step each of these centers are replaced by two new centers with relative displacements $\frac{1}{2}\mathbf{J}_4$ and $-\frac{1}{2}\mathbf{J}_4$. This can be continued for any number of passive spins each doubling the number of centers. Note that the multiplet patterns are unchanged under a sign change of $\mathbf{J}_i$ reflecting the fact that they are sensitive only to the *relative* sign of the couplings involved.

The power of this type of techniques lies in the fact that an otherwise invisible coupling J_{2i} can be measured very accurately in, for example, COSY-type spectra if the coupling J_{1i} exceeds the linewidth. In large proteins with broad lines the experimental arrangements with J_{1i} equal to a one-bond $^{13}C - ^1H$ or $^{15}N - ^1H$ coupling allows the measurement of most valuable homo- and hetero-nuclear long-range couplings.

When the passive spins are of a different isotope than I_1 and I_2 the I_1, I_2-selective mixing sequence does not pose any problems and the desired cross-peak multiplet patterns are generated automatically. The situation is different when passive spins are of the same isotope as I_1 and I_2 and selective pulses are not possible or would be too time consuming because of a large number of interesting cross peaks.

If only nonselective pulses shall be applied in homonuclear spin systems there are two possibilities to obtain the desired cross-peak multiplets: (i) Suppression of contributions where the spin states of the passive spins have changed by elaborate linear combination of spectra recorded with different rotation angles β or $\pi J\tau$. In the planar and bilinear COSY experiments this method is only feasible for special cases. (ii) Application of one short rotation angle β or $\pi J\tau$ (typically $<40°$) such that the cross-peak contributions corresponding to unchanged spin states of passive spins dominate in amplitude those with changed spin states.

CONCLUSIONS

In this paper we have described the principles of techniques which are gaining widespread recognition as the methods of choice for measurement of J coupling constants in proteins. Technically, the experiments rely on small or no perturbation of passive spins. The structural features of the cross-peak multiplets resulting from these experiments are essentially identical and only differ in the basic patterns. The most important experimental variants are of the COSY type where pure absorption peakshapes are obtained easily in contrast to the multiple-quantum, bilinear, and planar experiments which also may suffer from spin system dependent delays. A heteronuclear $^{13}C - {}^1H$ or $^{15}N - {}^1H$ one-bond coupling constant must be involved in order to make homo- or hetero-nuclear ^3J or ^4J measurable in ^{13}C and ^{15}N enriched proteins.

The interested reader can find more details about measurement of J coupling constants from E.COSY type multiplet patterns[2], about rotation angle cycles for suppression of undesired multiplet components[3], about multiple-quantum spectroscopy[4,5], and about bilinear and planar COSY experiments[6,1], respectively.

REFERENCES

1. T. Schulte-Herbrüggen, Z. L. Mádi, O. W. Sørensen, and R. R. Ernst, *Mol. Phys.* **72**, 847 (1991).
2. C. Griesinger, O. W. Sørensen and R. R. Ernst, *J. Magn. Reson.* **75**, 474 (1987).
3. C. Griesinger, O. W. Sørensen, and R. R. Ernst, *J. Chem. Phys.* **85**, 6387 (1986).
4. L. Braunschweiler, G. Bodenhausen, and R. R. Ernst, *Mol.Phys.* **48**, 535 (1983).
5. R. R. Ernst, G. Bodenhausen, and A. Wokaun, "Principles of Nuclear Magnetic Resonance in One and Two Dimensions", Clarendon Press, Oxford (1987).
6. M. H. Levitt, C. Radloff, and R. R. Ernst, *Chem. Phys. Lett.* **114**, 435 (1985).

COMPARISON OF THE NMR AND X-RAY STRUCTURES OF HIRUDIN

G. Marius Clore and Angela M. Gronenborn

Laboratory of Chemical Physics, Building 2
National Institute of Diabetes and
Digestive and Kidney Diseases
National Institutes of Health
Bethesda, MD 20892, U.S.A.

ABSTRACT

The solution structure of hirudin determined by NMR is compared to that of the subsequently determined crystal structure. The well-defined region common to both structures is the core of the N-terminal domain which comprises residues 2-30 and 37-48. The backbone conformation of the two structures is very similar with an atomic rms difference of <1 Å. A number of side chains have essentially identical conformations in the NMR and crystal structures. The majority of side chains, however, are highly surface exposed and disordered both in solution and in the crystal.

INTRODUCTION

Hirudin, a small protein of 65 residues from the leech Hirudo medicinalis, is the most powerful natural anticoagulant known and acts by binding specifically to thrombin, thereby inhibiting the thrombin catalysed conversion of fibrinogen to fibrin (see Markwardt[1,2] and Magnusson[3] for reviews). High resolution solution structures of both recombinant wild type and a Lys-47→Glu mutant were initially solved by NMR in our laboratory[4]. This was followed by an independent NMR structure determination of wild type hirudin[5]. Qualitatively, the two sets of NMR structures appear to be very similar but no quantitative comparison is feasible at this stage as the coordinates from this second NMR structure determination are not available. Eighteen months after publication of the original NMR structure, a 2.3 Å resolution X-ray structure of the complex of hirudin and thrombin was solved by molecular replacement using a highly refined X-ray structure of a-thrombin complexed with D-Phe-Pro-Arg-chloromethyl ketone[6] as a starting model

Computational Aspects of the Study of Biological Macromolecules by Nuclear Magnetic Resonance Spectroscopy, Edited by J.C. Hoch *et al.,* Plenum Press, New York, 1991

57

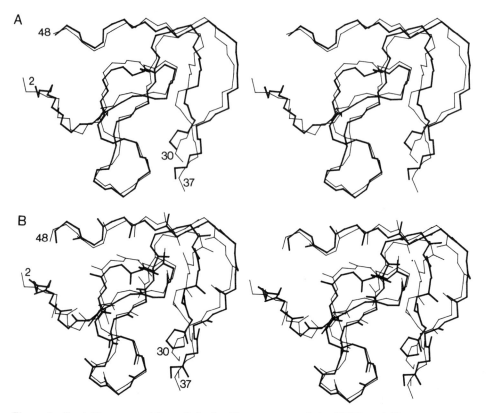

Figure 1. Best fit superposition of the backbone atoms of the NMR and X-ray structures of hirudin. The restrained minimized mean structure $(\overline{SA}_{wt})r$ of wild type hirudin is represented as thick lines, while the X-ray structure is shown as thin lines. N, C^a and C atoms are displayed in (A), while the backbone carbonyl atoms are also shown in (B).

and was partially facilitated by the use of the NMR structure[7]. The recombinant hirudin used for the X-ray work is a variant which exhibits the following changes relative to the wild type hirudin used in the NMR studies: Val-1→Ile, Val-2→Thr, Asp-33→Asn, Glu-35→Lys, Lys-36→Glu, Gln-49→Glu and Asp-53→Asn. In this paper we present a comparison of the NMR and X-ray structures of hirudin.

Hirudin is composed of essentially two domains, an N-terminal globular portion comprising residues 1-48 and a C-terminal tail from residue 49 to 65. The C-terminal domain is disordered in solution as evidenced by the absence of any non-sequential interresidue NOEs. In contrast, in the X-

Figure 2. Best fit superposition of all atoms for two segments of the NMR and X-ray structures of Hirudin. (A) and (C) show superpositions of the 32 individual simulated annealing NMR structures, while in (B) and (D) the restrained minimized mean structure $(\overline{SA}_{wt})r$ of wild type hirudin is represented as thick lines, while the X-ray structure is shown as thin lines. Residues 2-14 and 44-47 are shown in (A) and (B), and residues 16-30 and 37-43 in (C) and (D).

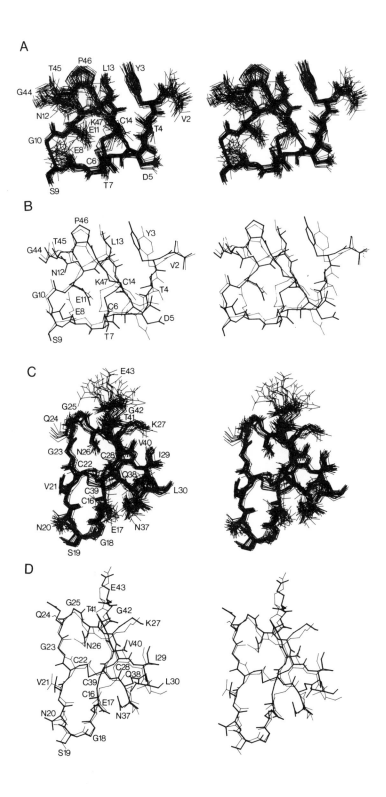

59

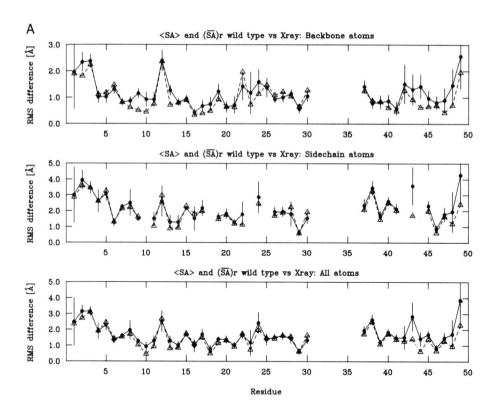

Figure 3. Atomic RMS differences between the X-ray structure and the wild type (A) and Lys-47→Glu mutant (B) NMR structures, and between the two NMR structures (C). In (A) and (B) the triangles represent the RMS difference between the restrained minimized mean NMR structures and the X-ray structure, while the closed circles are the average RMS differences between the individual simulated annealing (SA) NMR structures and the X-ray structure, with the bars representing the standard deviations in these values.

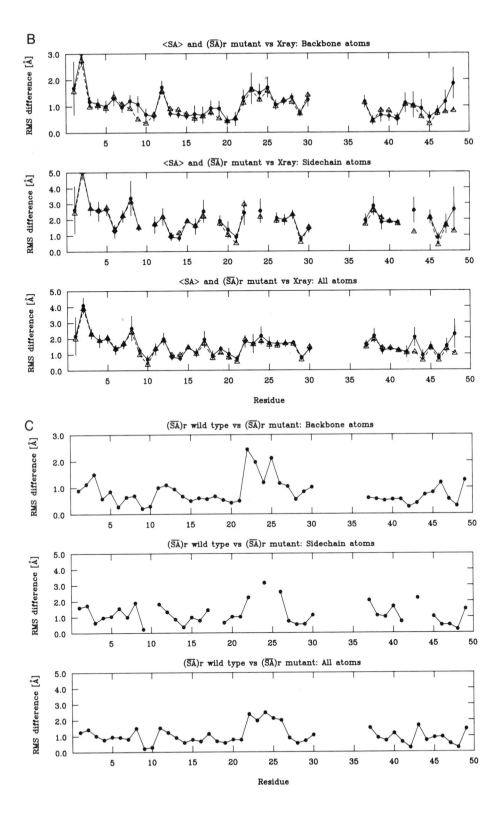

61

ray structure of the hirudin-thrombin complex, the C-terminal tail is well ordered as a result of a large number of interactions with the fibrinogen exosite on thrombin. In the X-ray structure the C-terminal tail consists of two stretches of polypeptide chain, residues 49-54 and 55-60, which are approximately 18 and 9 Å long, respectively, terminating with a helical type III reverse turn (residues 61-64). However, the C-terminal domain has no stabilizing interactions back to the N-terminal domain so that in the absence of fibrinogen, it is hardly surprising that the C-terminal tail is disordered in solution.

In solution, the N-terminal domain (residues 1-48) comprises a compact core stabilized by three disulfide bridges (residues 2-30, 16-28, and 37-48) and an exposed finger of anti-parallel β-sheet (residues 31-36). The orientation of this finger with respect to the core could not be determined in solution due to the absence of any long range NOEs between the exposed finger and the core. In the crystal structure, residues 31-36 are disordered and, with the exception of the backbone nitrogen atom of residue 31 and the backbone C^{α}, C and O atoms of residue 36, are not visible in the electron density map. In addition, residue 1 is disordered in solution but well defined in the crystal as a result of interactions with the primary specificity pocket of thrombin. Consequently, we have restricted our quantitative comparison between the NMR and X-ray structures to residues 2-30 and 37-48.

Backbone and side chain atom superpositions of the NMR wild type and X-ray structures of hirudin (residues 2-30 and 37-48) are shown in Figs. 1 and 2, respectively, while a plot of the atomic rms differences as a function of residue number is depicted in Fig. 3 and a summary of the atomic rms differences between the structures is given in Table 1. The overall atomic rms difference between the X-ray structure and the NMR restrained minimized mean structures for the wild type $(\overline{SA}_{wr})$r and Lys-47$\rightarrow$Glu mutant $(\overline{SA}_{mut})$r hirudins are between 0.9 and 1.0 Å for the backbone atoms, comparable to the backbone atomic rms difference between the two NMR hirudin structures, and about 1.6 Å for all atoms. The deviations between the NMR and X-ray structures of hirudin are also comparable to those reported previously for tendamistat (1.05 Å for the backbone atoms and 1.8 Å for all atoms[8]). Thus, overall, we can conclude that there is little difference in the backbone conformation. Nevertheless, it is evident from the data in Table 2 which presents a comparison of the experimental NMR distance restraints with their respective calculated values in the X-ray structure, that there must be some differences as there are quite a few distances which are violated in the X-ray structure. Interestingly, the large majority of these involve at least one backbone proton in addition to side chain protons. From Fig. 3, one can ascertain that the largest differences between the NMR and X-ray structures involve the N-terminus (residues 2-4), Asn-12 in the loop

Table 1. Atomic Rms Differences between NMR and
X-ray structures [a]

	Atomic Rms Difference For Residues 2-30 and 37-48 (Å)	
	Backbone atoms	All atoms
$< SA_{wt} >$ vs. X-ray	1.07±0.07	1.83±0.08[b]
$\overline{SA}_{wt}$ vs. X-ray	0.88	1.53[b]
$(\overline{SA}_{wt})$r vs. X-ray	0.91	1.63[b]
$< SA_{mut} >$ vs. X-ray	1.08±0.08	1.75±0.08[b]
$\overline{SA}_{mut}$ vs. X-ray	0.92	1.45[b]
$(\overline{SA}_{mut})$r vs. X-ray	0.96	1.58[b]
$< SA_{wt} >$ vs. $<SA_{mut} >$	1.13±0.20	1.62±0.20
$(\overline{SA}_{wt})$r vs. $(\overline{SA}_{mut})$r	0.84	1.18

[a] The notation of the NMR structures is as follows:
$< SA_{wt} >$ are the final 32 final dynamical simulated annealing structures and $< SA_{mut} >$ are the 32 final dynamical simulated annealing Lys-47→Glu mutant hirudin structures obtained by NMR (Folkers et al., 1989); $\overline{SA}_{wt}$ and $\overline{SA}_{mut}$ are the mean structures obtained by averaging the coordinates of the respective individual SA structures best fitted to residues 2-30 and 37-48; $(\overline{SA}_{wt})$r and $(\overline{SA}_{mut})$r are the structures obtained by restrained minimization of $\overline{SA}_{wt}$ and $\overline{SA}_{mut}$, respectively [b] Also excludes Tyr-3.

Table 2. Comparison of the experimental distance restraints with the calculated interproton distances in the X-ray structure

RMS Difference	Violations of Experimental NMR Restraints		
	0.5-1.0 Å	1.0-2.0 Å	>2.0 Å
A. X-ray vs experimental NMR data for wild type hirudin Interresidue			
short ($|i - j| \leq 5$) (187)	0.593	16	27	1
long ($|i - j| >5$) (179)	0.984	22	9	2
Intraresidue (209)	0.415	8	1	1
H-bond restraints (12)	0.091	0	0	0
B. X-ray vs experimental NMR data for Lys-47→Glu mutant hirudin Interresidue			
short ($|i - j| \leq 5$) (174)	0.509	10	18	1
long ($|i - j| > 5$) (172)	0.896	26	15	6
Intraresidue (204)	0.419	7	1	1
H-bond restraints (12)	0.091	0	0	0

[a] The experimental NMR restraints are those of Folkers et al., (1989) and the numbers in parentheses refer to the number of restraints. The 12 hydrogen bonding restraints relate to six hydrogen bonds identified on the basis of a qualitative interpretation of the NMR data. The experimental NMR interproton distance restraints are classified into three distance ranges, 1.8-2.7, 1.8-3.3 and 1.8-5.0 Å, corresponding to strong, medium and weak NOEs. In the case of the NMR structures, there are no violations of the experimental interproton distance restraints larger than 0.5 Å.

closed off at its base by the disulfide bridge between Cys-6 and -14, and the C-terminus (residues 47-49). The differences at the N- and C-termini are hardly surprising as these residues are involved in interactions with thrombin. The large backbone atomic rms difference for Asn-12 is due to a different orientation of the backbone carbonyl group such that it hydrogen bonds with the backbone NH of Cys-22 in the NMR structures and with the backbone NH of Lys-47 in the X-ray structure. This arises from large differences in the ψ angle of Asn-12 and the ϕ angle of Leu-13 which have values of 160°and 65°, respectively, in the X-ray structure compared to values of -60°and -79°, respectively, in the NMR structures. In this respect it is worth noting that the positive ϕ angle for a Leu residue found in the X-ray structure is somewhat unusual, and ϕ angles in this region of conformational space are generally found only for Gly or Asn residues at position 2 of a type II turn. The reason for this difference is not clear. In the NMR structure, the ϕ angle of Leu-13 is fixed by the strong NOE between the Thr-7 $(C^{\alpha}H)$ and Asn-12$(C^{\beta}H)$, while in the X-ray structure, these two protons are 11.7 Å apart.

The conformations of two of the disulfide bridges (Cys-6-Cys-14 and Cys-22-Cys-39) are very similar in the X-ray and NMR structures (Fig. 2). The third disulfide bridge between Cys-16 and Cys-28, appear to be mirror images in the X-ray and NMR structures (Fig. 2D). However, this does not reflect a real difference, as the conformation of this particular disulfide bond is poorly determined in the NMR structures and examination of the individual NMR structures reveals the presence of both conformations (Fig. 2C).

Because the majority of residues have high surface accessibilities[4], it is not surprising that a number of side chains at the surface are found to have different conformations in the crystal and in solution. For example, the side chains of Asp-5 and Thr-7 are clearly well defined from the NMR data (cf. Fig. 2A) and in distinctly different orientations from those present in the crystal structure (Fig. 2B). The same is true of Tyr-3 and the difference there is directly attributable to specific interaction between Tyr-3 and thrombin in the X-ray structure. Another class of side chains have rather large B-values (in the range 55-65 Å^2) in the X-ray structure and are correlated with residues whose conformation is ill-defined in solution (*e.g.*Glu-8, Glu-17, Asn-20, Gln-24, Lys-27, Gln-38, Glu-43). Finally, there are a number of side chains whose conformations are essentially identical in the crystal and NMR structures (*e.g.*Ser-9, Glu-11, Leu-13, Val-21, Asn-26, Ile-29, Leu-30, Pro-46, Lys-47).

In conclusion, the NMR and crystal structures of hirudin are in very close agreement, and most, but not all, the differences are small and due to the uncertainties in both the NMR and X-ray coordinates. However, some of the larger differences can be attributed to conformational changes resulting from the interaction of hirudin with thrombin in the X-ray structure.

ACKNOWLEDGEMENTS

This work was supported by the Intramural AIDS Targeted Antiviral Program of the Office of the Director of the National Institutes of Health. We thank Dr. Alexander Tulinsky for the coordinates of the hirudin crystal structure.

REFERENCES

1. F. Markwardt, *Methods Enzymol.* **19**, 924-932 (1979).
2. F. Markwardt, *Biomed. Biochim. Acta* **44**, 1007-1013 (1985).
3. S. Magnusson, *Enzymes (3rd Ed.)* **3**, 277-321 (1972).
4. P. J. M. Folkers, G. M. Clore, P. C. Driscoll, J. Dodt, S. Köhler, and A. M. Gronenborn, *Biochemistry* **28**, 2601-2617 (1989).
5. H. Haruyama, and K. Wüthrich, *Biochemistry* **28**, 4301-4312 (1989).
6. W. Bode, I. Mayr, U. Baumann, R. Huber, S. R. Stone, and J. Hofsteenge, *EMBO J.* **8**, 3467-3473 (1989).
7. T. J. Rydel, K. G. Ravichandran, A. Tulinsky, W. Bode, R. Huber, C. Roitsch, and J. W. Fenton, *Science*, **249**, 277-280 (1990).
8. M. Billeter, A. D. Kline, W. Braun, R. Huber, and K. Wüthrich, *J. Mol. Biol.* **206**, 677-687 (1989).

THE APPLICATION OF THE LINEAR PREDICTION PRINCIPLE TO NMR SPECTROSCOPY

H. Gesmar and J. J. Led

University of Copenhagen
Department of Chemistry
The H.C. Ørsted Institute
5, Universitetsparken
DK-2100 Copenhagen Ø, Denmark

ABSTRACT

The deviations of the discrete Fourier transform from the theoretical spectrum are shortly reviewed based on the exact analytical expression of the discrete Fourier transform of an exponentially decaying FID. They are aliasing, truncation errors, non-linear phase distortions, and pseudo baseline levels. These discrepancies can be reduced significantly by linear prediction extrapolation of the FID prior to the Fourier transformation. Frequencies, relaxation rates, intensities and phases can also be estimated directly from the FID.

INTRODUCTION

Since the introduction of high resolution pulse NMR[1], spectral estimation of the free induction decay has been based on the discrete Fourier transform (DFT). This method is computationally very efficient, numerically stable, and it preserves all information present in the original free induction decay (FID). In spite of these qualities, a number of new spectral methods has been introduced into the analysis of NMR data over the last years. The reason for this is the deviation of the DFT from the theoretical spectrum: aliasing, truncation errors, non-linear phase distortions, and the appearance of pseudo base line levels. Based on the exact analytical expression of the DFT of an exponentially decaying FID, these deviations are briefly reviewed in the following and it is described how the linear prediction (LP) principle has been applied to NMR data in order to avoid these discrepancies.

Computational Aspects of the Study of Biological Macromolecules by Nuclear Magnetic Resonance Spectroscopy, Edited by J.C. Hoch *et al.*, Plenum Press, New York, 1991

67

THE DISCRETE FOURIER TRANSFORM OF THE FREE INDUCTION DECAY

When a signal $F(t)$ is sampled over a finite period of time only the discrete Fourier transform can be calculated as already mentioned in the introduction, i.e.,

$$\hat{S}(\frac{m}{N\,\Delta t}) = \sum_{k=0}^{N-1} F(\Delta t\,k)\exp(-i2\pi mk/N) \tag{1}$$

Δt being the sampling interval and N the number of sampled points. In the case of exponential decay, the sampled time domain signal has the following form

$$F_k = F(\Delta t\,k) = \sum_{j=1}^{p} I_j\exp((i2\pi\nu_j - R_{2j})(k\,\Delta t + T_{in}) + i\varphi_j) \tag{2}$$

where I_j is the amplitude of the j'th resonance, ν_j the chemical shift, R_{2j} the transverse relaxation rate, and φ_j the phase. T_{in} is the initial delay between the rf pulse and the first point, $F(0)$. Introducing Eq. 2 in Eq. 1, the discrete NMR spectrum $\hat{S}$ can be calculated as the finite sum of a geometrical progression (see ref. 2).

$$\hat{S}(\nu) = \sum_{j=1}^{p} A_j \frac{1 - \exp((i2\pi(\nu_j - \nu) - R_{2j})T_{aq})}{1 - \exp((i2\pi(\nu_j - \nu) - R_{2j})/SW)} \tag{3}$$

where $\frac{m}{N\Delta t}$ has been replaced by the frequency, ν, $N\,\Delta t$ by the acquisition time, T_{aq}, and $\frac{1}{\Delta t}$ by the sweep width ,SW. The symbol A_j is defined as

$$A_j = I_j\exp((i2\pi\nu_j - R_{2j})T_{in} + i\varphi_j) \tag{4}$$

Obviously,the periodic function in Eq. 3 differs significantly from the theoretical spectrum, i.e., the well known sum of complex Lorentzians

$$s(\nu) = \sum_{j=1}^{p} I_j\exp((i\varphi_j)/R_{2j}\frac{1}{1 - i2\pi(\nu_j - \nu)/R_{2j}} \tag{5}$$

However, as demonstrated in ref. 2, the theoretical spectrum in Eq. 5 can be approximated by the discrete spectrum in Eq. 3 under certain ideal conditions: If the acquisition time, T_{aq}, is large, compared to the relaxation time of the slowest decaying component of Eq. 2. ($T_{aq} \gg 1/R_{2j}$) and if the sweep width, SW, is large as compared to the relaxation rates, R_{2j}, as well as to the involved $(\nu_j - \nu)$, Eq. 3 reduces to

$$\hat{S}(\nu) = SW \sum_{j=1}^{p} A_j/R_{2j}\frac{1}{1 - i2\pi(\nu_j - \nu)/R_{2j}} + \sum_{j=1}^{p} A_j/2 \tag{6}$$

Eq. 6 represents the optimum discrete Fourier transform spectrum, i.e., a frequency sampled sum of complex Lorentizians. However, even this optimum differs from the theoretical spectrum in Eq. 5 on two particular points: a constant with the value of $\sum_{j=1}^{p} A_j/2$ has been added to each point, and each of the p Lorentzians has been multiplied by the factor $\exp((i2\pi\nu_j - R_{2j})T_{in})$.

The former of the above mentioned deviations from the theoretical spectrum results in a pseudo baseline in the DFT spectrum. As the value of the baseline level corresponds to half the value of the first point in the FID this pseudo baseline can be removed by multiplying the first point in the FID by 1/2 before performing the Fourier transformation. This is particularly important in the case of two dimensional spectra where the pseudo baselines will lead to F_1 and F_2 ridges[3].

The latter deviation, i.e., the initial factor of $\exp((i2\pi\nu_j - R_{2j})T_{in})$, gives rise to a frequency dependent phase distortion of the DFT spectrum. It is worthy of note that this phase factor does not depend linearly on the frequency, ν. It depends on the center frequencies of the individual Lorentzians. In most cases, when the spectrum does not contain severe overlap, a linear phase correction procedure is able to solve the problem. However, if the involved Lorentzians are broad and poorly separated, or if small signals are situated on the tail of a large solvent signal, the spectrum is only correctly phased at the center of each line, and baseline undulation will occur[4]. Thus, the true spectrum cannot be produced even under ideal sampling conditions, i.e., even when the signal is allowed to relax completely during the sampling period and when the sweep width is large. Obviously, this effect is more pronounced when the initial delay is long. This explains why a prolonged initial delay is rarely used, even though the quality of the first points of the FID is often reduced by overflow in the AD converter, non-linear effects of the amplifier, and distortion due to filtering.

It should be emphasized that the Lorentzian approximation to a resonance signal only holds near the center frequency of this signal (i.e., $(\nu_j - \nu)/SW \ll 1$). In the general case, described by Eq. 3, $\hat{S}$ is a periodic function with the period $SW = \Delta t^{-1}$, i.e., $\hat{S}(\nu) = \hat{S}(\nu + SW)$. This periodicity describes the well known effect of aliasing or zoning[5].

In many cases the DFT spectrum is even far from the optimum given by Eq. 5. When the FID is truncated (i.e., $T_{aq} \simeq 1/R_{2j}$) the j'th Lorentzian is overlaid by a new periodic expression with a period of one point (T_{aq}^{-1}). This will lead to leakage in the spectrum[6], if the frequency is not an integer part of the sweep width (i.e., $\nu_j \neq m_j T_{aq}^{-1}$, m_j being an integer) resulting in an apparently misphased signal (Fig. 1). Furthermore, the relative signal intensity cannot be measured as the area between the curve and the baseline because the pseudo baseline level is altered by factors $\exp((i2\pi(\nu_j - \nu) - R_{2j})T_{aq})$ depending on the individual resonances. The general rule that

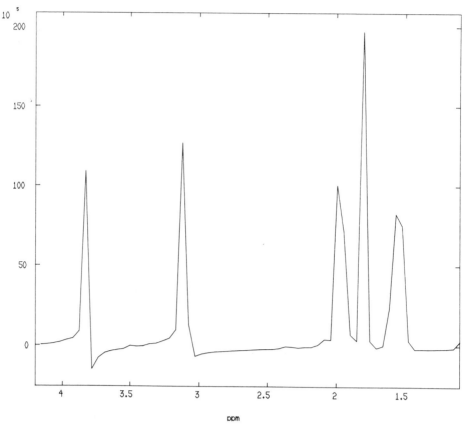

Figure 1. 500 MHz ^{1}H DFT NMR spectrum of lysin. Only the first 512 points of the original 16K FID were included in the DFT. No 'zero filling' was applied.

the first point of the FID equals the sum of the points in the spectrum (except for a constant of proportionality) is still valid of course, but it is of no practical value because frequency and relaxation dependent amounts of intensity have leaked into the pseudo baseline level. Thus the relative areas of the Lorentzian like signals have been altered. This is another important divergence from the theoretical spectrum.

In case of zero filling before the discrete Fourier transformation, the exponential term in the numerator of Eq. 3 causes even more pronounced distortions in the resulting spectrum. This can be seen by letting m assume fractional values in $\nu = mT_{aq}^{-1}$. 'Wiggles' with a period of T_{aq}^{-1} will then appear in the spectrum in connection with the signals whose components in the FID have been truncated (Fig. 2).

As described above the DFT spectrum differs from the theoretical spectrum to a large extent, even in cases where the presence of noise can be neglected. Against these deficiencies a number of alternative spectral methods have been applied to NMR data[7] and among these the LP principle has

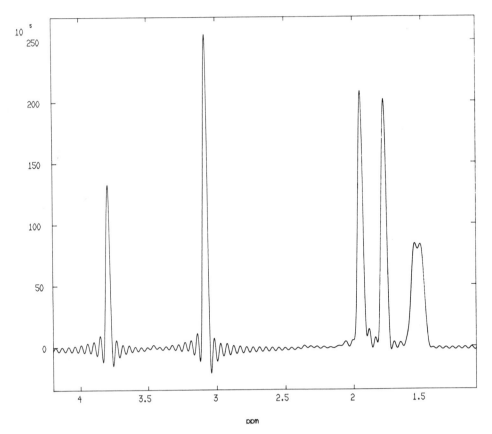

Figure 2. 500 MHz ^{1}H 'zero filled' DFT NMR spectrum of lysin. Only the first 512 points of the original 16K FID were included in the DFT, but it was 'zero filled' until 16K.

proven to be very useful[8-12]. In the following two different applications of the LP principle are described: a qualitative method that focuses on reducing the above mentioned deficiencies of the DFT by manipulating the FID prior to Fourier transformation and a quantitative method that allows the spectral parameters to be determined directly from the FID.

THE QUALITATIVE LP METHOD

When a FID conforming to Eq. 2. is sampled, the resulting function F_k has the following property according to Kay and Marple[13].

$$F_k = \sum_{m=1}^{p} f_m F_{k-m} \qquad (7)$$

disregarding the effect of the noise. Here f_m is the m'th forward prediction coefficient, which is independent of k and therefore can be determined from

71

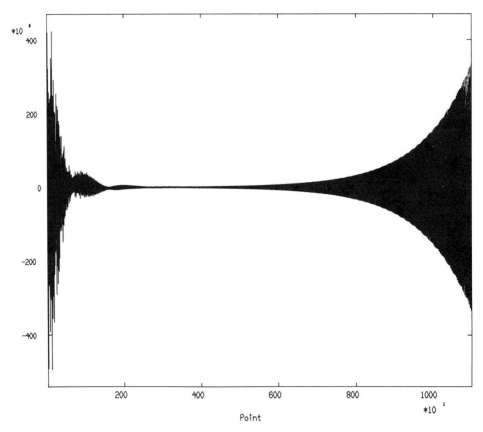

Figure 3. 16K FID, LP extrapolated from the 512 points, the DFT of which is shown in Fig. 1. 150 LP coefficients, determined from 353 equations, were applied (see text). Only the first 11,000 points are shown.

the data points by some linear calculation. The prediction coefficients will be related to the frequencies, ν_j, and the relaxation rates, R_{2j}, through the characteristic polynomial

$$z^p - \sum_{m=1}^{p} f_m z^{p-m} = P(z) \qquad (8)$$

whose roots, C_j, according to Kay and Marple[13], are given by the equation

$$C_j = \exp((i2\pi\nu_j - R_{2j})\Delta t) \qquad (9)$$

When the value of p is unknown the number of prediction coefficients must be increased to a value larger than p in order to assure the determination of all the involved complex exponentials. This results in an excess of roots of the characteristic polynomial in Eq. 8. As shown by Kumaresan[14], however, this set of roots is divided into two classes: one containing p signal roots conforming with Eq. 9 and another containing the extraneous roots, the latter situated inside the unit circle. In the case of forward prediction as in Eq. 7 the signal roots, C_j, are also situated inside the unit circle due to positive values of R_{2j} rendering a separation of the two classes impossible. Only in the case of backward prediction

$$F_k = \sum_{m=1}^{p} b_m F_{k+m} \qquad (10)$$

is a separation of the two classes possible because the signal roots fall outside the unit circle. When noise is present, as also stated by Kumaresan[14], the extraneous roots fall close to the unit circle or even outside.

Tirendi and Martin[11] have demonstrated that the truncation errors described in the preceding section can be reduced considerably by applying the LP principle as an alternative to 'zero filling'. Having calculated the LP coefficients, f_m, the truncated FID is extrapolated until the point of total transverse relaxation, simply by direct application of Eq. 7. This, however, must be done with some care as can be seen in Fig. 3. In this case 150 LP coefficients were determined from 353 equations of the type of Eq. 7, k ranging from 159 to 512. One of the 150 roots of the characteristic polynomial, Eq. 8, and its complex conjugate, fell outside the unit circle. By increasing the amount of averaging, i.e., by increasing the number of equations of the type of Eq. 7 as compared to the number of forward prediction coefficients a solution can be found having all of the characteristic roots inside the unit circle. Thus, when only 130 LP coefficients where determined from 373 equations no zeros were outside the unit circle. Part of the resultant 16K extrapolation is shown in Fig. 4 and the corresponding DFT spectrum in Fig. 5. Comparing Figs. 1, 2 and 5 with the 'genuine' spectrum in Fig. 6 it is seen that by extrapolating the FID before the DFT, the artifacts described in the previous section have been reduced considerably. Furthermore the resolution seems to be enhanced.

The first data points of a FID are often corrupted by the electronic filter of the spectrometer and ought not to be included in the DFT. However, the initial delay T_{in} resulting from this operation would lead to severe phase distortions as pointed out previously. Also, certain experimental conditions, e.g., selective pulses, can introduce significant initial delays. As shown by Marion and Bax[12] the initial data points can be recalculated or even extrapolated back to zero if a number of backward prediction coefficients are determined from equations like Eq. 10.

In Figs. 7 and 8 an example of recalculation of initial time domain data points is shown. The insulin spectrum in Fig. 7 shows a significant curved baseline, although the FID was oversampled, the sweep width being 40 ppm.

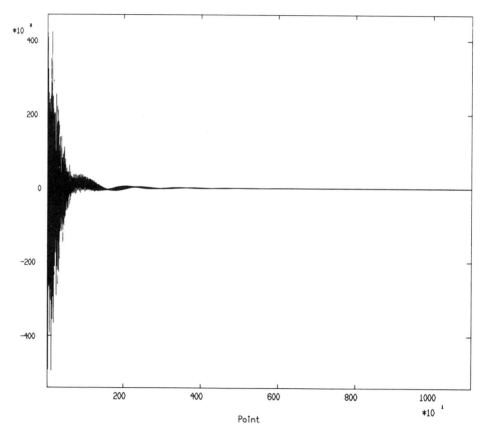

Figure 4. 16K FID, LP extrapolated from the 512 points, the DFT of which is shown in Fig. 1. 130 LP coefficients, determined from 373 equations were applied (see text). Only the first 11,000 points are shown.

In order to correct this baseline, 18 backward LP coefficients were determined from 36 equations of the type of Eq. 10, k ranging from 7 to 42. The initial 6 points of the FID were thus recalculated by backward extrapolation prior to the DFT, resulting in the spectrum in Fig. 8. Obviously the baseline curvature has been removed.

The above example serve to show that deviations of the DFT from the theoretical spectrum can be reduced significantly by performing backward and forward linear prediction extrapolation on the FID, prior to the DFT.

THE QUANTITATIVE LP METHOD

The aim of the quantitative LP approach is to estimate the spectral parameters directly from the FID. The resonance frequencies, ν_j, and the relaxation rates, R_{2j}, are determined from the LP coefficients through Eq. 8 and Eq. 9 and the intensities, I_j, and phases, φ_j, are found by a linear least-squares

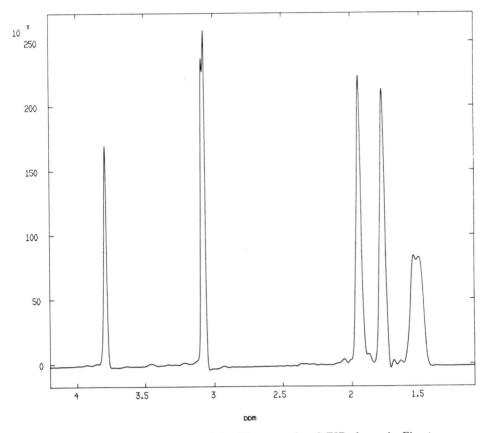

Figure 5. DFT spectrum of the LP extrapolated FID shown in Fig. 4.

(LSQ) calculation based on Eq. 2. This method is based on the work of Baron de Prony[15], further developed by Kumaresan and Tufts[16], and introduced into the field of NMR spectroscopy by Barkhuijsen et al.[8] In the following it is described how the quantitative LP method has been applied to two-dimensional (2D) NMR data. A more extensive description of the quantitative LP method is given by Gesmar et al.[2].

In the two-dimensional case, the free-induction decay is assumed to have the form

$$f(t_1, t_2) = \sum_{j,h} I_{hj} \exp(i\varphi_{hj}) \exp((i2\pi\nu_h^{(1)} - R_{2h}^{(1)})t_1) \exp((i2\pi\nu_j^{(2)} - R_{2j}^{(2)})t_2)$$

(11)

where $\nu_j^{(2)}$, $\nu_h^{(1)}$, $R_{2j}^{(2)}$, and $R_{2h}^{(1)}$ are the frequencies and transverse relaxation rates in the t_2 and t_1 dimensions. I_{hj} is the corresponding amplitude and φ_{hj}, the phase. When the signal is sampled at regularly spaced intervals in

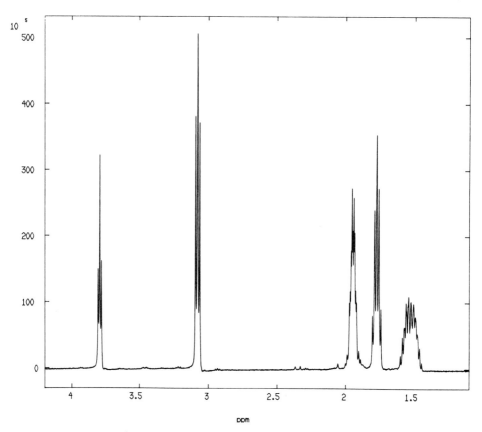

Figure 6. 500 MHz ^{1}H 16K DFT NMR spectrum of lysin. The entire 16K FID was included in the DFT.

the two dimensions, the following discrete function is obtained.

$$F_{l,k} = \sum_{j=1}^{p_2} D_{lj} \exp((i2\pi\nu_j^{(2)} - R_{2j}^{(2)})\Delta t_2 k) \qquad (12)$$

where

$$D_{lj} = \sum_{h=1}^{p_j} I_{hj} \exp(i\varphi_{hj}) \exp((i2\pi\nu_h^{(1)} - R_{2h}^{(1)})\Delta t_1 l) \qquad (13)$$

Δt_1 and Δt_2 are the sampling intervals in the t_1 and t_2 dimensions, respectively. The value of p_2 is the number of resonances in the t_2 dimension, and p_j is the number of correlations of the j'th resonance. For each value of l, the linear-prediction principle, as stated in Eq. 10, can be applied to Eq. 12. Since, furthermore, $\exp((i2\pi\nu_j^{(2)} - R_{2j}^{(2)})\Delta t_2)$ does not depend on l, neither do the backward prediction coefficients. Therefore, the following equation

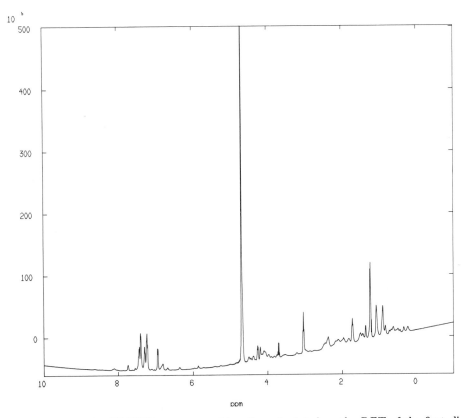

Figure 7. 500 MHz ^{1}H NMR spectrum of insulin, calculated as the DFT of the first slice of a HOHAHA FID. The curved baseline is due to corruption of the first points of the FID in the t_2 dimension.

is fulfilled for all k and l:

$$F_{lk} = \sum_{m=1}^{p_2} b_m F_{l,k+m} \qquad (14)$$

Eq. 14 shows that the entire two-dimensional FID can be included in the determination of the backward coefficients belonging to the t_2 dimension. The linear-prediction order is chosen to be larger than the expected number of resonances in the t_2 dimension. As in the one-dimensional case, described by Gesmar and Led[10], the b_m's are found by Cholesky decomposition of the normal equations, constructed from the overdetermined system of linear equations (Eq. 14). The matrix corresponding to the total set of normal equations is formed by summing the normal equation matrices corresponding to each value of l in Eq. 14 (or t_1 in Eq. 11),i.e., the normal equation matrix corresponding to each individual FID in the t_2 dimension. Also the right hand side of the normal equations is formed by adding the individual

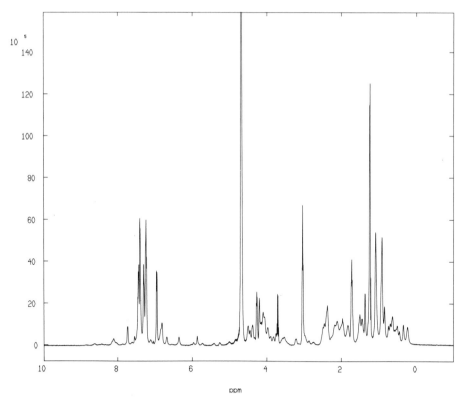

Figure 8. Same spectrum as shown in Fig. 7. The curved baseline has been removed by a LP recalculation of the initial time domain data points, prior to the DFT.

right hand sides. Having found the b_m's, the values of the frequencies, $\nu_j^{(2)}$, and transverse relaxation rates, $R_{2j}^{(2)}$, connected with the t_2 dimension, are determined from

$$C_j = \exp((-i2\pi\nu_j^{(2)} + R_{2j}^{(2)})\Delta t_2) \tag{15}$$

the backward equivalent of Eq. 9. As mentioned in the preceding section the extraneous roots, originating from the excess of linear-prediction coefficients have a tendency of falling inside the unit circle. Because backward prediction is used, the signal roots, C_j, are situated outside the unit circle, due to the positive values of $R_{2j}^{(2)}$ in Eq. 15. Contrary to the case where forward prediction is used, the two classes of roots are thus easily separated, and the extraneous roots are excluded from the following part of the calculation. It should be emphasized that the large-order Cholesky decomposition, which together with the polynomial rooting is the time-consuming part of the procedure (see Gesmar and Led[17]), is only performed once for a given 2D data set.

Since the C_j's are known at this point, the D_{lj}'s in Eq. 12 can be evaluated from another overdetermined system of linear equations, viz.,

$$F_{lk} = \sum_{j=1}^{p_2} D_{lj} C_j^{-k} \qquad (16)$$

The calculation must be carried out for each value of l, but as the coefficients C_j are independent of l, the triangularization of the coefficients matrix is only performed once, thus reducing the number of operations considerably.

According to Eq. 13, D_{lj} itself is a sum of decaying complex exponentials, and as such the linear-prediction principle applies once again

$$D_{lj} = \sum_{m=1}^{p_j} B_{jm} D_{l+m,j} \qquad (17)$$

B_{jm} representing the t_1-backward coefficients belonging to the j'th resonance. For each value of j, a complete linear-prediction calculation and subsequent LSQ intensity determination are applied, and thus estimates of the frequencies, $\nu_h^{(1)}$, and relaxation rates, $R_{2h}^{(1)}$, in the t_1 dimension are produced together with the value of intensity, I_{hj}, and phase, φ_{hj}, connecting the frequencies in the two dimensions. *A complete determination of the spectral parameters of the two-dimensional free-induction decay has thus been achieved.*

In connection with the final t_1 calculation, a few details should be noted. As in the former applications of the linear-prediction principle, the prediction order must be chosen somewhat larger than the expected number of resonances, since D_{lj} is corrupted by noise. The finite variance of D_{lj} is not only caused by the white noise contribution to the individual FIDs in the t_2 dimension, but it is also affected by the t_1 noise in the general case. Because of the limited number of frequency correlations for each t_2 resonance, the determination of the spectral parameters can be accomplished involving only few t_1 values. This is an important advantage over the DFT spectrum in which 'wiggles' and 'ripples', originating from the truncation, would reduce the value of the spectral estimate. However, the resolution of the two-dimensional linear-prediction estimator is enhanced when the number of applied t_1 values is increased.

The applicability of the quantitative 2D LP procedure has been demonstrated by Gesmar and Led[18]. The spectral parameter pairs, $(\nu_j^{(2)}, \nu_h^{(1)})$, $(R_{2j}^{(2)}, R_{2h}^{(1)})$, and the corresponding intensities, I_{hj}, and individual phases, φ_{hj} were estimated from the ^{1}H COSY free-induction decay of threonine in D_2O and DMSO, the Fourier transform of which is shown in Fig. 9a. In the t_2 dimension, 800 real backward prediction coefficients, b_m, were applied,

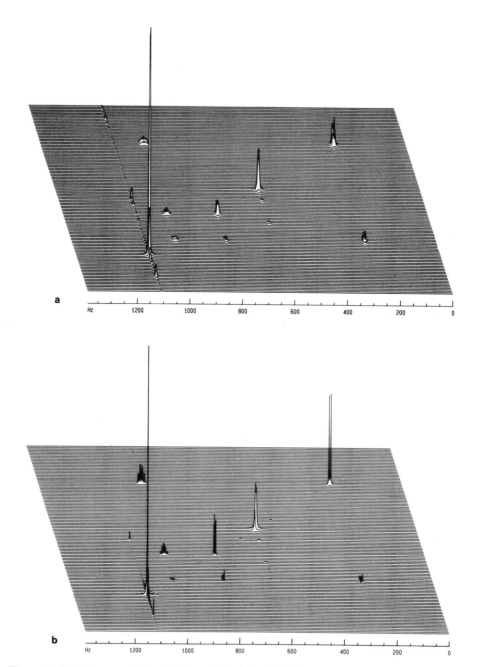

Figure 9. (a) Phase modulated 250.13 MHz ^{1}H 2D DFT COSY power spectrum of threonine in D_2O and DMSO recorded in 172×1024 data points, applying a sweep width of 1392.76 Hz in both dimensions. The FID was premultiplied by an unshifted 2D sine bell before DFT in 128×1024 points. The resonance at 616.5 Hz is due to DMSO (2.5 ppm). (b) 2D LP spectrum of the FID whose DFT spectrum is shown in (a), recalculated from the estimated parameters. (Reproduced by permission of Academic Press from ref. 18.)

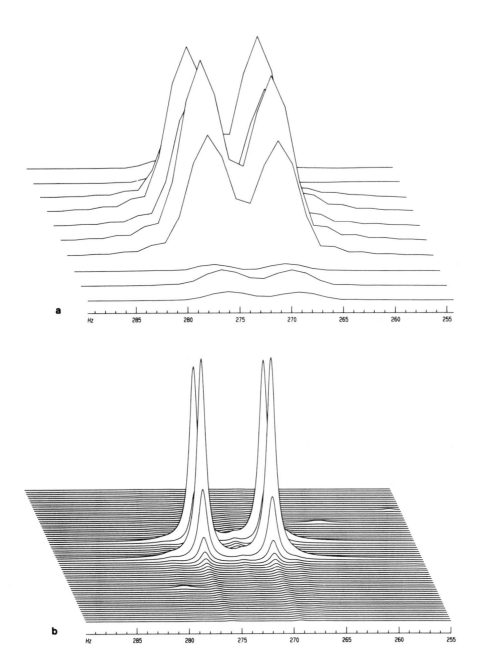

Figure 10. (a) Expansion of the region between 255.0 and 290.0 Hz in both dimensions of the DFT spectrum in Fig. 9a. In this case the FID was 'zero filled' to 250 points in the t_1 dimension after the sine bell filtering. The expansion shows the (CH_3, CH_3) multiplet. (b) LP spectrum of the same region, calculated from the estimated parameters (see ref. 18). (Reproduced by permission of Academic Press from ref. 18.)

and all 172 rows of the data matrix were included in the normal equations corresponding to Eq. 14. The first 16 points of each row were deleted in order to suppress possible fast, nonexponential decaying transients. Only real numbers were used in this part of the calculation due to the sequential quadrature detection in the t_2 dimension[19]. The solution of the characteristic polynomial Eq. 8 resulted in 114 roots outside the unit circle. As the backward coefficients are real valued, these roots fall in pairs of complex conjugates, thus representing 57 potential signals, i.e., 57 values of $\nu_j^{(2)}$ and $R_{2j}^{(2)}$. The normal equations corresponding to Eq. 16 were also formed on the basis of the entire two-dimensional free-induction decay, and consequently 57 times 172 complex values of D_{lj} were evaluated. For each of the 57 values of j, a complete linear-prediction calculation was performed on the 172 complex values of D_{lj} applying 64 complex backward prediction coefficients. Thus parameters corresponding to 206 signals were determined by the calculation. The apparent excess consists mostly of small, insignificant signals that can be classified as artifacts.

To demonstrate the general quality of the estimated parameters a two-dimensional Lorentzian lineshaped spectrum was calculated directly from these values. This spectrum is shown in Fig. 9b. A more detailed description of this type of spectrum is given by Gesmar and Led[18].

It is important to emphasize that the LP spectrum described above is only a practical way of visualizing a large amount of data. Any specific spectral information must be taken from the parameter tables provided by the LP procedure. In Figs. 10 and 11 smaller sections of the DFT spectrum and the LP spectrum are expanded. The corresponding parameters are listed in ref. 18. As can be seen, the resolution of the LP method is beyond what can be obtained by inspection of the DFT spectrum, although the latter has been zero-filled in both dimensions and premultiplied by a two-dimensional sine bell. It is also worthy of note that the phase estimates are exactly as one should expect: The signals on the diagonal are in-phase while the off-diagonal signals are pairwise in counterphase.

As mentioned above, a number of small extraneous signals is found by the procedure, particularly in the vicinity of intense signals. This is can also be seen in Fig. 10. It should be emphasized that these are not removed by the procedure because they are real signals, in the sense that they conform with Eq. 12. Only physical knowledge of the system under study can separate them from the relevant signals.

CONCLUSION

The qualitative LP method, i.e., backward and forward LP extrapolation of the time domain signal before the DFT, is able to improve the quality of

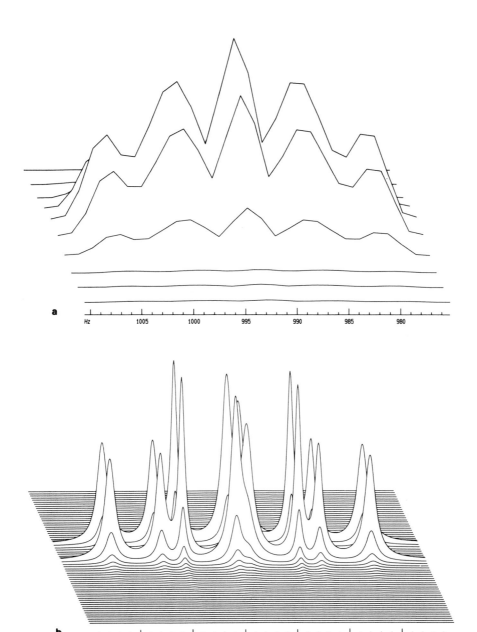

Figure 11. (a) Expansion of the region between 255.0 and 290.0 Hz in the F_1 dimension and 975.0 and 1010.0 Hz in the F_2 dimension of the DFT spectrum in Fig. 9a. In this case the FID was 'zero filled' to 256 points in the t_1 dimension after sine bell filtering. The expansion shows the (β-CH, CH$_3$) multiplet. (b) LP spectrum of the same region, calculated from the estimated parameters (see ref. 18). The signals are pairwise in counterphase, but for reasons of clearness the phases have all been set to zero in the spectrum. (Reproduced by permission of Academic Press from ref. 18.)

NMR spectra considerably. Intensity distortions, non-linear phase problems, 'ripples', and pseudo baselines are of much less significance than in the normal DFT spectra.

The quantitative LP method seems to be able to estimate reliable values of frequencies,transverse relaxation rates, intensities, and phases from both one- and two-dimensional NMR data. It does not suffer from the previously described deficiencies of the DFT. In the 2D case there is no need for phase-cycling procedures in order to avoid dispersion-like lineshapes. Finally, it produces tables of spectral parameters, making further automated analyses possible without the need for peak finding procedures.

The major drawback of the LP method is the extensive computing time. However, development of more dedicated numerical procedures as well as faster computers can undoubtedly reduce this disadvantage considerably in the future.

As a general conclusion, it is fair to say that the LP principle has proven its applicability to spectral analysis of NMR data.

ACKNOWLEDGEMENTS

This work was supported by the Danish Technical Research Council, J.No. 16-3922H, and 16-4679H,the Ministry of Industry, J.No. 85886, and Julie Damm's Studiefond. Free access to the RC 8000 minicomputer at the H.C. Ørsted Institute is also acknowledged. We are grateful to Ms. Majbritt Jørgensen for her care and patience when preparing the typewritten manuscript.

REFERENCES

1. R. R. Ernst, *Adv. Magn. Reson.* **2**, 7 (1966).
2. H. Gesmar, J. J. Led, and F. Abildgaard, *Prog. NMR Spectrosc.* **22**, 255 (1990).
3. G. Otting, H. Widmer, G. Wagner, and K. Wütrich, *J. Magn. Reson.* **66**, 187 (1986).
4. P. Plateau, C. Dumas, and M. Guéron, *J. Magn. Reson.* **54**, 46 (1983).
5. D. I. Hoult and R. E. Richards, *Proc. R. Soc. London, Ser. A* **344**, 311 (1975).
6. E. O. Brigham,*The Fast Fourier Transform*, Chap 6, Prentice-Hall Inc., Englewood Cliffs, NJ (1974).
7. D. S. Stephenson, *Prog. NMR Spectrosc.* **20**, 515 (1988).
8. H. Barkhuijsen, R. de Beer, W. M. M. J. Bovée, and D. van Ormondt, *J. Magn. Reson.* **61**, 465 (1985).
9. J. Tang, C. P. Lin, M. K. Bowman, and J. R. Norris, *J. Magn. Reson.* **62**, 167 (1985).
10. H. Gesmar and J. J. Led, *J. Magn. Reson.* **76**, 183 (1988).
11. C. F. Tirendi and J. F. Martin, *J. Magn. Reson.* **81**, 577 (1989).
12. D. Marion and A. Bax, *J. Magn. Reson.* **83**, 205 (1989).
13. S. M. Kay and S. L. Marple, *Proc. IEEE* **69**, 1380 (1981).
14. R. Kumaresan, *IEEE Trans.* **ASSP-31**, 217 (1983).
15. G. R. B. Prony, *J. de L'Ecole Polytechnique, Paris* **1**, 24 (1795).
16. R. Kumaresan and D. W. Tufts, *IEEE Trans.* **ASSP-30**, 833 (1982).

17. H. Gesmar and J. J. Led, *J. Magn. Reson.* **76**, 575 (1988).
18. H. Gesmar and J. J. Led, *J. Magn. Reson.* **83**, 53 (1989).
19. A. G. Redfield and S. D. Kunz, *J. Magn. Reson.* **19**, 250 (1975).

NMR DATA PROCESSING AND STRUCTURE CALCULATIONS USING PARALLEL COMPUTERS

Wayne Boucher, Andrew R. C. Raine and Ernest D. Laue

Cambridge Centre for Molecular Recognition
Department of Biochemistry
Tennis Court Road
Cambridge CB2 1QW
Great Britain

ABSTRACT

The determination of protein structures from NMR data involves a considerable amount of computer time, both for the calculation and interpretation of spectra. Parallel computers offer individual laboratories super-computer performance at reasonable cost.

Reconstruction of NMR spectra using the maximum entropy method involves a large number of Fourier transforms. We have ported a conventional 2-D maximum entropy program, written in FORTRAN 77, to our Meiko Computing Surface (a transputer-based parallel computer). The major modification necessary was the implementation of an efficient data communication scheme. This is particularly important for the 2-D Fourier transform, which involves each processor communicating with every other during the transpose stage. We have also implemented a stand-alone parallel 2-D Fourier transform program which minimizes communication time by overlapping this with the computation.

Obtaining the 3-D coordinates of a protein structure from 2-D spectra is commonly performed using a combination of distance-geometry and molecular dynamics calculations. We have implemented an efficient systolic loop algorithm for parallel molecular dynamics simulations on the Meiko. The program has been modified to include the distance constraints that are derived from the NMR data.

INTRODUCTION

The study of proteins, their structure, and their interactions with other molecules is fundamental to our understanding of biology at the molecular level. In our laboratory we use two-dimensional and three-dimensional nuclear magnetic resonance (2-D and 3-D NMR) experiments to determine the structure of macromolecules of biological importance. We are developing software for a Meiko Computing Surface, a parallel processing machine designed around the transputer, to process NMR data and to compute structures consistent with those data.

Computational Aspects of the Study of Biological Macromolecules by Nuclear Magnetic Resonance Spectroscopy, Edited by J.C. Hoch *et al.*, Plenum Press, New York, 1991

PARALLELIZATION OF FOURIER TRANSFORMS

Taking the Fourier transform of a 2-D NMR data set is relatively inexpensive computationally, but there are various, more expensive, schemes to facilitate the interpretation of a spectrum from a noisy or incomplete data set. For example, maximum entropy[1] is an iterative algorithm that generally needs to compute the transform hundreds of times in order to produce an "optimal" spectrum. In 3-D NMR there is so much data (in practice 8 megawords or more) that even the direct Fourier transform is computationally expensive. In principle, it is sensible to parallelize both the 2-D maximum entropy algorithm and the 3-D direct Fourier transform because of the amount of computation. Although having several processors might be useful in principle, there are several numbers that should be considered before embarking on massive parallelization of computer programs. We can split the total time to run a multi-processor program into three parts:
(i) the time to read and write the data to and from the disk, T_{io};
(ii) the time to move data between processes, T_{comm};
(iii) the time to do the calculation, T_{calc}.

Times (i) and (iii) are familiar from sequential computers. Now, if T_{io} or T_{comm} is much larger than T_{calc} then using the given parallel processor is not an efficient way to solve the desired computational problem. If T_{calc} dominates, then parallelization is useful. Of course, these times depend on the type of parallel machine being used, as well as the number of processors, the amount of data and the algorithm under consideration. T_{io} scales approximately linearly with the amount of input/output data. T_{calc} scales approximately inversely with the number of processors and in a problem-dependent way with the amount of processed data. T_{comm} scales in a complicated way depending on the problem and the number of processors and how they are connected together (their "topology"). Transputers can have at most four links to other transputers, and hence the number of links grows linearly with the number of processors. Thus for the Meiko Computing Surface we find that T_{comm} scales approximately linearly with the amount of communicated data and approximately inversely with the number of processors except for a multiplicative factor due to the average number of links an average inter-processor message must use (this depends on the algorithm, the number of processors and their topology).

There will always be a maximal number of processors that it is sensible to use for a problem of a given size. If the problem size is assumed to grow linearly with the number of processors, then there should be no such limit for applications with only local (e.g., nearest neighbor) patterns of inter-processor communication. However for applications with global communi-

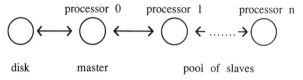

processor 0   processor 1   processor n

disk   master   pool of slaves
Figure 1. Schematic view of processors.

cation requirements (such as the 2-D Fourier transform) there will always be a limit on the number of processors that can be used, due to the multiplicative factor mentioned above. The question of whether to parallelize reduces to whether the maximal number of processors that should be used is more than 1!

We have found it useful to have one processor, which we call the master, to run a program dedicated to reading and writing the data to and from disc, sending out the data and receiving it back from the other processors, and possibly doing some bookkeeping. In the maximum entropy program the master also does the optimization calculations. The rest of the processors, called the slaves, run a program that does the main part of the computation. Fig. 1 gives a schematic view of this picture. We can then split T_{comm} into two parts, $T_{comm}^{(1)}$, which is the amount of time to get the data back and forth between the master and the slaves, and $T_{comm}^{(2)}$, which is the amount of time for inter-slave (and possibly intra-slave) communication that is needed as part of the computation. For a 2-D Fourier transform program written in Fortran and run on the Meiko we find that $T_{comm}^{(1)}$ dominates $T_{comm}^{(2)}$, but for our maximum entropy program the opposite is true.

The 2-D Fourier transforms we study have both row and column lengths that are powers of 2. It is traditional, although not necessary, for the data from NMR experiments to satisfy this requirement (data that is not is often zero-filled out to the next higher power of 2). This allows the standard fast Fourier transform algorithm (see for example, W.H. Press, et al.[2]) to be used. There are fast Fourier algorithms for transforms that are not powers of 2, but the size of data from NMR experiments that is not a power of 2 is so varied that the simplest recourse is to stick with tradition. Considerations of load balancing and symmetry then imply that the number of processors doing the calculations should also be a power of 2.

The communication time for fixed number of processors and topology scales as N, where N is the amount of data. The computation time of the fast Fourier transform algorithm scales as $N \log N$, independently of the dimension of the transform. $\log N$ can be taken to be approximately constant, with value 20, for the range of N that occur in NMR data sets. Thus the computation time and communication time scale approximately the same way. For the 2-D Fourier transform we then find that T_{calc} and

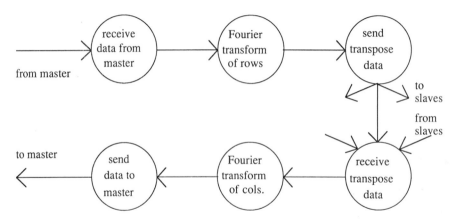

Figure 2. Schematic view of process.

$T_{comm}^{(1)}$ are of the same order of magnitude when the number of processors is of the order of 32. By extrapolating our results we expect that T_{calc} and $T_{comm}^{(2)}$ are of the same order of magnitude when the number of processors is of the order of 128. It is this latter figure which is relevant for the maximum entropy program. Naturally, the particular hardware and software available determine the precise figures.

Disc input/output also needs to be discussed. For the Meiko we find that T_{io} dominates T_{calc} for the program that just does direct Fourier transforms, and the same is true for maximum entropy if the number of processors is more than of the order of 8. Thus for our system there is no point in using more than one processor for direct Fourier transforms. We believe that in the future, disc input/output (which we have found to be limited by system software rather than hardware) will be improved by more than an order of magnitude, and this will then make multi-processing of direct Fourier transforms sensible. Keeping these future developments in mind, we will ignore T_{io} for now and concentrate instead on T_{comm} and T_{calc}

Transputers separate the task of communication from computation in hardware and thus allow the efficient parallelization of communication with computation. Occam is the natural language for writing parallel programs for transputers. However, the existing maximum entropy code is in Fortran. On the Meiko the parallel facilities for non-Occam languages are primitive (and this is perhaps just as well, since any advanced facilities would probably not be portable). Thus we have found it useful to explicitly set up in software a paradigm for the processes of a program.

At the simplest level, we can view a process as in Fig. 2 (a similar viewpoint is expressed, for example in K.C. Bowler, et al.[3]). There are three stages. Data is input from some other process or processes, some calculations are done, and data is output to some other process or processes.

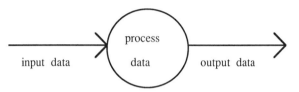

Figure 3. The six processes running on each slave.

In general, one or two of the three stages may be left out, and processes that are sequential can share memory. Each processor has one or more processes running on it. On the Meiko, intra-processor communication is much faster than inter-processor communication, hence it is the latter which needs to be minimized (although the former should not be allowed to get out of hand).

We now concentrate on 2-D Fourier transforms (a brief discussion of multi-dimensional Fourier transforms in the context of parallel processing is provided in G. Fox, et al.[4]). We assume that the number of slave processors (as well as the rows and columns) is a power of 2, and is no more than the number of rows and columns (this is always the case in practice). Even with this assumption, there are still several possible ways to decompose the calculation of a 2-D Fourier transform among the slaves. We have chosen for each slave first to do the 1-D Fourier transforms of its share of the rows, then send the relevant column data to the other slaves and receive its column data in return, then do the 1-D Fourier transforms of its share of the columns. Thus, we can identify six processes that need to run on each slave:
(1) input the data from the master processor;
(2) do the Fourier transforms for the rows;
(3) package and send the data to the other slaves;
(4) receive and unpackage the data from the other slaves;
(5) do the Fourier transforms for the columns;
(6) output the data to the master processor.

Processes (3) and (4) together must transpose the 2-D data, and are special in that if they are run "sequentially", this should be taken to mean that the sending and receiving are intertwined in one process. Some of the six processes can be run in parallel, but processes (5) and (6) have to wait until process (4) is done. Thus no more than four of the processes need to be run in parallel at any given time. A sequential view of the six processes is given in Fig. 3. It should be remembered that for two processes that are sequential and residing on one processor, the data does not actually need to be moved physically from the first process to the second.

Parallelization using Occam would be straightforward, using the PAR construct. Parallelization using Fortran (or C) is more difficult. We have designed our own parallelization by allowing each process to be split into a number of cycles. During each cycle the three stages of the process are carried out (if they are carried out at all), but only partially. For example, with two cycles the process would: input the first half of the data; process the first half of the data; output the first half of the data; input the second half of the data; process the second half of the data; output the second half of the data. If the number of cycles is only one then the process is sequential with respect to any other process which must either send input or receive output from this one. If the number of cycles is greater than one then, for example, the process to which output is being sent (each cycle) can start doing useful work before the original process is done. Thus it is by arranging for the number of cycles to be greater than one that some degree of parallelization is introduced into the program above and beyond that due simply to having more than one processor available. Since there is only one CPU per transputer this is a sensible strategy on the Meiko only if it is communication that is being parallelized with computation. We see that four of the six processes for the 2-D Fourier transform program involve communication. We analyzed which groupings of the six processes should be parallelized. Before stating the results we will more fully describe how the program is structured.

The master processor, as well as transmitting the data to and from the disk and the slave processors, is used as a control center, instructing the slaves when to begin each of the six processes that they carry out. Letting n be the maximum number of the six processes running in parallel at any time, each of the slaves has n processes running in parallel all the time (some of them perhaps idle now and again). Each of the processes has exactly the same program running (this feature could easily be modified for more sophisticated requirements). Let P be the number of slave processors. It is convenient to split the nP processes running on the slaves into n groupings of P processes. Each task that is to be carried out by the slaves is done by the processes of a particular grouping.

Each of the processes has exactly the same program running, which includes the following Fortran-like code (leaving out a few details):

```
command= go
while (command .ne. stop)
        call receive_from_master(command_info)
        call request(command_info)
end while
```

Command_info is an array of length twenty words (either integers or, via an "equivalence" statement, reals). The first word of command_info is an integer specifying the next command to be carried out by the process. The rest of command_info gives additional information including: whether there is input from another of the n processes (of this slave and possibly also of the other slaves), and if so which one; whether any calculations need to be done, and if so which one; whether there is output to another process, and if so which one; the number of cycles for the process. When the master is finished with its entire program it sends a command to the slaves telling them to stop (when they are done with the current command). The request subroutine is a long "if-then-else" sequence. Each "if" block looks like:

```
if (command .eq. some_command) then
        call wrapper(some_in, some_process, some_out)
endif
```

where some_in, some_process and some_out are the subroutines for the input, processing and output stages of the process determined by some_command. The subroutine wrapper is a coded version of the simple picture of a process given in Fig. 2, with the possibility of more than one cycle allowed:

```
subroutine wrapper(in_sub, process_sub, out_sub)
call wrapper_info
do i = 1, number of cycles
        if (in_flag) call in_sub
        if (process_flag) call process_sub
        if (out_flag) call out _sub
end do
```

The subroutine wrapper_info translates between the information in command_info that the master sends and the variables that the slave process uses, e.g., the number of cycles. The three flags are Boolean variables that indicate which of the three stages this particular process does. Finally, the time it takes to execute each command is counted.

This modular approach to parallel processing allows for the easy incorporation of already existing sequential code.

COMMUNICATION

It is worthwhile taking some time to consider the communication between processes. There is a fundamental limitation in the current generation of transputer chips in that they can only be connected to at most four other transputers and the sending of messages is by physical naming. The latter condition means that if one transputer needs to send a message to another transputer that is not connected to it, it is up to the software programmer to explicitly route the message through intermediate transputers. Fortunately, there is a program (that runs, for example, on the Meiko) called Tiny, available from the University of Edinburgh and written by Lyndon Clarke, which allows the routing of messages using logical names. That is, a message can be sent between two transputers without reference to any transputers which the message might have to pass through on its way. This has the obvious advantage of relieving the programmer of having to explicitly route messages, but it also has the advantage of allowing the effect of different topologies on program speed to be easily determined. (The "topology" of a group of transputers is the equivalence class of the graph which describes the way the transputers are connected together.)

Transputers pass messages synchronously, that is, if one process is trying to send data to another process then the first process is prevented from continuing until the second one has acknowledged the reception of the data. This has the advantage of not requiring extra memory to store unread messages, but at the cost of possible delays in processing. On the other hand, Tiny allows messages to be buffered, which can create its own problems. For example, with Tiny, if a message arrives and the buffer into which it should be put is full then the system freezes (an alternative would be for some message to be discarded, but that is not any better). If this happens in a particular program, then either the number of buffers has to be increase (which costs memory) or the rate of message passing has to be decreased (which costs time).

Tiny uses a number of buffers of fixed size. This can be wasteful of memory, because lots of small messages (such as we can get with the transpose step of the 2-D Fourier transform) take up the same amount of buffers as an equal number of larger messages. Another option would have been to use one larger buffer with bits of it allocated as needed. This is slower in implementation since some calculation would need to be done for every request, to determine which part of the buffer to allocate.

There is one more feature of Tiny which proves to be very convenient. Communication between processes can be made selective, through the use of "channels". For each process the programmer specifies a list of channels which it can use for receiving and sending messages. With the use of only one channel the messages arriving at a process are forced to be read in a first come, first served basis. With more than one channel, the process can choose which channel to read from first. For example, if the master processor sends out a message on one channel, the slaves can finish reading data on another channel before having to respond to the master. As long as there are enough buffers, this is a very convenient way to set up communication. For the 2-D Fourier transform program we have one "global" channel on which all processes can send and receive messages, and which is used for sending the Fourier transform data back and forth from the master to the slaves. Letting n be the number of groupings of processes, as above, we have n channels, one for each group, which the master

Table 1

Number Of Slaves	Master	Relative Efficiency	Receive Data	Send Data	Fourier Transform	Transpose	Sum For Slaves
32	4.15	43%	1.59	0.19	1.41	0.40	3.59
16	5.85	61%	1.40	0.48	2.81	0.35	5.04
8	8.92	79%	1.41	0.68	5.63	0.60	8.32
4	15.14	94%	1.41	0.92	11.26	1.17	14.76
2	28.36		1.41	1.01	22.51	3.10	28.03

uses to broadcast the information indicating which task is to be carried out next and how. Finally, we have $n(n + 1)/2$ channels, one for each pair of groups. This allows each group to send messages exclusively to any other group, and is used for all the communication between the slaves, including the transpose. It should be noted that although there are a total of $1 + n + n(n + 1)/2$ channels, the master process uses only $1 + n$, and each slave process only $2 + n$ of the channels. Using this many channels means that it is not necessary to synchronize all the message sending between the various processes.

FOURIER TRANSFORM RESULTS

In order to test which of the six processes above should be parallelized we used a 512×512 (real) data set, running on up to 32 slaves. It is convenient to introduce a shorthand notation to indicate which of the six processes is running on which of the n available processes. For example, (1 2 3 3 3 1) indicates that the first three processes (inputting data from the master, doing the row Fourier transforms and sending the transpose data) are done in parallel; the sending and receiving of the transpose data is done "sequentially" (which in this case means that they are intertwined); the column Fourier transforms, as always, wait for the transpose to finish, and are done in parallel with outputting data to the master. The fourth and fifth process both use process 3 in order to avoid having to move the data between them, which there is no advantage doing since they are sequential. For this example $n = 3$. Other examples of the shorthand notation are (1 1 1 1 1 1), which is a totally sequential program $(n = 1)$, and (1 2 3 4 4 1), which is the limit of parallelization $(n = 4)$.

Having a higher n suffers from the disadvantages that more memory is required and each of the n processes is slowed down if they need to share the available computational resource. Countering these two disadvantages is the ability to speed up the n processes if they can overlap computation with communication.

Table 2

Number Of Slaves	Number of Cycles	Master	Relative Efficiency	Sequential Multiple	Useful Computation
32	16	2.52	69%	61%	56%
16	16	3.81	91%	65%	74%
8	16	6.98	99%	78%	81%
4	32	13.57	102%	90%	83%
2	64	27.77		98%	81%

In Table 1 we list the times to run the program sequentially, i.e., for (1 1 1 1 1 1). The second column gives the time (in seconds) it took the master to run and the third column gives the efficiency relative to the time for two slaves. The subsequent columns give the times (in seconds) it took the slaves *on average* to complete the indicated tasks. The master times are longer than the sum of the average slave times because some slaves take longer to send the data to the master than others. The times depend on the topology and those given are the best found (there is less than 1% difference between the various topologies for no more than eight slaves, less than 5% for sixteen slaves, but up to 30% for thirty-two slaves). The times for the Fourier transforms halve when the number of slaves is doubled, as it should. The efficiency drops markedly when the computation time is of the same order of magnitude as the communication time, as expected.

The sequential program (1 1 1 1 1 1) provides a benchmark against which parallelization may be tested. At the opposite end of the spectrum is the program (1 2 3 4 4 3). Results for this are given in Table 2. The second column gives the (determined) optimal number of cycles for parallelization. The third column gives the time (in seconds) for the master to run. The fourth column gives the efficiency relative to two processors. The fifth column gives the percentage of time for the master to run relative to the sequential results given above. The last column gives the percentage of time spent by the slaves on the actual Fourier transform (using the times given above). It can be seen that the relative efficiency figures have improved compared with the sequential program. This, along with the sequential multiple figures indicate that communication is indeed being successfully overlapped with computation.

Many other programs have been tried besides the above two. For more than eight slaves, the best times were achieved for (1 2 3 4 4 3). However for eight slaves, for example, the program (1 2 2 2 2 1) with 8 cycles was faster than the best (1 2 3 4 4 3) result by about three percent. Similar results hold for less than eight slaves. The problem seems to lie with the transpose step, which involves some indexing computation in addition to the communication.

MOLECULAR DYNAMICS

Molecular dynamics simulation is a powerful computational technique which allows the calculation of structural, dynamic and thermodynamic properties of atomic and molecular ensembles. It involves the integration of the equations of motion of the particles comprising the system of interest, generating a trajectory of the conformation of the system through time.

Simulations of protein dynamics allows the investigation of phenomena not accessible to conventional experimental techniques. In addition, molecular dynamics can be used as a tool to aid the determination of protein structure from x-ray diffraction[5] and two-dimensional nuclear magnetic resonance data[6].

Although molecular dynamics simulation can tell us a great deal about a system, it is an expensive technique. The equations of motion are integrated using a finite difference method such as that of Verlet[7], where the coordinates (r_i) and velocities (v_j) of the particles are stepped through time using the equations:

$$r_i(t + \Delta t) = r_i(t) + \Delta t v_i(t + 1/2 \Delta t)$$
$$v_i(t + 1/2 \Delta t) = v_i(t - 1/2 \Delta t) + \Delta t a_i(t)$$

and

$$m_i a_i = -\partial/\partial r_i E \qquad (1)$$

where E is a function representing the potential energy of the system, and m_i is the mass of particle i.

In order for this approach to be valid, the time-step (Δt) must be shorter than the period of the fastest motion present in the system, usually this is between 1-10 fs. This means that very many time-steps will be required to bring the system into thermal equilibrium. Furthermore, if the inter-particle interactions are significant at separations comparable to the size of the system, the number of pair-wise interactions that have to be evaluated at each time-step will be $1/2N(N-1)$. Thus an efficient parallelization of the molecular dynamics algorithm, to run on a transputer based parallel computer would be of great value.

In an earlier paper[8], efficient methods for performing molecular dynamics simulations of simple fluids on a ring of transputers was described. This paper describes how these methods have been generalized to accommodate the features of larger molecules, and hence allow efficient simulation of protein dynamics.

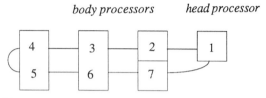

Figure 4. Distribution of particles over processors for the SLS method.

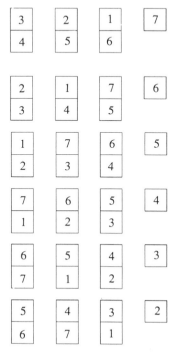

Figure 5. Data movements for the SLS method.

SYSTOLIC LOOP METHODS

As an example of the systolic loop methods, one (termed SLS) is described. Consider a system of seven particles, interacting via pair-wise forces. The data representing the coordinates, and the forces acting upon each particle can be distributed over four processors in the following manner (Fig. 4):

The pair-wise forces 4-5, 3=6, and 2-7 can be evaluated in parallel. If the coordinates and forces are rotated around the processor ring, then interactions 3-4, 2-5 and 1-6 can be evaluated. If the data movements are repeated then, after seven such systolic pulses, all the pair-wise interactions will have been evaluated, and the data will have returned to their original processors (Fig. 5). The time taken for the calculation will be a third of the time that a single processor would take, but the total elapsed time will include the time taken to rotate all the data once round the loop.

```
      12 11 10           9 8 7           6 5 4           3 2 1

      13 14 15        16 17 18        19 20 21


         9 8 7           6 5 4           3 2 1        21 20 19

      10 11 12        13 14 15        16 17 18
                          etc
```

Figure 6. Distribution of data, and data movements for the SLS-G method.

This scheme can be extended for the case where there are many more particles than processors. Consider a system of twenty one particles, distributed over the same four processors (Fig. 6).

If the data are moved as groups of particles rather than individually, then the body processors can evaluate forces between groups of particles, while the head processor can evaluate the forces between particles within each group. Furthermore, as the time taken for the calculation is proportional to the square of the size of the groups, while the communication time depends only on the total number of particles in the system, the calculations become more efficient as the size of the system increases. In practice it was found that, for a simulation of the Lennard-Jones fluid, optimum efficiency (close to 100%) was obtained when the group size exceeded 16 particles. For a simulation of 256 particles, a systolic loop program running on 30 T800 processors was equivalent to a conventional program running on a CRAY 1S.

PROTEIN STRUCTURE

Proteins are long chain molecules, having a polymeric backbone with different side-chains branching from each backbone unit (Fig. 7). In their biologically active form most proteins fold into a compact globular conformation. Some side-chains can then form additional chemical bonds, and thus cross-link the backbone. Flexible covalent bonds are customarily modelled using two- three- and four-body forces (Fig. 8).

The systolic loop methods only guarantee that every pair-wise interactions taken into account, although some multi-body interactions could be evaluated within and between some groups. In order to use a systolic loop algorithm for protein simulation, some way must be devised to ensure that all the three- and four-body interactions occur between atoms within at most two groups. In practice, a simple rule can be applied when dividing the atoms into groups, which will enforce this condition.

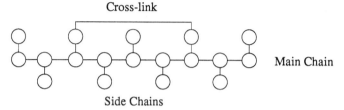

Cross-link

Main Chain

Side Chains

Figure 7. Cartoon of protein topology. Proteins have a long chain backbone, with different side-chains branching from it. Some side-chains can form covalent bonds with one another, creating closed cycles.

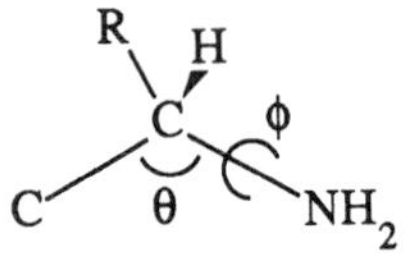

$$R \quad H$$
$$C \quad \phi$$
$$C \quad \theta \quad NH_2$$

Figure 8. Bond geometry is defined by 2 (bond length b), 3 (bond angle θ) and 4-body (torsion angle ϕ) relationships. $E_{bond} = \sum_{bonds} \frac{1}{2}(r - r_0)^2$ $E_{bond} = \sum_{angles} \frac{1}{2}(\theta - \theta_0)^2$ $E_{bond} = \sum_{torsions} k_\phi\{1 + cos(n\phi - \phi_0)\}$

ATOM DIVISION RULE

Group boundaries will necessarily cut atom-atom bonds (protein molecules are so large that there will always be more processors than molecules). If, however, atoms which participate in cross-boundary bonds are chosen such that none contributes to more than one such bond, then all the three- and four-body interactions will be limited to at most two groups as required (Fig. 9). This allows the systolic loop methods to be used unchanged to simulate any flexible macromolecular system.

BOND LENGTH CONSTRAINTS

As mentioned above, the time-step of a simulation is limited by the period of the fastest characteristic motions of the system being investigated. In systems with flexible chemical bonds, the fastest motions will be bond stretching vibrations. As these are not closely coupled to the other degrees of freedom of the molecule, an increase in the usable time-step can be achieved by fixing the lengths of bonds, without significantly affecting the dynamics of the system as a whole. For large molecules this is most often done using the SHAKE method[9], in which the coordinates of atoms are adjusted to satisfy the bond-length constraints after an unconstrained time-step has been made (Fig. 10). Using SHAKE to constrain bond-lengths allows the time-step to be increased by a factor of 2-3, so an efficient implementation within the systolic loop scheme is desirable.

100

$$E_{bond} = \sum_{bonds} \frac{1}{2} k_b (r - r_0)^2$$

$$E_{angle} = \sum_{angles} \frac{1}{2} k_\theta (\theta - \theta_0)^2$$

$$E_{torsion} = \sum_{torsions} k_\phi \{1 + \cos(n\phi - \phi_0)\}$$

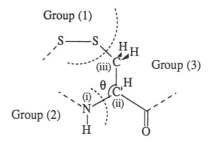

Figure 9. Division of atom into groups. No atom is allowed to contribute bonds to more than one group. Thus all the three- and four-body interactions are accommodated within the original systolic loop scheme.

Because satisfying one constraint may well cause another to be violated, and because some bonds are split over processor boundaries, each iteration of the SHAKE algorithm is potentially as expensive in communication time as a full force evaluation. However, as was noted[9], the shortest period motions in protein structures will be those of bonds involving hydrogen atoms. Constraining just these terms using SHAKE allows an increase of the time-step by a factor of two. Hydrogen atoms are monovalent, and therefore bonds involving hydrogens need never cross group boundaries. If this is the case, then the constraints within one group will be independent of those in other groups, and so SHAKE can be applied to all groups in parallel with no communication penalty at all.

IMPLEMENTATION

A systolic loop program (SLS-PRO) for molecular dynamics simulation of proteins has been written, in Occam 2, for MEIKO and INMOS hardware. SLS-PRO uses the potential function from the established protein simulation suite GROMOS[10], and reads and writes GROMOS format files. Thus input data can be prepared using the appropriate programs from GROMOS on a conventional computer, and simulations run using SLS-PRO on transputers.

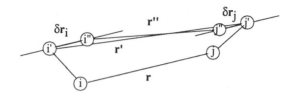

Figure 10. Vector relationships involved when satisfying a bond-length constraint using the SHAKE method. i and j are the atoms before the time-step, i' and j' are the atoms after the unconstrained time-step and i'' and j'' are the atoms after the SHAKE correction has been applied. The corrections dr_i and dr_j are calculated:

$$dr_1 = -g\frac{\mathbf{r}_{ij}}{m_i}$$

$$dr_j = g\frac{\mathbf{r}_{ij}}{m_j}$$

$$g = \frac{\left(|r_0|^2 - |r'_{ij}|^2\right)}{2M\mathbf{r}_{ij}\cdot\mathbf{r}'_{ij}}$$

$$M = \frac{1}{m_i} + \frac{1}{m_j}$$

where m_i and m_j are the masses of atoms i and j respectively. This process is iterated over all constrained ij pairs until all the constraints are satisfied.

PERFORMANCE

The performance of SLS-PRO was assessed by comparing simulations of crambin (a small protein, $\sim$ 400 atoms per molecule) on three, five and twelve processors, with equivalent simulations run using GROMOS on a microVAX 3600. The results are shown in Table 1.

As can be seen, for the simulations taking all pair-wise interactions into account, the speed-up is more or less linear. When a 1.0 nm cutoff is applied to the interactions (reducing the number of terms evaluated by a factor

Table 3. Performance of SLS-PRO for a simulation of crambin

	Time Per 100 Steps (1.0 nm Cutoff)	Speedup Relative To VAX	Time Per 100 Steps (All Pairs)	Speedup Relative To VAX	Average Group Size
microVAX360	705	1.0	2045	1.0	400
3 T800s	215	3.3	450	4.5	80
5 T800S	165	4.4	260	7.9	44
12 T800s	60	11.8	125	16.4	17

of three), the the time taken for inter-processor communication is more significant, and the efficiency drops. Overall, a T800 transputer gives a realizable performance equivalent to 1-1.5 microVAX 3600 processors for these calculations.

CONCLUSION

The work described in this paper demonstrates that the use of parallel computers can be a cost-effective route to obtaining super-computer performance in a research group environment for some of the typical calculations involved in the determination of protein structure from NMR data. Furthermore, the algorithms developed are more generally applicable to a wide range of scientific computations.

ACKNOWLEDGEMENTS

The authors acknowledge the support of Meiko Scientific, Polygen, the SERC and its Cambridge Centre for Molecular Recognition. ARCR would like to thank Dr. R.E. Hubbard, Dr. D. Fincham and Dr. W. Smith for many useful discussions.

REFERENCES

1. S. Sibisi, J. Skilling, R. G. Brereton, E. D. Laue, and J. Staunton, *Nature* **311**, 446-447 (1984).
2. W. H. Press, B. P. Flannery, S. A. Teuklosky, W. T. Vetterling, "Numerical Recipes," Cambridge University Press, Cambridge (1986).
3. K. C. Bowler, R. D. Kenway, G. S. Pawley and D. Roweth, "An Introduction to Occam 2 Programming", Chartwell-Bratt, Lund (1987).
4. G. Fox, M. Johnson, G. Lyzenga, S. Otto, J. Salmon, and D. Walker, "Solving Problems on Concurrent Processors, Vol. I: General Techniques and Regular Problems", Prentice-Hall, New York (1988).
5. A. T. Brunger, M. Karplus, and G. A. Petsko *Acta Cryst. Sect A* **45**, 50 (1989).
6. G. M. Clore and A. M. Gronenborn *Crit. Rev. Biochem.* **24**, 479-564 (1989).
7. L. Verlet *Phys. Rev.* **159**, 98-103 (1967).
8. A. R. C. Raine, D. Fincham, and W. Smith, *Comput. Phys. Commn.* **55**, 13-30 (1989).
9. J. P. Ryckaert, G. Ciccotti, and H. J. C. Berendsen, *J. Comput. Phys.* **23**, 327 (1977).
10. W. F. van Gunsteren, H. J. C. Berendsen, J. Hermans, W. G. J. Hol, and J. P. M. Postma, *Proc. Natl. Acad. Sci. USA* **80**, 4315 (1983).

SOFTWARE APPROACHES FOR DETERMINATION OF 3-DIMENSIONAL MOLECULAR STRUCTURES FROM MULTI-DIMENSIONAL NMR

George C. Levy, Sophia Wang, Pankaj Kumar,
Gwang-woo Jeong, and Philip N. Borer

NMR and Data Processing Laboratory
Syracuse University
Syracuse, New York 13244-4100

ABSTRACT

Two and higher dimensional NMR spectroscopies offer extraordinary power for detailed structure elucidation of proteins, nucleic acids and other important biomolecules. The methodology of elucidating biopolymer structures at atomic resolution from NMR spectroscopic data incorporates primarily NOESY experiments, but also may add spin-spin coupling constants and other information measured from COSY and other NMR experimentation. There are several important challenges that must be overcome in order for this methodology to be generally applicable to a broad range of biomolecules. One of the most important long-term goals of this research arena is to be able to determine structures with a confidence level sufficient to allow utilization of the structural information without confirmatory experimentation such as single X-ray structures.

In order to achieve this long-term goal, a number of issues must be dealt with: 1) primary 2D (nD) data reduction must incorporate techniques to allow accurate determination of sufficient NOESY cross-peak volumes; 2) computational schemes must be developed which not only determine refined molecular structures from the experimental information, but which also reflect confidence levels in the determined structures based on intelligent error analysis through all procedures; 3) corrections for these calculations must include, at a minimum, correction for dynamics variations, correction and recovery for missed spectral assignments, and wide sampling of possible molecular geometries.

Development of automated and assisted multi-dimensional NMR spectral assignment techniques is critical for many of these studies, where hundreds or even thousands of cross peaks may be significant for analysis. Techniques incorporating automated NOESY walks, pattern recognition for identification of specific sites, and other techniques will have to be used together for optimal spectral assignment. Of course 3- and higher dimensional spectroscopy will also assist in this area.

At Syracuse University, one of the primary goals realized at this time, is optimal preparation of the data for analysis. Use of non-linear processing techniques based on the maximum likelihood method (MLM) and specialized protocols increases the number of cross peaks that can be used for 3D structure determination. Experiments and computation underway indicates that these non-linear techniques have broad applicability and

Computational Aspects of the Study of Biological Macromolecules by Nuclear Magnetic Resonance Spectroscopy, Edited by J.C. Hoch *et al.,* Plenum Press, New York, 1991

that, across a range of spectral conditions, they are robust and *quantitative* (or where dynamic range is too high, corrections may be possible to quantitate the smallest peaks). Preliminary results on synthetic and mixed data show superior quantification of cross-peak volumes over a range of peaks sizes exceeding 50:1.

A second area of investigation at Syracuse University involves utilization of parallel and distributed computing methods. These are initially being applied to two applications: 1) 3D NMR data processing and 2) using a genetic algorithm for NMR molecular modeling.

The basic idea is to utilize, in parallel, workstation and other computers coexistent on local and wide-area computer networks. In cases where specialized computing hardware such as MIMD parallel computers (examples: Alliant FX/80, Hypercubes, etc.) or SIMD architectures (example: Connection Machine) are available, an additional opportunity is present to dissect a computational application and allocate appropriate portions to that specialized hardware. This type of distribution of processing tasks is included in the work underway. Thus, on a computational network such as the one existing at Syracuse University which incorporates a large number of Sun work stations, IBM RISC System 6000's, and a large configuration Connection Machine 2, as well as an Alliant FX/80, more than an order of magnitude speedup in realization of compute and I/O applications such as 3D NMR data processing and matrix manipulation found in aspects of NMR molecular modeling.

INTRODUCTION

The opportunity presented by combining multi-dimensional NMR spectroscopy with determination at atomic resolution of solution structures of important biological molecules takes an extraordinary expenditure of effort in a growing number of laboratories at this time.

However, it should be noted that there are significant problems facing complex molecular modeling from NMR data. Also, while results from the methods being developed in this application are increasingly automated, there is still the difficult and as yet largely manual requirement of spectroscopic assignments. A number of groups are working on pattern recognition and other automated methods of spectral assignment which promise to facilitate this task, but as yet no general solution is in sight.

Ambiguous spectral assignments form one of the potential challenges to accurate determination of molecular structures from multi-dimensional NMR data. However, this source of potential error is not alone. Included as significant problems are: 1) the determination of sufficient, correct NOESY peak volumes, 2) delineate all necessary structural details, limitations or errors in application of theory to calculation, such as assumption of overall molecular dynamics, 3) inclusion of internal molecular dynamics models, 4) corrections for spin diffusion and other deviations from the isolated spin pair approximation, etc. Difficulties can also occur as a result of incomplete sampling of molecular geometries during reduction of hydrogen-hydrogen distance information into 3-dimensional crude molecular structures, and through the stages of molecular structure refinement. Exacerbating determination of accurate structures is the possibility of an equilibrium between structural isomers or conformers and resulting in inaccurate assessment of distances and consequently structural details.

Refinement of crude 3-dimensional solution structures determined through NMR NOESY experiments often incorporates NMR J-coupling and non-NMR experimental data, supplemented by energetics or molecular dynamics calculations. All of these important sources of information have their own potential error sources embedded within those techniques.

Having introduced many of the challenges and problems associated with NMR molecular modeling, it is important to reflect on the extraordinary opportunity presented by this new analytical approach. In spite of the various error sources, the quality of information that can be obtained from these studies, and their unique nature, argue strongly for an extraordinary effort to be made to develop reliable experimental methodology and computational protocols for NMR molecular modeling.

The Syracuse University effort in this area, thus far has largely been involved in preparation of 2-dimensional NOESY data for optimal quantification of the maximum number of cross peaks, with that preparation simultaneously incorporating significant resolution and contrast enhancement, on order to facilitate spectral assignments and the quantification procedures. The Syracuse group has also begun studies in parallel and distributed computing and application of these technologies to NMR data processing and NMR molecular modeling. Preliminary results from the laboratory are presented in this chapter.

CONTRAST ENHANCEMENT PROTOCOLS AND THEIR USE IN NMR MOLECULAR MODELING

Two-dimensional NMR spectra, particularly of complex biological molecules, often exhibit highly overlapped peak regions with insufficient signal-to-noise ratio for conventional resolution enhancement techniques such as Fourier deconvolution. Further, the limited attainable signal-to-noise often can restrict utilization of even isolated NOESY cross peaks for inter-proton distance determination. Usually, the longest distance measured in this approach is of order 5 Å, with the smallest cross peaks that are used corresponding to approximately 2% of the size of cross peaks of hydrogens that are in close proximity (ca. 2.5-2.7 Å).

Mathematicians have developed powerful tools for contrast enhancement (signal-to-noise contrast and resolution enhancement contrast) and their results have had considerable impact in astronomy, medical imaging, and other fields. A variety of approaches have been utilized, including maximum entropy[1-14], maximum likelihood[15-18], minimum-area reconstructed spectra[19], Jansson-Van Cittert deconvolution[20-26], signal restoration and image recovery techniques[25-26], iterative spectral deconvolution[27-32], linear prediction[33-39], and applications of Bayesian prediction techniques[40-42].

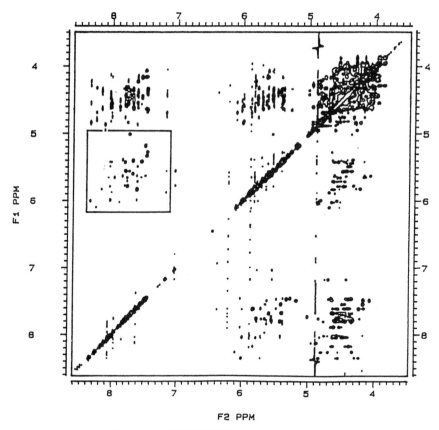

Figure 1. NOESY RNA 24-mer hairpin loop.

The problems are comparable in all application fields, including NMR: to enhance interesting features that occur in a "blurry" intensity pattern of a 2-dimensional "picture". When there is an appropriate functional form that represents the cause of the blurring, corrections can be made. For NMR spectroscopy, the proper functional form is a Lorentzian lineshape, and deconvolution can be performed according to well-tested and rigid protocols. It is possible with our current MLM and related protocols to effectively remove linewidth, resulting in resolution increases of 2-3 fold.

The MLM protocol, along with MEM and some of the other procedures, also dramatically reduces the apparent noise. The noise reduction is probably the least important aspect. MLM is not used to try to reduce the signal-averaging time. The apparent gain in signal-to-noise is almost wholly cosmetic — any signal that is at the level of the noise will not be enhanced. Likewise, any peak that is enhanced will also be apparent upon close examination of the original spectrum. The "deblurring" aspect of the deconvolution is incredibly useful, however.

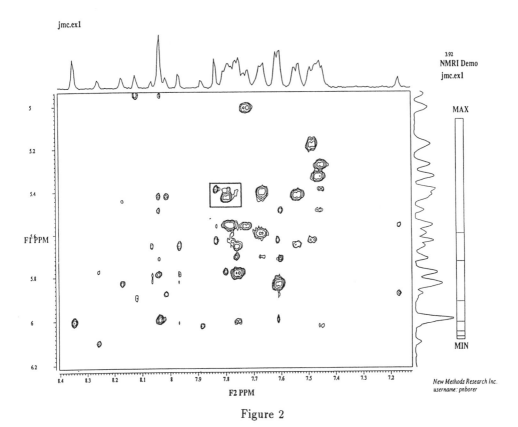

jmc.ex1

3.92
NMRI Demo
jmc.ex1

New Methods Research Inc.
username: pnborer

Figure 2

Fig. 1 illustrates the typical overlap problem in a NOESY spectrum of an RNA 24-mer. (We thank Drs. P. N. Borer, O. Uhlenbeck, and J. Gott for letting us use this spectrum). This spectrum, obtained on the S.U. GN-500 (1 k × 1 k spectrum shown, but in other cases up to 4 k × 4 k spectra are used) indicates one of the main problems of molecular structure determination from NOESY data. Most of the regions of the RNA spectrum are very heavily overlapped, although the H1'-base region outlined by the zoom box is relatively simple to interpret. Fig. 2 shows this H1'-base region in an expansion (ca. 7.1 to 8.4 ppm-F2; 5 to 6.2 ppm-F1). Fig. 2 is processed conventionally with shifted sign bell apodization and FFT processing. It is easy to contrast Fig. 2 to the results in Fig. 3, which show a Maximum Likelihood calculation, followed by a symmetrization operation. Peaks that are heavily overlapped in the conventionally processed Fig. 2 are easily separated in the MLS (Maximum Likelihood Symmetrized) spectrum. This can be further seen from the spin-spin multiplet structures appearing in the MLS spectrum, and is also visible in the projections given on the F1 and F2 axes

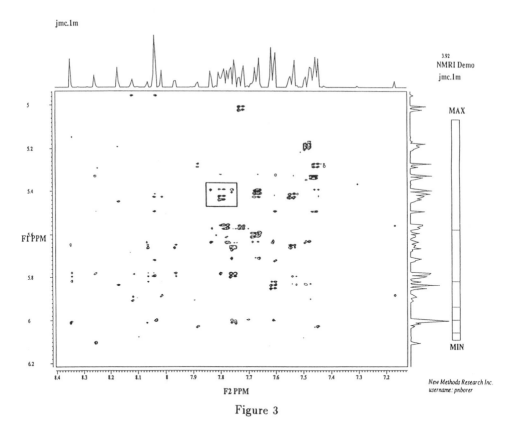

Figure 3

(compare with Fig. 2). For Figs. 1-3, 1 k × 1 k data were utilized; in fact, the most significant results on this molecule utilize a combination of zero-filling, symmetrization, and maximum likelihood calculation. Figs. 2 and 3 show a small box outlining a few cross peaks near 5.4 ppm, 7.8 ppm. Fig. 4 shows the peaks of this small region processed conventionally at 1 k × 1 k dataset size (not including imaginary points) and also at 4 k × 4 k dataset size after additional zero-filling. The effect of symmetrized ML deconvolution on the twice zero-filled spectrum, is dramatic. Note that all of the low S:N peaks in Fig. 4f are real and have been assigned to specific hydrogens.

Thus far we have developed two symmetrized MLM protocols, listed below with two conventional FFT processing schemes.

PROTOCOLS

A. Data ◇ Apodization ◇ Zerofill ◇ FFT ◇◇◇ DISPLAY

B. Data ◇ Apodization ◇ Zerofill ◇ FFT ◇ Symmetrize ◇◇◇ DISPLAY

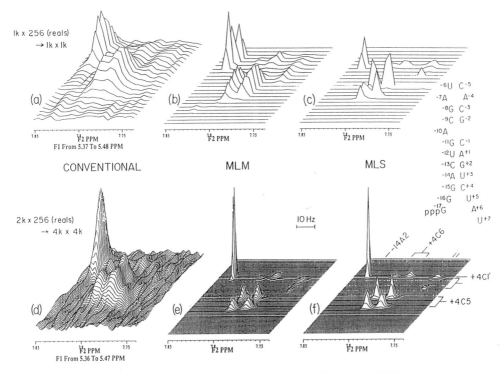

Figure 4. Effect of digital resolution on MLM and MLS.

C. Data ◇ Apodization ◇ Zerofill ◇ FFT ◇ MLM ◇ Symmetrize ◇◇◇ DIS-
   PLAY (MLS)

D. Data ◇ Apodization ◇ Zerofill ◇ FFT ◇ Symmetrize ◇ MLM ◇◇◇ DIS-
   PLAY (SML)

Of the four protocols listed here, C and D offer marked advantage with respect to both apparent signal-to-noise ratios and cross-peak separation. Noise is suppressed by the Maximum Likelihood procedure, and peak separation is improved by the deconvolution of cross-peak lineshapes, also as a result of the MLM procedure. *Symmetrization recovers part of the resolution typically limited by data acquisition.*

With small molecules, it is true that an experienced spectroscopist can construct NOESY walks for most Regions Of Interest, but there are often ambiguities, and as the molecular systems become more complex, there are real challenges in assignment and quantification of cross-peak volumes. Reliable quantitative methods for contrast enhancement will be *critical* for assigning such spectra and determining interproton distance constraints from the NOESY cross-peak volumes.

It is possible to wreak havoc by trying to over-enhance a spectrum. For instance, in a spectrum where the natural linewidths are ca. 6-8 Hz, application of 10 Hz line-sharpening, rather than the 5 Hz we would recommend causes the peaks to become distorted and lose intensity. With 20 Hz linewidth reduction in this case, many peaks can be completely suppressed.

We note here that a 256 × 256 point region can be enhanced by MLM in 15 minutes of cpu time on our current Stellar GS-1000.

MEASUREMENT OF NOESY CROSS PEAKS

In the final analysis, it is most important for the data processing to produce quantitative integrations for cross peaks over the maximum practical dynamic range extending, if possible, across all measurable small peaks (it is not necessary to quantitate on-diagonal peaks including strong solvent peaks). The MLM protocols, as in the case of maximum entropy and other non-linear techniques, can only have a certain, limited dynamic range. Our preliminary results indicate that we are able to control the algorithm to give quantitative results over at a factor of 25 of peak intensity (or integration). We believe that calibration curves can extend this to approximately a dynamic range of 100-500. If this proves possible, then it should prove possible to measure meaningful measured NOE cross-peak volumes to produce meaningful interproton distances exceeding 5 Å and approaching 6 to 6.5 Å. This alone would have significant affect on NMR molecular modeling.

RECENT RESULTS ON MLM QUANTITATION USING SYNTHETIC DATA MIXED WITH REAL DATA

Fig. 5 shows a synthetic spectrum consisting of 32 peaks covering a dynamic range, in this case, of 12.5:1. Fig. 6 shows the same spectrum combined with real data and covering now a total dynamic range of 150:1. Fig. 6a shows the conventionally processed spectrum and Fig. 6b the MLM spectrum. Synthetic peaks in both spectra are circled. It is easier to get some visual perspective using stack plots of the upper right-hand portion of these spectral regions and this is shown in Fig. 7. Fig. 7a and c show the synthetic data and combine synthetic/real data processed by conventional FFT methodology; the right-hand spectra b and d show the synthetic data and the combined synthetic/real data processed by MLM with 6 Hz line narrowing in each dimension. It is interesting to note both the quality of the spectral data in Fig. 7d and also particularly the quality of the baseplane and contrast enhancement that is observed even for small peaks in the real data such as seen near the bottom of the spectrum at approximately 3000 Hz (circled). Fig. 8 shows the result of quantifica-

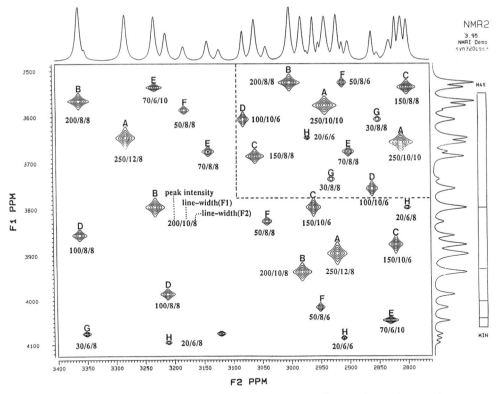

Figure 5. Synthetic spectrum consisting of 32 peaks (12.5:1 dynamic range).

tion of groups of peaks having the same integrated intensity, but varying linewidths, processed conventionally and by MLM. The MLM quantification was superior for 6 of the 8 groups of peaks with 4 times better overall precision.

A second combined synthetic/real dataset was constructed with a dynamic range of 20:1 and these data are shown in Fig. 9. In this test, different values of deconvolution were used to evaluate the line sharpening on quantification. Some results are shown in Fig. 10, again indicating significantly enhanced quantification with the MLM protocol vs. conventional FFT processing.

In order to reliably evaluate MLM and related quantification capabilities and limitations, we need to perform a range of computer experiments such as the two that have been reported here. It will also be important to use different real datasets as mixing data for the combination of real and synthetic marker data. Most important to the overall objective of extending 2-dimensional NOESY cross-peak measurements to add additional usable data for 2.3 to 5 angstrom internuclear distances and possibly to extend

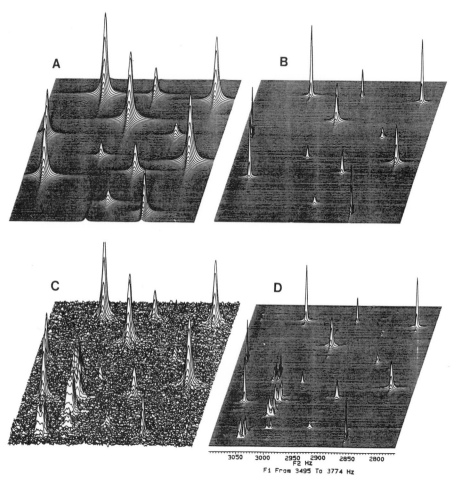

Figure 6. Conventional(A) and MLM(B) spectra of combined real and synthetic data (150:1 Dynamic range: MLM: -6Hz/100 iterations).

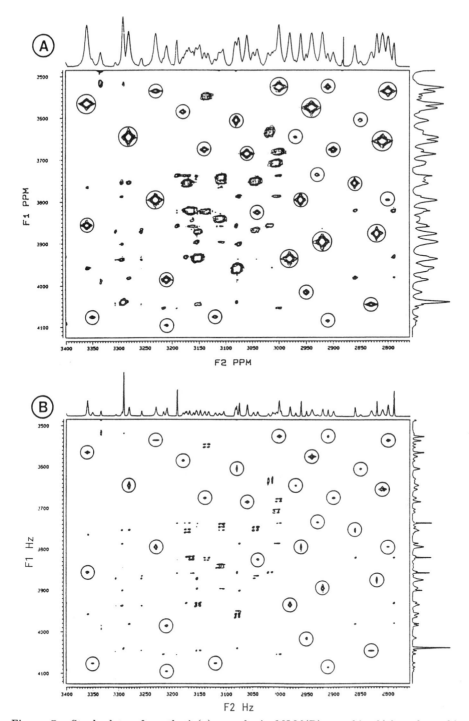

Figure 7. Stack-plots of synthetic(a), synthetic MLM(B), combined(c) and combined MLM(D) Data (MLM: -6Hz/100 iterations).

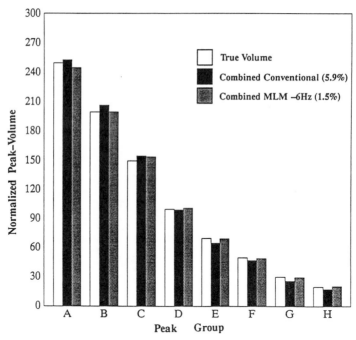

Figure 8. Volume deviation of synthetic conventional- and MLM-data in combined spectra: Each single alphabet code represents a group consisting of 4 synthetic peaks with same intensity but different line-width.

the distance range to 6 or slightly more angstroms, if we are able to obtain reasonable integrals on very low S:N cross peaks using MLM combined with techniques such as those described below. (Longer distance measurements using paramagnetic probes, has been reported. In that case the much larger magnetic moment of the unpaired electron supports the measurement).

PARALLEL AND DISTRIBUTED COMPUTING FOR NMR DATA PROCESSING AND NMR MODELING

Maximum likelihood processing is one example of the computationally intensive scheme for data reduction. For NMR spectra of 3- or higher dimensions, even standard Fourier transform data reduction can tax computational resources, and certainly limit the amount of interaction between the scientist and the data. Tomorrow's 3-dimensional NMR spectra will consist of 10^8 to 10^9 data elements (or more) and computation of 3-dimensional Fourier transform for such a large array requires both cpu intensive and disk I/O intensive operations for sustained calculation. A maximum likelihood calculation on a data array of this size would be prohibitive on today's computational machines.

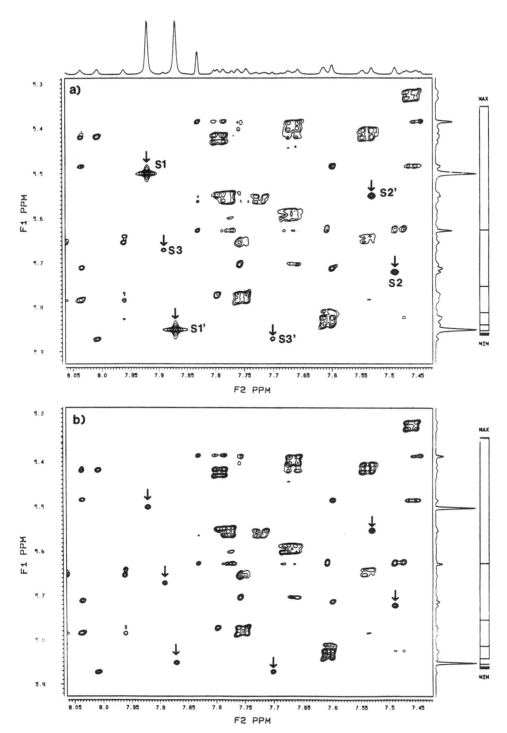

Figure 9. MLM spectra of combined real and synthetic data
a) MLM (-4 Hz/100 iterations).
b) MLM (-6 Hz/100 iterations).

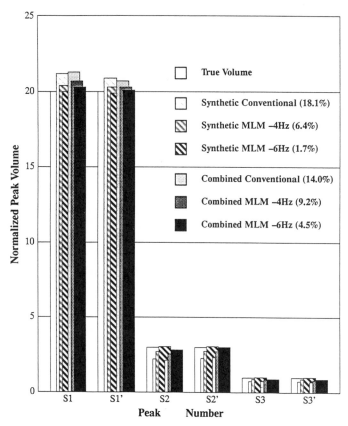

Figure 10. Volume deviation of synthetic conventional and MLM data in combined spectra.

This situation will undoubtedly change over the next few years. Computing technology is advancing at such a rapid and accelerating rate, that it is anticipated that by 1992 a high-end desktop workstation (cost $30,000 ?) will have compute power for these types of applications equivalent to a Cray XMP; by 1996-98, a single user workstation may approach gigaflop speeds. It is widely predicted that by the year 2000, the larger supercomputers will be teraflop machines (over 1,000 times more powerful than today's supercomputers); those machines will be massively parallel architectures and an exciting aspect of that fact is that the parallel architectures will probably be fully scalable. That is, a desktop workstation in 2000 might consist of approximately 1% of the resources present in a year 2000 supercomputer, and thereby offer 10 gigaflops of compute power.

All of this science fiction is a few years away. Nevertheless, utilization of massive parallelism and distributed network computing is not automatically achieved and it is for this reason that our laboratory is spending considerable effort to develop new computing paradigms to take advantage of the expected computing environments.

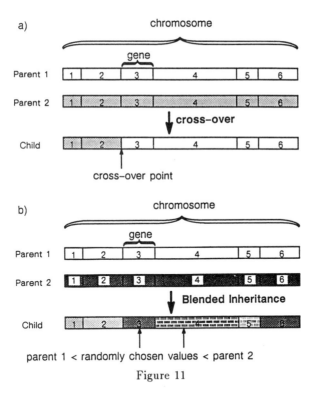

a)

chromosome

gene

Parent 1

| 1 | 2 | 3 | 4 | 5 | 6 |

Parent 2

| 1 | 2 | 3 | 4 | 5 | 6 |

↓ **cross-over**

Child

| 1 | 2 | 3 | 4 | 5 | 6 |

cross-over point

b)

chromosome

gene

Parent 1

| 1 | 2 | 3 | 4 | 5 | 6 |

Parent 2

| 1 | 2 | 3 | 4 | 5 | 6 |

↓ **Blended Inheritance**

Child

| 1 | 2 | 3 | 4 | 5 | 6 |

parent 1 < randomly chosen values < parent 2

Figure 11

The Syracuse laboratory is currently developing computing methods to exploit use of distributed and cooperative network computing, with additional fine-grain parallel optimization for problems requiring and supporting those capabilities (including large array Fourier transform and MLM processing as well as aspects of NMR molecular modeling).

Much of the success of using distributed computing depends on the existence of *efficient* distributed parallel algorithms. These algorithms must be scalable, i.e., with an increase in the size of the problem they must be able to use more machines to keep the overall processing time low. This means that they must exhibit linear or non-linear speedups.

Currently, much of the work on parallel distributed programming has been focused on homogeneous distributed systems. In practice, computer systems available form a heterogeneous distributed system. Heterogeneity can occur in a variety of ways, and must be taken into account for the efficient processing.

The above NMR related applications offer a high degree of parallelism, ranging from coarse-grained to fine-grained. Since our distributed system comprises of virtually all kinds of architectures such as SIMD (Connection Machine), MIMD (8 processor Alliant FX-80), and a network of high performance workstations, our goal is to develop the applications in a manner

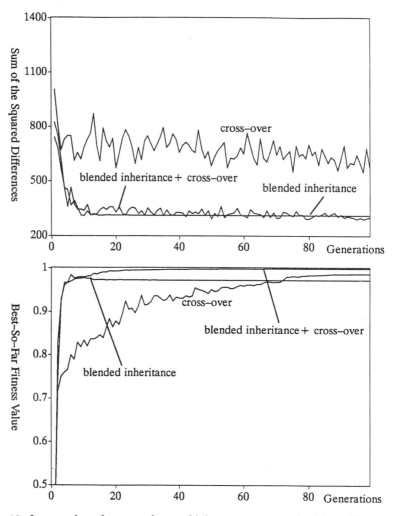

Figure 12. Improved performance by combining cross-over and mixing. Parameters: 500 chromosomes' pool, 100 generations, categorized searching space (BIG, SMALL, NULL and NO-INFO; see text for detail).

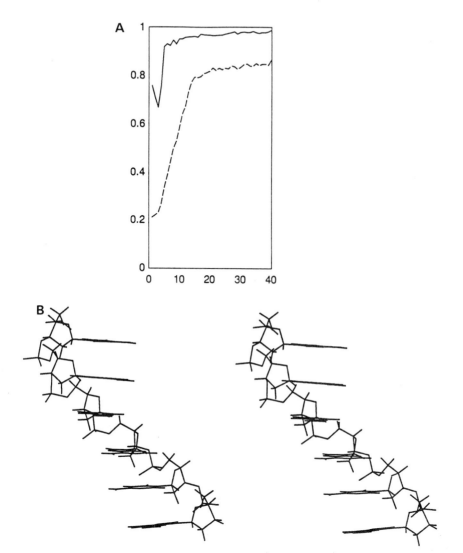

Figure 13. (Top) The solid line shows the BSF value (see text) and the dash line shows the avearage fitness value of the chromosome pool. The results were obtained from step 3 (structure refinement) of a DNA hexamer d(CA-CACG). The pool size is 200. The number of generation is 40.

(Bottom) The stereoview of the DNA d(CGCGCA) structure. See Fig. 6 for parameters. The structure was obtained by weighted average from the 40th generation chromosome pool using GA step 3 refinements.

that parallelism is exploited at all levels. This approach would allow the different sections of the applications to be mapped onto the most suitable architecture, e.g., fine-grained parallel execution can be performed most efficiently on Connection Machine, while those computations which are not suitable for Connection Machine can be performed on the Alliant or Stellar.

Developing distributed programs with linear speedups has until now been considered a difficult task. This is because it requires a proper decomposition strategy and elimination of as much synchronization and communication overhead as possible.

In practice, a problem can be decomposed into subproblems by a variety of methods, with the degree of parallelism ranging from very coarse-grained to fine-grained. Until now, performance results for parallel algorithms on loosely-coupled distributed systems, has been limited only to a small number of processors, based on very coarse-grained parallelism. Utilizing a larger number of processors for the same problem size has often required more finer-grained parallelism, but the resulting higher synchronization and communication cost, limits the performance.

The situation gets complicated when efforts are made to extend parallelism from one algorithm to another at the application level. To reduce the effect of synchronization, the computation must be restricted to only a small number processors. However, to reduce the execution time, more processors must be used. To achieve both of these conflicting goals, we outline a hierarchical scheme, where initially the problem is divided into very coarse-grained subproblems. Thus synchronization is limited only to a small number of processors. However, each of the subproblems is then recursively subdivided into finer grained subproblems to utilize any remaining idle processors. *The size and the number of these finer-grained subproblems is decided dynamically at run time and is a function of the size of the problem, number of processors available, and communication latency.* Since our aim is to distribute the processing to as many processors as possible, and at the same time *keep most of them busy,* we use a dynamic load balancing scheme to achieve the optimal decomposition strategy and size. The method is especially attractive for heterogeneous distributed systems, where computing resources may greatly vary from node to node, thus requiring irregular decomposition strategies.

One of the problems now at hand is to provide an efficient dynamic load balancing scheme. The goal of any load balancing scheme is for each processor to perform an equitable share of the total work load. At the operating system level, dynamic load balancing is achieved by migrating the task at run time from a heavily-loaded machine to another, idle, machine. However, in parallel programming environments this approach is not appropriate for a number of reasons. Firstly, moving a task at run time from

Table 1

Data	Configuration	Time[a]	Speedup
128×128×128	SPARC 1+	2400 secs	1
128×128×128	12 SPARCstations (heterogeneous)	236 secs	10.16
256×256×256	SPARC 1+	18816 secs	1
256×256×256	12 SPARCstations (heterogeneous)	1873	10.04

[a]3D FFT algorithm uses small RAM configuration and is not yet optimized.

one machine to another involves high overhead in terms of setting up the new execution environment. Secondly, migration can take place only in a homogeneous environment, moving tasks from one machine architecture to a different one is often not possible. Thirdly, the migration process involves high communication overhead as the data required by the task must also be retransmitted with each migration. Fourthly, it requires each processor to broadcast its load from time to time, thus generating additional network traffic. Finally, the system cannot use some application-specific information for higher efficiency.

PERFORMANCE RESULTS

As shown in Table 1, near linear speedups were obtained for our implementation of 3D FFT. Much of the success can be attributed to the simple but effective load balancing scheme. Due to load balancing, communication was done only when a request was made, which was initiated before the requesting processor became idle. Thus communication was able to keep up with the data requirements of the workers. Also, since each packet of work contained the sender's address, each worker was able to anticipate the needs of other workers, thus communication was initiated even before the other worker had requested for data. Thus we avoided transmitting data to machines where it was never going to be used. As a result of this load balancing and the paradigm used for the algorithm most communication took place in parallel to the computation, without affecting overall processing time.

Multi-dimensional Fourier transform processing is certainly not the only application for distributed network processing, or for fine-grained parallel processing. The brief section below describes our initial work in utilizing the genetic algorithm approach for parallel implementation of NMR molecular modeling.

A NEW PROJECT: PARALLEL MOLECULAR MODELING USING A GENETIC ALGORITHM

Twenty-five years ago, John Holland pioneered the field of genetic algorithms[43]. These algorithms simulate natural evolution, with each species evolving according to its "fitness" to an environment. As in natural evolution, each species searches for beneficial adaptation during its evolution. The genetic algorithm is a "weak" method that makes no assumptions about the system at hand. It is a search method based on the localization of solutions in hyperplanes and the search space. The Syracuse laboratory is using the genetic algorithmic approach in an attempt to develop a robust NMR molecular modeling system which will be able to generate and evaluate complex molecular structures. It is hoped that this approach may also produce an adaptive modeling scheme which extends itself, providing confidence in later results.

Our implementation of the genetic algorithm is envisioned to proceed in several steps. Initially, the molecular correlation time is evaluated. The second step is a genetic optimization procedure for interproton distances determined directly from the experimental NOE cross-peak data. In the third step, interproton distances are used as a target in a crude structure determination. Finally, the relaxation matrix of the structures are calculated, diagonalized and compared with the experimental data; additional experimental data can also be incorporated at this step. The genetic algorithm in step 2 optimizes distances but in steps 3 and 4 we plan to experiment with genetic optimization of the dihedral angle variable.

Implementation of the genetic algorithm for this study utilizes standard techniques and adds a blended inheritance mechanism to facilitate rapid evolution (convergence) as well as utilization of both vector target functions and a sharing target function.

Fig. 11 shows the mechanisms of crossover and blended inheritance; mutation is not shown but is represented by a single change of value for an individual gene of the chromosome. Fig. 12 shows the advantage of blended inheritance utilized with crossover to assist rapid convergence in early generations and to allow the final "fit" to be optimal. (Fig. 12 shows the results of genetic operations for determination of interproton distances; step 2 as described above). Finally, Fig. 13 shows preliminary results obtained for step 3 (of 4) modeling, from a test case of a DNA hexamer which has about 50 hydrogens. The pool size was 200 and it ran for 40 generations.

SUMMARY

This snapshot view of the research at Syracuse University on NMR data processing and its utilization for determination of solution structures of complex molecules is a status report as of mid-1990. We expect to see many advances (and experience some inevitable setbacks) during the next several years. The potential for reliable and validated NMR molecular modeling is so important, that we plan to proceed with as rapid a pace as we can.

REFERENCES

1. B. R. Frieden, *J. Opt. Soc. Am.* **62**, 511 (1972).
2. B. R. Frieden, in "Picture Processing and Digital Filtering", T. S. Huang, ed., 177-248, Springer-Verlag, Berlin/New York (1975).
3. S. F. Gull and G. J. Daniell, *Nature* (London), **272**, 686 (1978).
4. R. K. Bryan, Ph.D. thesis, University of Cambridge (1981).
5. S. Sibisi, *Nature* **301**, 134 (1983).
6. J. Skilling, *Nature* **309**, 748 (1984).
7. S. Sibisi, J. Skilling, R G. Brereton, E D. Laue, and J. Staunton, *Nature* **311**, 446 (1984).
8. J. Skilling and R. K. Bryan, *Mon. Not. R. Astron. Soc.* **211**, 111 (1984).
9. J. C. Hoch, *J. Magn. Reson.* **64**, 436 (1985).
10. E. D. Laue, J. Skilling, J. Staunton, S. Sibisi, and R. B. Brereton, *J. Magn. Reson.* **62**, 437 (1985).
11. P. J. Hore, *J. Magn. Reson.* **62**, 561 (1985).
12. J. F. Martin, *J. Magn. Reson.* **65**, 291 (1985).
13. E. D. Laue, J. Skilling, R. B. Brereton, S. Sibisi, and J. Staunton, *J. Magn. Reson.* **62**, 446 (1985).
14. E. D. Laue, M. R. Mayger, J. Skilling, and J. Staunton, *J. Magn. Reson.* **68**, 14 (1986).
15. J. Capon, *Proc. IEEE* **57**, 1408 (1969).
16. F. Ni and H. A. Scheraga, QCPE Documentation, 573 (1988).
17. F. Ni and H. A. Scheraga, *J. Magn. Reson.* **82**, 413-418 (1989).
18. R. E. Hoffman, A. Kumar, K. D. Bishop, P. N. Borer, and G. C. Levy, *J. Magn. Reson.* **83**, 586-594 (1989).
19. R. H. Newman, *J. Magn. Reson.* **79**, 448 (1988).
20. P. A. Jansson, R. H. Hunt, and E. K. Plyler, *J. Opt. Soc. Am.* **60**, 596 (1970).
21. W. E. Blass and G. W. Halsey, "Deconvolution of Absorption Spectra," Academic Press, New York (1981).
22. P. A. Jansson, in "Deconvolution with Applications in Spectroscopy", (P.A. Jansson, ed.), 96-132, Academic Press, New York/Orlando (1984).
23. G. W. Halsey and W. E. Blass, in "Deconvolution with Applications in Spectroscopy, (P.A. Jansson, ed.), 188-225, Academic Press, New York/Orlando (1984).
24. G. J. Thomas, Jr. and D. A. Agard, *Biophys. J.* **46**, 763 (1984).
25. B. P. Medoff, "Proceedings, IEEE International Conference on Acoust., Speech, Signal Processing", Tampa, FL, 1073-1076 (1985).
26. B. P. Medoff, in "Image Recovery: Theory and Application", H. Stark, ed., 321-368, Academic Press, New York/Orlando (1987).
27. F. Ni and H. A. Scheraga, *J. Raman Spectrosc.* **16**, 337 (1985).
28. F. Ni, G. C. Levy, and H. A. Scheraga, *J. Magn. Reson.* **66**, 385 (1986).

29. A. R. Mazzeo and G. C. Levy, *Comput. Enhanced Spectrosc.* **3**, 165, (1986).
30. A. A. Bothner-By and J. Dadok, *J. Magn. Reson.* **72**, 540 (1987).
31. M. A. Delsuc and G. C. Levy, *J. Magn. Reson.* **76**, 306 (1988).
32. A. R. Mazzeo, M. A. Delsuc, A. Kumar, and G. C. Levy, *J. Magn. Reson.* **81**, 512-519 (1989).
33. H. Barkhuusen, R. De Beer, W. M. M. J. Bovée, and D. Van Ormondt, *J. Magn. Reson.* **61**, 465 (1985).
34. J. Tang and J. R. Norris, *J. Magn. Reson.* **69**, 180 (1986).
35. J. Tang and J. R. Norris, *J. Chem. Phys.* **84**, 5210 (1986).
36. A. E. Schussheim and D. Cowburn, *J. Magn. Reson.* **71**, 371 (1987).
37. H. Gesmar and J. J. Led, "Spectral Estimation of Two-dimensional NMR Signals Applying Linear Prediction to Both Dimensions", Thesis, Univ. of Copenhagen (1987).
38. F. Ni and H. A. Scheraga, *J. Magn. Reson.* **70**, 506 (1987).
39. M. A. Delsuc, F. Ni, and G. C. Levy, *J. Magn. Reson.* **73**, 548 (1987).
40. G. L. Bretthorst, "Bayesian Spectrum Analysis and Parameter Estimation", Ph.D. thesis, Department of Physics, Washington University, St. Louis, Missouri, August (1987).
41. E. T. Jaynes, in "Maximum-Energy and Bayesian Spectral Analysis and Estimation Problems", C. R. Smith and G. J. Erickson, eds., p. 1, Reidel, Dordecht, Holland (1987).
42. G. L. Bretthorst, C. C. Hung, D. A. D'Avignon, and J. J. H. Ackerman, *J. Magn. Reson.* **79**, 369-376 (1988).
43. J. H. Holland, K. J. Holyoad, R. E. Nisbett, and P. R. Thagard, *Induction: Process of Inference, Learning and Discovery* The MIT Press (1986).

APPLICABILITY AND LIMITATIONS OF THREE-DIMENSIONAL NMR SPECTROSCOPY FOR THE STUDY OF PROTEINS IN SOLUTION[1]

Rolf Boelens[a], Christian Griesinger[b], Lewis E. Kay[c], Dominique Marion[d], and Erik R.P. Zuiderweg[e,f]

[a]Department of Chemistry
University of Utrecht
Padualaan 8
3584 CH Utrecht, The Netherlands
[b]Institute for Organic Chemistry
Johann Wolfgang Goethe University
Niederurseler Hang
6000 Frankfurt-am-Main 50
Federal Republic of Germany
[c]Laboratory of Chemical Physics
Bld. 2, NIDDK, National Institutes of Health
Bethesda, MD 20892, USA
[d]Centre de Biophysique Moleculaire
C.N.R.S.
F-45071 Orleans Cedex 2, France
[e]Pharmaceutical Discovery
Abbott Laboratories
Abbott Park, IL 60064, USA
[f]Present address:
Biophysics Research Division
University of Michigan
2200 Bonisteel Blvd.
Ann Arbor, MI 48109, USA

ABSTRACT

Assignment procedures for backbone resonances of proteins are outlined for homonuclear and heteronuclear three-dimensional NMR spectroscopy. Phasecycling procedures in fast 3D experiments are discussed. Examples of data space reducing processing techniques are given. A new triple resonance based assignment procedure for protein resonances is outlined. The general advantages and disadvantages of homonuclear and heteronuclear three-dimensional NMR spectroscopy are discussed.

INTRODUCTION (E.R.P. Zuiderweg)

The determination of the three-dimensional structures of small biomolecules (MW <10,000) in solution by two dimensional (2D) NMR spectroscopy has become a well established technique (for a review see ref. 1). The protocol consists of the assignment of the resonances of the proton NMR spectrum by scalar-correlated spectroscopy (COSY, DQF-COSY, TOCSY/HOHAHA), followed by the identification of nuclear Overhauser effect cross peaks between the resonances of sequentially adjacent structural units (amino acids, bases, or monosaccharides) from 2D NOE spectra. This assignment forms the basis of the subsequent search for NOE distance constraints between sequentially remote protons using 2D NOE spectra. The obtained NOE constraints are then used in structure calculation programs to obtain an ensemble of structures compatible with the NMR data.

This approach breaks down for larger systems, mainly because of the overlap problems in the 2D data sets caused by the larger number of resonances. This overlap interferes with unambiguous analysis. At the time two-dimensional NMR was developed, it was already realized that extension of the methods into more dimensions would lead to additional resolution[2]; just as two-dimensional spectra are better resolved than one-dimensional spectra, so are three-dimensional (3D) spectra better resolved than two-dimensional datasets. Thus, the above mentioned problem of overlapping resonances for larger systems can in principle be resolved by three dimensional (3D) spectroscopy. 3D NMR experiments are constructed by recording a series of 2D experiments as a function of another 2D experiment[3] as shown in Fig. 1.

The first reports on 3D NMR combined two homonuclear (proton) experiments such as COSY with COSY, NOESY with COSY, COSY with J-spectroscopy[3,4,4b]. These pioneering experiments showed that 3D NMR was practical and that the large amount of data involved could be handled properly; early applications of these homonuclear 3D experiments (NOESY with HOHAHA[5,6]) on proteins were published in 1988. A second approach to 3D NMR was demonstrated quickly thereafter, by combining a heteronuclear correlation experiment such as HMQC with a homonuclear experiment such as NOESY[7]. These experiments have the advantage that only large inter-

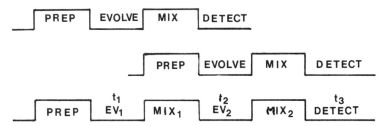

Figure 1. The construction of a three dimensional NMR experiment from two two-dimensional NMR experiments. The data are acquired during t_3 as a function of a matrix of t_1 and t_2 time variables, which are incremented independently.

actions are involved in the coherence transfer steps, which make them very suitable for larger systems, but have the disadvantage that the biomolecules need to be isotopically enriched. The application of heteronuclear 3D experiments to proteins (^{15}N labeled) was demonstrated in the next year[8,9]. A homonuclear 3D experiment suitable for the studies of larger proteins (NOE-NOE) was developed in the same year[10]. In the first half of 1990, the problem of the effect of the larger linewidth for larger systems on coherence transfer efficiency was addressed by redirecting these transfers over the ^{13}C backbone[11,12], protein NOE spectra were resolved with respect to ^{13}C[13,14], the first 4D experiment appeared[15], and an alternative assignment protocol for ^{13}C, ^{15}N labeled proteins was described[16,17].

To date, 3D NMR methods have been applied to Staphylococcal Nuclease, C5a, T4-lysozyme, Calmodulin, Interleukin 1 β, Rnase, Cyclophylin, Purotoxin, pike parvalbumin, Carbohydrates and DNA fragments. These rapid applications were possible because several problems associated with the recording, processing and interpretation of 3D NMR data could be solved (more versatile pulse-programmers, larger disks, faster computers, and advances in the biotechnology of isotopic labeling). The following reports describe some of these aspects, several issues that have not been solved to-date, and give an overview of the current state-of-the-art in 3D spectroscopy.

PROTEIN SEQUENTIAL ASSIGNMENT PROCEDURES IN HOMO- AND HETERONUCLEAR 3D NMR SPECTROSCOPY (C. Griesinger)

Homo- and heteronuclear 3D NMR spectroscopy[18] of biomacromolecules relies mainly on the fact that the resolution due to the display of three chemical shifts is increased compared to the resolution of 2D spectra. One prominent application of 3D spectra so far concerns sequential assignment of resonances. For this purpose it is of interest how the increased resolution of 3D spectra translates into higher reliability and less ambiguity in the sequential assignment process.

The assignment procedure in 2D spectra can be described in the following way. Given a chain of spins A,B,C,D,E which pairwise interact either via scalar or dipolar coupling, we will find by recording the appropriate 2D spectra the following cross peaks: $(\omega_1, \omega_2) = (\Omega_A, \Omega_B), (\Omega_B, \Omega_C), (\Omega_C, \Omega_D)$, and (Ω_D, Ω_E). These successive cross peaks have each one chemical shift in common. Thus they are connected in the 2D spectrum/spectra by a one dimensional search. The procedure becomes ambiguous as soon as there is degeneracy in the 1D spectrum.

In three dimensions, the optimal combination of 3D experiments will produce the following chain of cross peaks: $(\omega_1, \omega_2, \omega_3) = (\Omega_A, \Omega_B, \Omega_C)$, $(\Omega_B, \Omega_C, \Omega_D), (\Omega_C, \Omega_D, \Omega_E)$. Again, successive cross peaks can be connected by a one-dimensional search, however now keeping two frequency coordinates fixed when going from one cross peak to the next. Therefore this 3D assignment procedure only becomes ambiguous if there is overlap between pairs of resonances which is less likely than overlap between resonances themselves. Thus this is the optimal 3D assignment procedure. It is more robust against overlap than 2D assignment procedures.

Analysis of some often used 3D sequences in application for proteins reveals the following: The homonuclear sequence NOESY-TOCSY[5,6] can be applied in the optimal way for sequential assignment both for β sheet (relying on the strong NOE between H_i^α/NH_{i+1} and the intraresidual NH, H_α coupling) or α helical (relying instead on the NH_i, NH_{i+1} NOE) secondary structures. The chain of cross peaks is constituted for β-sheet by: $(\omega_1, \omega_2, \omega_3) = (NH_i, H_{i-1}^\alpha, NH_{i-1}), (H_{i-1}^\alpha, NH_i, H_i^\alpha), (NH_{i+1}, H_i^\alpha, NH_i)$ and for α-helix by: $(\omega_1, \omega_2, \omega_3) = (NH_i, NH_{i-1}, H_{i-1}^\alpha), (NH_{i-1}, NH_i, H_i^\alpha), (NH_{i+1}, NH_i, H_i^\alpha)$.

For heteronuclear sequences[7,18] like NOESY-hetero-COSY and TOCSY-hetero-COSY with $^{15}N/^{13}C$ enriched proteins the situation is more complicated. If ^{15}N enrichment alone is used, only for α-helical secondary structures the optimal assignment procedure can be applied: $(\omega_1, \omega_2, \omega_3) = (NH_i, ^{15}N_{i-1}, NH_{i-1}), (NH_{i-1}, ^{15}N_i, NH_i), (NH_{i+1}, ^{15}N_i, NH_i)$. This is demonstrated by a chain of three cross peaks substantiating the Val54-Glu55-Ala56 moiety contained in one of the α-helices of ribonuclease A in Fig. 2. If heteronuclear sequences should be applied, β-sheet secondary structures need for the ideal assignment procedure in addition to ^{15}N labeling also ^{13}C labels at least in the C_α-positions to get the following chain of resonances from the altogether four combinations of NOESY or TOCSY with C,H-hetero-COSY or N,H-hetero-COSY: $(\omega_1, \omega_2, \omega_3) = T(H_i^\alpha, ^{15}NH_i, NH_i), T(H_i^\alpha, ^{13}C_i, NH_i), N(H_i^\alpha, ^{13}C_i, NH_{i+1}), N(H_i^\alpha, ^{15}N_{i+1}, NH_{i+1}), T(H_{i+1}^\alpha, ^{15}NH_{i+1}, NH_{i+1})$. (TOCSY peaks are designated with a T, NOESY peaks with an N). These experiments have been successfully carried out for Ribonuclease H[19].

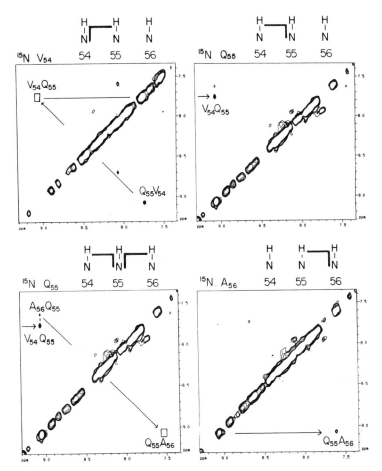

Figure 2. The sequential walk through the tripeptide moiety Val54-Glu55-Ala56 of ribonu-clease A by a one-dimensional search is demonstrated in a NOESY-^{15}N,H-hetero-COSY experiment. ^{15}N planes are shown together. The assignment of the respective cross peaks is given by a bar connecting the three correlated spins on the top of each spectrum. The NH resonance assignment is given in the spectra. Rectangles in the spectra designate po-sitions of the type $(NH_i, {}^{15}N_i, NH_{i-1})$ which are obtained by reflection of the observable $(NH_{i-1}, {}^{15}N_i, NH_i)$ peaks at the $\omega_1 = \omega_3$ plane (C. Griesinger).

DISCUSSION OF DR. GRIESINGER'S PAPER
(E.R.P. Zuiderweg)

The general question of the sensitivity of homonuclear 3D NMR spectroscopy as compared to heteronuclear 3D NMR spectroscopy was raised by the moderator. The workers who co-authored the first (homonuclear) 3D experiment on a small protein (Drs. H. Oschkinat, C. Griesinger and G. M. Clore (ref. 5)), indicated that despite the relatively high concentration (6.8 mM), the NOESY-HOHAHA 3D spectrum of the small purotoxin protein was not entirely complete.

HOMONUCLEAR 3D NMR OF BIOMOLECULES (R. Boelens)

Structure determination of biomolecules by NMR relies strongly on the observation of proton-proton NOE's[1]. First, they are essential in the so-called sequential assignment methods where neighbouring proton spin systems (often derived from proton J-coupling networks) are connected via unique NOE's at the polymer backbone. Secondly, most NMR methods for structure determination are based on proton-proton distance constraints, which can be derived from the NOE intensities. However, for large molecules (with a molecular weight above 10 kDa) the overlap of cross peaks in 2D NOE spectra even at high magnetic fields is already considerable, which complicates the assignment process and reduces the amount of observable distance constraints. Recently, both homo- and heteronuclear 3D NMR methods have been proposed to increase the resolution of the NMR spectra[4-9,20]. An advantage for the homonuclear technique is that no special isotope labeling of the biomolecular material is required. This makes it suitable for a broad range of biomolecules. Furthermore a single homonuclear 3D experiment can in principle contain all information required for both sequential and structural analysis. However, a clear disadvantage is that the mixing processes of homonuclear 3D, i.e., homonuclear J-coupling and cross relaxation, can cause inefficient magnetisation transfer. In addition, sequential assignment strategies based on only the proton-proton NOE can be ambiguous. Previously applications of homonuclear 3D NMR with proteins, oligosaccharides and DNA fragments have been given[21] and sequential assignment procedures based on 3D NOE-HOHAHA spectra have been developed[22,23].

A Non-Selective Homonuclear 3D Experiment

For homonuclear 3D ^{1}H NMR it seems very logical to combine a NOE and a HOHAHA experiment, since two important proton-proton interactions are now measured in one experiment. Fig. 3a shows the 3D NOE-HOHAHA pulse sequence, where the FID in the time domain t_3 is recorded as a function

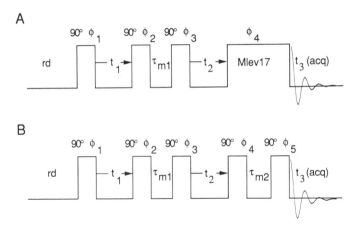

Figure 3. Pulse sequence for a 3D NOE-HOHAHA (a) and a 3D NOE-NOE experiment (b). MLEV-17 indicates the HOHAHA mixing sequence and τ_m corresponds to the NOE mixing time. Usually experiments with rf pulses with different phases are added (R. Boelens).

of two variable times t_1 and t_2. After Fourier transformation in the three dimensions a 3D spectrum is obtained with three independent frequency axes. A cross peak in this spectrum arises when magnetization of one proton is transferred in the first (NOE) mixing period to a second proton and then in the second (HOHAHA) mixing period to a third one. Since both the t_1 and t_2 time domains have to be incremented independently, a large number of FIDs are recorded in order to obtain a sufficiently high resolution for these two domains. In order to reduce measuring times and the amount of collected data it has been proposed to use semi-selective pulses which limit the spectral width to be sampled for the evolution domains[5,20]. However, in many cases such semi-selective or 'soft' pulses require additional phase cycling on a 180° echo pulse in order to obtain absorptive spectra. In that case a non-selective 3D NMR experiment which samples the full proton spectral width in all domains can be obtained in almost the same time and is therefore to be preferred. Measuring times can be reduced by minimal phase cycling schemes for reduction of artifacts. The HOHAHA mixing does not require any phase cycling to select the proper magnetization transfer and is therefore very suitable for 3D NMR. The NOE mixing requires no special phase cycling either, at least if a homospoil pulse can be given in the mixing time without interference with the spectrometer stability. If not, a choice must be made for SQC suppression (in a two-step phase cycle and generally sufficient for large biomolecules) or additional DQC suppression (in a four-step phase cycle). Suppression of axial peaks which develop in the evolution periods, can be accomplished by inversion of the first rf pulse combined with receiver switching. Often, imperfections of the instrument demand dummy scans and additional phase cycling. In many cases these phases can be combined with the coherence suppression scheme. Thus, one FID in the 3D NOE-HOHAHA experiment can be recorded with one to 8 phases plus zero to two dummy scans. With 256 increments of both the t_1 and t_2 period and with a repetition rate of 1 s, this results in 18.2 to 182 hrs of measuring time.

A Single-Scan 3D NOE-NOE Experiment

The miminal amount of scans per FID required for a non-selective 3D experiment is one. In order to test this we recorded a non-selective 3D NOE-NOE experiment of the protein parvalbumin (109 a.a.) in ^1H_2O with only a single scan. Coherences were suppressed by homospoil pulses in both NOE mixing times of 150 ms. The last rf pulse of the NOE-NOE pulse sequence was a semi-selective 45-τ-45 pulse, which would not excite the ^1H_2O resonance. The quadrature image was eliminated from the 3D spectrum by positioning the rf carrier to the left of the amide region and discarding one half of the spectrum. Axial peaks were not suppressed, but there was sufficient space between the carrier and the spectrum to eliminate overlap. The loading of our instrument's pulse programmer together with the storage of the FID on disk caused an effective relaxation delay of 2.3 s. With 160*256 increments this resulted in 30 hrs measuring time. Fig. 4 shows a cross section perpendicular to ω_3 through this 3D spectrum at 10.33 ppm. Comparison with a more traditional 3D NOE-NOE experiment recorded with 8 scans per FID and phase cycling, indicated that many 3D cross peaks can still be observed in this shorter experiment cf. Fig. 6). In fact, the S/N ratio was surprisingly good, probably due to limited sampling of t_1/t_2 noise in the short experimental time.

Analysis of Sequential and Medium Range Connectivities of a Protein by 3D NOE-HOHAHA and 3D NOE-NOE Spectroscopy

Recently, we have analyzed the amide ω_3 cross sections of the 3D HOHAHA-NOE spectrum of the protein parvalbumin in ^1H_2O and compared the sequential and medium range connectivities, which can be obtained from the 3D spectrum, with those extracted from 2D spectra[23]. The 3D spectrum allowed the observation of 455 3D cross peaks involving short and medium range NOE's, on which the assignment of 108 amino acids could be based. For the sequential contacts this is comparable to what was obtained previously from a whole series of 2D NOE spectra. In general, there were less d_{NN} based NOEs observed, first because of the lower digital resolution which makes closely resonating protons indistinguishable and secondly because of the short T_1 and T_2 relaxation times of amide protons in the mixing and evo-

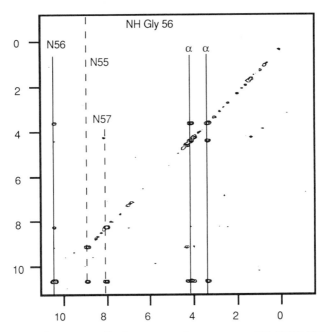

Figure 4. Cross section perpendicular to ω_3 of a one-scan 3D NOE-NOE experiment of parvalbumin at the NH frequency of Gly 56 ($\omega_3 = 10.33$ ppm). The spectrum was recorded at 500 Mhz with a 3D NOE-NOE sequence (*cf.* Fig. 3b) with a semi-selective 45-τ-45 detection pulse and the rf carrier positioned to the left of the spectrum (R. Boelens).

lution periods which makes 3D cross peaks involving two amide protons most vulnerable. Fig. 5a summarizes the sequential and medium range connectivities which could be observed in a 3D HOHAHA-NOE experiment recorded with a 'clean' MLEV17 sequence[24] which gives a more efficient HOHAHA magnetization transfer. The 3D dataset allowed even a better definition of the secondary structure than was previously possible with 2D, since a series of new medium range NOE's were observed, mainly because the medium range $d_{\alpha\beta}$ and $d_{\alpha N}$ NOEs can be detected in non-overlapping regions. Furthermore, a preliminary analysis of the 3D 'clean' HOHAHA-NOE spectrum demonstrated that most long range NOE connectivities as obtained from 2D spectra could be found in the 3D spectrum as well (A. Padilla and R. Boelens, unpublished results). Of course, it should be realized that the analysis of the 3D HOHAHA-NOE spectra of parvalbumin was not an ab initio assignment, but was based on previous assignments obtained from 2D spectra. Until now we only showed that one 3D HOHAHA-NOE spectrum in principle contains all the information needed for assignment, secondary structure analysis and probably tertiary structure determination.

For large proteins the HOHAHA magnetization transfer in the 3D HO-HAHA-NOE experiment could become inefficient in case of weakly coupled protons. Therefore, we have explored the feasibility of a homonuclear 3D

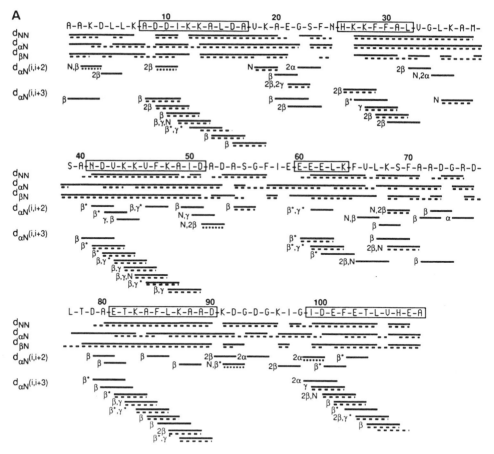

Figure 5. (a) Sequential and medium range NOEs in parvalbumin obtained from a 'clean' 3D HOHAHA-NOE spectrum (——) and those obtained from a set of 2D NOE spectra (- - - - - - -).

technique that uses only NOE mixing periods[25]. We recorded a 3D NOE-NOE spectrum of parvalbumin in ^1H_2O. Fig. 6 shows the corresponding cross sections from Ser 55 to Phe 57 in a 3D 'clean' HOHAHA-NOE spectrum and in a 3D NOE-NOE spectrum, demonstrating that sequential connectivities can be observed in both type of experiments. In fact most 3D connectivities identified in the 3D NOE-HOHAHA spectrum (Fig. 5a) can also be found in the 3D NOE-NOE spectrum (Fig. 5b). Although more complex for analysis, the 3D NOE-NOE spectrum can be used also for unraveling NOE patterns, as encountered in the sequential assignment of protein spectra. Since for large molecules cross relaxation becomes increasingly more efficient, the 3D NOE-NOE technique seems very suitable for large molecules.

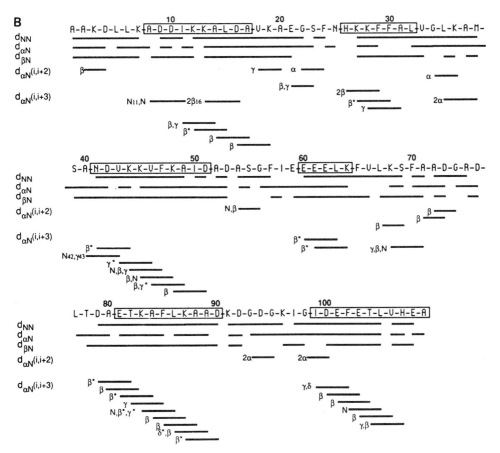

Figure 5 (continued). (b) Sequential and medium range NOEs in parvalbumin obtained from a 3D NOE-NOE spectrum (R. Boelens).

DISCUSSION OF DR. BOELENS' PAPER (E.R.P. Zuiderweg)

It was re-emphasized that the first 2D spectrum of the single scan 3D NOE-NOE spectrum contained hardly any peaks, while the sensitivity of the resulting 3D spectrum was quite acceptable. This illustrates the multiplexing procedure in 3D NMR very nicely; the sensitivity of the experiment is determined by the total number of scans. Furthermore, it was emphasized that it is therefore hazardous to compress data before all processing is done; when doing so information can be discarded by that process.

Dr. Boelens indicated that the protein concentration used for the non-selective homonuclear 3D experiments is around 5 mM.

Dr. Boelens made the general point that the labeling necessary for heteronuclear NMR is expensive, and that in many cases it is not a straightforward process. Especially in eukaryotic systems, where post-processing may be necessary, the labeling becomes even more of a challenge. Dr. Boelens also indicated that very good homonuclear 3D NMR spectra were obtained from DNA and carbohydrates; labeling of such systems is extremely difficult.

A general discussion developed about molecular weight limitations for NMR structure determination. Dr. Kay indicated that the question revolves around the necessity of experiments; if one can do without correlated information through the backbone, very high molecular weight proteins may be tackled; it is too early to come to a conclusion on these matters. Dr. Jardetzky made the comment that larger systems have been successfully studied with NMR (Trp repressor). Trp repressor dimer is over 25 kDa and has been studied using selective deuterium labeling. Dr. Jardetzky anticipated that much larger systems can be studied if these labeling methods are combined with heteronuclear 3D NMR.

Smaller systems may also benefit from 3D NMR and isotopic labeling since much overlap occurs in the spectra of such molecules as well. Dr. Zuiderweg indicated that a large amount of overlap became apparent in the 3D spectrum of C5a (a small protein, 8.5 kDa) once it was labeled with ^{15}N. It was also brought up by Dr. Kay that a project involving a 39 a.a. protein fragment would have been easier if ^{15}N labeling would have been available.

Dr. Boelens was asked how one could tell that cross peaks in the 3D NOE-NOE spectra are due to sequential interactions. Dr. Boelens indicated that a statistical approach is needed to do a reliable assignment from NOE-NOE spectra, and that work along those lines is in progress.

Figure 6. Cross sections through a 'clean' 3D HOHAHA-NOE and a 3D NOE-NOE spectrum of parvalbumin. Corresponding cross sections are shown from Ser 55 to Phe 57. Solid lines indicate the spin system (NH,C$^{\alpha}$H,C$^{\beta}$H frequencies) linked to the NH frequency of the cross section. Dashed lines indicate frequencies of neighbouring (sequential and medium-range) spin systems. The 3D HOHAHA-NOE spectrum was recorded with 4 scans at 500 MHz with a 'clean' HOHAHA pulse sequence of 44 ms including trim pulses and a NOE mixing time of 150 ms. The 3D NOE-NOE spectrum was recorded with 8 scans and 2 dummy scans at 500 MHz with two identical mixing times of 150 ms (R. Boelens).

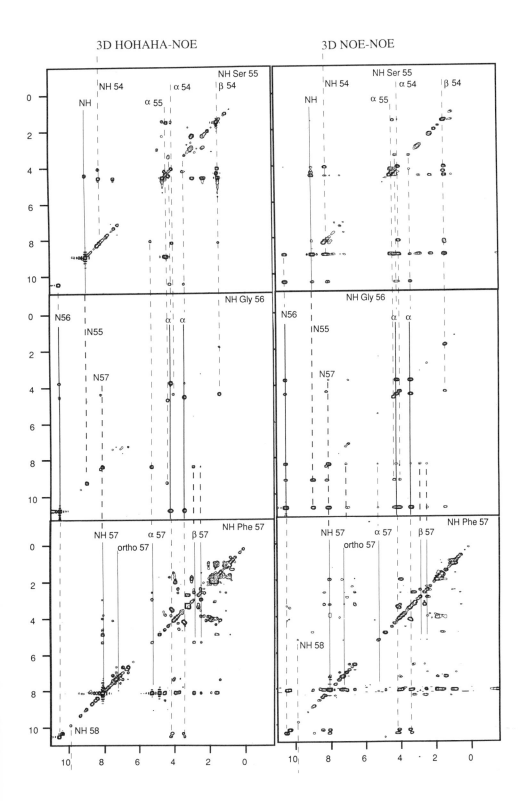

3D HOHAHA-NOE

3D NOE-NOE

139

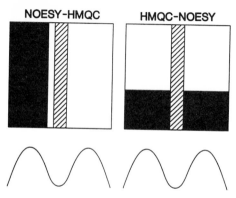

Figure 7. Comparison of the experimental artifacts observed in the NOESY-HMQC and HMQC-NOESY experiment. In NOESY-HMQC, the magnetizations, beared by any kind of protons (during t_1) are transferred by n.O.e. to the NH (detected during t_3), after its ^{15}N labelling during t_2. In HMQC-NOESY, the NH magnetizations, identified during t_1, are ^{15}N labelled during t_2, and then transferred to all protons. The regions of interest (NH-C$^\alpha$H, NH-C$^\beta$H correlation are shown as a black rectangle, and the noise artifacts during to the intense water correspond to the hatched area. In this respect, the NOESY-HMQC experiment is more suitable for protein studies. Furthermore, a frequency-selective excitation such as a 1-1 pulse (lower part of the figure) can be conveniently implemented for this option (D. Marion).

PRACTICAL CONSIDERATIONS FOR OPTIMIZING ^{15}N NOESY- OR HOHAHA-HMQC SPECTRA (D. Marion)

Optimization of Spectral Windows

3D experiments are made up from the combination of two 2D pulse sequences: for instance, in ^{15}N-resolved NOESY or HOHAHA 3D experiments, a homonuclear experiment is merged with a HMQC. As a result, two schemes can be devised, with either the HMQC or the homonuclear part (NOESY or HOHAHA) at the beginning of the sequence[7,8,26]. Whereas NOESY-HMQC and HMQC-NOESY are theoretically strictly equivalent, they differ in practice with regard to the location of various experimental artifacts, such as (t_1, t_2) noise and water residual signal. In fact, ^{15}N-resolved NOESY or HOHAHA spectra of ^{15}N-labelled proteins can only be recorded in H_2O, and one has therefore to cope with the suppression of the large water signal. For these reasons, the optimal compromise for recording these spectra make use of the NOESY- or HOHAHA-HMQC pulse sequence, where the amide proton chemical shift is sampled during the t_3 detection period (*cf.* Fig. 7).

Whereas this choice permits the easy observation of NH to C$_\alpha$H connectivities in an artifact-free area of the spectrum, it exhibits severe drawbacks, as far as digital resolution is concerned. During t_3 (where a increase of the digital resolution is almost free of charge), only a limited range of chemical shift (say from 5.5 to 10.5 ppm) has to be detected, as opposed to t_1 where

140

the entire ^{1}H shift range has to be sampled. Consequently, various tricks have been implemented in order to maximize the digital resolution. During the t_1 period, the carrier has to be in the middle of the proton spectrum, and during the t_3 period, in the middle of the NH region. This carrier shifting can be achieved either by an actual change of the synthesizer frequency (with phase coherence) or by more sophisticated NMR tricks. If the TPPi method[27] is used for the quadrature detection during t_1, the usual $\pi/2$ phase increment can be changed into a different value, leading to a shift not equal to SW/2. For hypercomplex data[28], the same result can be achieved using a linear time-dependent phase correction of the free induction decay, before FT. Moreover, the second solution is more convenient, because the amount of shift can anyhow be adjusted after the data collection, as opposed to the TPPi case.

These methods make it possible to center the spectral windows during t_1 and t_3 independently, and the spectral width can be squeezed to a maximum. Let us point out an additional way of increasing the digital resolution, only for hypercomplex data: If the first increment in t_2 (the ^{15}N dimension) is set in order to yield a 180° linear phase correction across the spectral width, the folded resonances will appear with opposite phase relative to the unfolded one making them easily identified. However, use of extensive folding may lead to signal cancellation of resonances with opposite sign. This method is not advisable in any of the proton dimensions, since baseline distortion will show up due to the immense diagonal peaks.

Processing of 3D NMR Spectra

Using spectrometers designed for 2D NMR, a 3D data set appears as several files (typically 64 or 128) recorded with different timing. For instance, for each t_2 value, a pair of (t_1, t_3) files are collected for a NOESY-HMQC 3D spectrum. Therefore, a straightforward manner of processing these data can rely on any commercially available software, for the F_1 and F_3 Fourier transform, supplemented by additional software for the F_2 FT. A prerequisite for this simple approach is an adequate format for the data storage. A software package has been written in a modular form by Kay, Marion, and Bax[26] for the F_2 Fourier transformation, as well as for the pre- and post-processing routines: all manipulations, that would normally be applied to a single data point in a one dimensional process, have been converted for plane processing.

The very first step involves the windowing of the data — each (t_1, t_3) plane is scaled by a constant factor — and the window function is designed to avoid too low a weighting factor for the last t_2 increments, at the expense of a rather small increase of truncation artifacts. Zero-filling is merely achieved by the creation of files, which only contain zeroed data points. The Fourier transform is a modified version of the 1D FFT for a set of N points, where N is a power of two. During its first step (the bit reversal routine), the reshuffling amounts to renaming the planes, without moving the data. The second step (the so-called butterfly algorithm) is a series of linear combination of planes: two planes are loaded from the disk, combined and then written again. As a result, the FT is done in place and the file structure (dictated by the choice of commercial software) is preserved.

Due to their size, 3D data cannot be interactively phased. In fact, with the exception of the detection dimension (t_3) where the hardware interferes with the phase problem (the zero order term originates from hardware phase shift, and the first order from filter time response), the required phase correction can be easily computed from the pulse sequence timing. The linear phase corrections are exclusively due to the finite duration of the pulses, during which the spins start processing. As a matter of fact, during a 90° pulse, the spins exhibit the same dephasing as during a delay of $2/\pi$ of its duration[2]. This simple rule can be used in order to compute the first-order term of the phase correction (the zero-order, as defined in the center of the spectrum, is zero). Let us point out that the time-domain phase correction (for carrier shifting) must be taken into account for computing the phase correction parameters. Here again, the t_2 phase correction amounts to a combination of planes, and is performed in place. The flexibility of this software has been shown recently on 4D NMR, where the 2 additional dimensions have been processed with no modification: files have to be renamed between the two processings.

Display of 3D NMR Data

^{15}N-resolved NOESY or HOHAHA spectra are conveniently displayed as ^{1}H-^{1}H contour plots, since the ^{15}N chemical shift is of no major concern, except for resolving overlaps. In fact, due to the large number of planes, there is a need for automation, and a laser printer can be used overnight for plotting the data. Handling of 64 or 128 sheets of paper for data interpretation is rather tedious, although most of the information is contained in rather limited areas of the spectrum. The 3D data can be compressed back into a 2D spectrum, by extracting strips along F_1 at the various chemical shifts of the amide pairs ($F_3 = {}^1$H and $F_2 = {}^{15}$N) (see Fig. 8 and Driscoll et al.[29]). Once the strips have been sequentially ordered, this data representation permits to back up the assignment by comparison of the peak patterns, principally of the C^β protons.

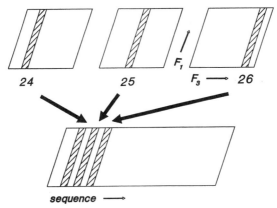

Figure 8. Data reduction for 3D NMR. Once the ^{15}N and ^{1}H frequencies of each amide group have been tabulated, strips running parallel to F_1 can be extracted by software in order to build a 2 spectrum. The strips are ordered arbitrarily before assignment and according the sequence after assignment (D. Marion).

DISCUSSION OF DR. MARION'S PAPER (E.R.P. Zuiderweg)

The question was raised how maximum entropy processing compared with linear prediction. Dr. Marion indicated that in his hands maximum entropy processing was more robust, and that he successfully applied it to process the indirect proton domain in a 3D NOESY-HMQC experiment.

The techniques described by Dr. Marion are geared towards minimizing the required data space. The question was raised by the moderator whether the acquisition of more data points at a higher sampling rate during detection (oversampling[30]) would not be advantageous (for dynamic range). Dr. M. Delsuc, co-author on a paper on this subject, indicated that the impact of oversampling is dependent on the make of the spectrometer and that it has to be determined case-by-case whether it is worth the burden of additional data.

Data table compression was discussed. Dr. Marion indicated that he processed with MEM only the (proton) columns containing signal for the last transform of the NOESY-HMQC data. This could be done because the diagonal shows where the cross peaks are to show up.

A NOVEL APPROACH FOR SEQUENTIAL ASSIGNMENT OF LARGER PROTEINS: TRIPLE RESONANCE THREE-DIMENSIONAL NMR SPECTROSCOPY (L.E. Kay)

Complete assignment of the backbone proton resonances of proteins forms the basis for both detailed structural and dynamical studies by NMR. Traditionally, this is carried out by relying on both through-space and through-bond connectivities provided by homonuclear NOESY and HOHAHA/TOC-

SY spectra[1]. While feasible for small proteins (< 10 kD), this approach does not yield unambiguous assignments for larger molecules due to extensive overlap and decreasing sensitivity of experiments relying on through bond proton magnetization transfer.

The limitations described above can be overcome by novel 3D heteronuclear triple resonance experiments which exploit the relatively large one-bond J couplings between the backbone ^{13}C and ^{15}N nuclei and between the backbone protons and the ^{15}N and ^{13}C nuclei to which they are directly attached[16,17]. The new experiments yield spectra almost free of overlap that provide sufficient information to obtain complete sequential backbone resonance assignments in a manner completely independent of the secondary structure of the protein under investigation. Because magnetization transfers often involve carbonyl or nitrogen spins having linewidths of only 4-8 Hz for proteins in the 15-20 KD molecular weight regime the sensitivity of the experiments is high.

Fig. 9 shows the four novel 3D pulse schemes developed to assign backbone resonances in proteins. The first 3D experiment, HNCO, correlates NH and ^{15}N chemical shifts of an amino acid with the carbonyl chemical shift (C') of the preceeding residue, thereby providing valuable sequential connectivity information. The second experiment, HNCA, correlates the NH and ^{15}N chemical shifts with the intra-residue C^{α} chemical shift. In addition, a correlation between the NH and ^{15}N spins and the C^{α} spin of the preceeding residue is often observed due to the two-bond ^{15}N-$^{13}C^{\alpha}$ coupling. In these favorable cases, a second source of sequential connectivity information is obtained. The third experiment, HCACO provides information relating to intra-residue connectivity by correlating the H^{α}, C^{α} and C' chemical shifts. Finally, the HCA(CO)N experiment provides sequential connectivity information by relating the H^{α} and C^{α} shifts with the ^{15}N shift of the subsequent residue.

Analysis of a HOHAHA-HMQC[31] spectra can also be extremely useful in the sequential assignment process. When combined with the HNCA experiment, the HOHAHA-HMQC experiment firmly establishes intraresidue correlations between pairs of ^{15}N-NH and C^{α}-H^{α} backbone resonances, despite significant overlap present in the 1D ^{15}N, ^{13}C and ^1H spectra. Once C^{α}-H^{α} pairs are obtained, the intra-residue C' chemical shift is determined from the HCACO experiment. The NH, H^{α}, ^{15}N, C^{α} and C' shifts from a

---

Figure 9. Pulse sequences of (a) the HNCO experiment, (b) the HNCA experiment, (c) the HCACO experiment and (d) the HCA(CO)N experiment. For schemes a and b, H_2O presaturation is used and in addition, in (a) low-power water irradiation is used during the intervals delta (solid bars). The mechanism of the experiments as well as details such as the phase cycling employed and the implementation of the methods on modern spectrometers can be found in ref. 17 (L.E. Kay).

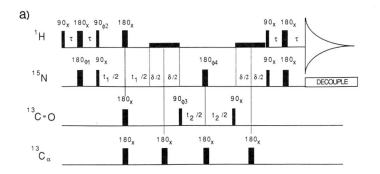

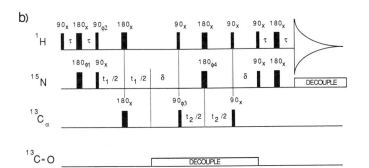

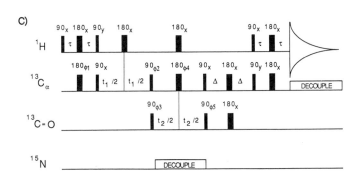

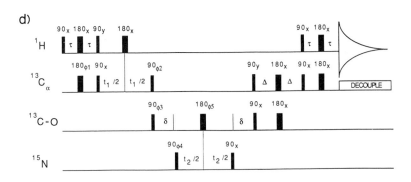

particular residue are subsequently linked using the HCA(CO)N and HNCO spectra together with the HNCA spectrum in those cases where a sequential correlation is observed.

Fig. 10 illustrates the resolution and sensitivity that can be obtained by these methods. Several slices from five separate 3D NMR spectra recorded for the protein calmodulin (MW 16.7 KD) are presented. Fig. 10A is a region of the (C′,NH) slice taken from the HNCO spectrum at an ^{15}N frequency of 117.4 ppm, establishing a correlation between the NH and ^{15}N chemical shifts of Lys-21 and the C′ shift of Asp-20. Fig. 10B is part of the corresponding slice of the HNCA spectrum, identifying the Lys-21 C$^\alpha$ chemical shift. The corresponding slice from the ^{15}N HOHAHA-HMQC spectrum identifies the H$^\alpha$ chemical shift of Lys-21 (Fig. 10C). A knowledge of the H$^\alpha$ and C$^\alpha$ chemical shifts establishes the intra-residue C′ shift using the HCACO spectrum shown in Fig. 10D. The ^{15}N chemical shift of the subsequent residue, Asp-22, is obtained from the HCA(CO)N spectrum (Fig. 10E). Finally, a knowledge of the C′ chemical shift of Lys-21 and the ^{15}N shift of Asp-22 enables the assignment of the HN shift of Asp-22 from the HNCO spectrum of Fig. 10F. In this fashion all of the backbone resonances of Lys-21 have been assigned and the connection between Lys-21 and Asp-22 is thoroughly established. Note that the slice of Fig. 10G shows a weak two-bond connectivity between the Asp-22 amide nitrogen and the C$^\alpha$ carbon of Lys-21, providing yet another confirmation of the connectivity between Lys-21 and Asp-22.

The sequential assignment approach outlined above provides a very powerful and direct method for analyzing the NMR spectra of proteins that can be isotopically enriched with ^{15}N and ^{13}C. The sensitivity of the new pulse schemes presented is sufficient to permit the recording of a 3D spectrum with a high signal to noise ratio in less than 2 days for protein concentrations in the 1 mM range. Because the spectral regions of interest are usually rather limited in any given dimension, data matrices are quite small, typically between 2-8 Mword, depending on the type of experiment. The approach described, together with 4D NMR spectroscopy[15], should enable the determination of high resolution solution structures for proteins up to at least 200 amino acids in an automated fashion.

DISCUSSION OF DR. KAY'S PAPER (E.R.P. Zuiderweg)

The question was asked whether it would be desirable to go to even higher dimensions in these experiments. Dr. Kay indicated that that depends on the system; if lower dimensionality yields the necessary resolution, it would be easier to pick the peaks and integrate. For systems of higher molecular weight than 17 kDa, four-dimensional NMR may be necessary, in conjunc-

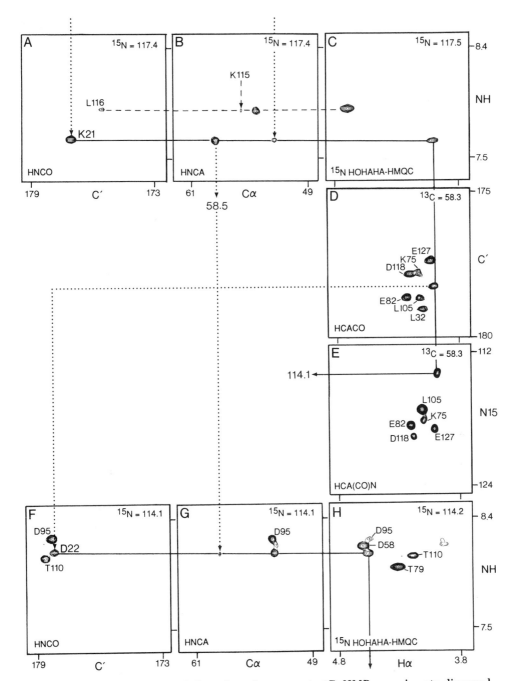

Figure 10. Selected regions of slices from five separate 3D NMR experiments discussed in the text. These regions illustrate the J correlations between Lys-21 and Asp-22. Solid and dotted lines trace the connectivity patterns for these two residues. The broken lines correspond to parts of the connectivity patterns observed for other residues (L.E. Kay).

tion with automated analysis. The problem of resolution in 4D NMR was addressed. Dr. Kay suggested to use multiple folding of the spectra in the different domains and to use a computed linear phase shift (180°) to be able to distinguish between folded and non-folded resonances (see also Dr. Marion's paper). Furthermore, it is extremely useful to use linear prediction; in fact, Dr. Kay has been experimenting with predicting 8 complex points out to 16. The experience is that sometimes negative T_2 values were created (interferogram increases in amplitude); but that these effects were dealt with by the apodization function and by "root reflection". The question was raised why it would be necessary to transform the predicted data; why is it not sufficient to use a table of coefficients. Dr. Kay indicated that that was a very interesting possibility. However, the approach of transforming was chosen because it would then be possible to go back and inverse transform another indirect dimension and repeat the linear prediction process.

Upon question, Dr. Kay indicated that it was necessary to have at least three channels on a spectrometer which can be phase cycled independently. More channels would give more capacities ($C=O$ treated independently from the rest of the carbon spectrum), and the number of experiments would certainly be expanded.

SUMMARY AND SOME UNDISCUSSED POINTS
(E. R. P. Zuiderweg)

The contributors to the session have addressed several aspects of 3D NMR. It was shown that homonuclear 3D experiments can be quite powerful for solving assignment problems in not too large proteins. The cost and availability aspect of heteronuclear 3D NMR was brought forth, but consensus was that the applicability of these experiments is such that people, time and money investments for the labeling are well worth it. It was indicated that not all labeling can be easily carried out in E. coli; plasmid rejection in minimal media as well as folding problems of proteins in these organisms necessitate the development of labeling procedures in other organisms.

Consensus was that the novel heteronuclear techniques would in principle enable biomacromolecules up to 25,000 kDa to be studied successfully; for larger systems additional dimensions (with a concommitant loss of approximately a factor of two in sensitivity per dimension when transfers are 100% effective) or additional selective labeling need to be considered to handle the broader lines. More dimensions or higher resolution in 3D NMR imply that hardly any phase cycling can be done per indirect time increment combination. The feasibility of that approach is demonstrated in Dr. Boelens paper. In order for these fast experiments to work properly, very high standards of rf reproducibility are required from the current and future equipment, and the unavailabilty of time for phase cycling of e.g., 180° pulses necessitates short pulses. Clearly, many channels are needed on the new spectrometers and a need for faster and/or higher resolution digitizers is emerging to cope with the large dynamic range associated with the spectra of large molecules. 3D NMR is also useful for the study of smaller molecules; it is anticipated that a significant increase in the precision of 3D structure determinations can be achieved for such molecules when overlap problems in the spectra are resolved by 3D NMR for such systems as well. No consensus was reached on the best protocol for the NMR study of larger systems. Homonuclear experiments as NOE-HOHAHA[5,6], NOE-NOE[10], heteronuclear experiments as ^{15}N and ^{13}C resolved NOESY and HOHAHA[7-9,13,14,31], ^{13}C-^{13}C coherence transfer[11,12], as well as the through-bond assignments[16,17] are all useful.

ACKNOWLEDGEMENTS

R.B. is thankful for the collaboration with Drs. G.W. Vuister, G.J. Kleywegt, R. Kaptein (U. Utrecht) and Dr. A. Padilla (CNRS, Montpellier, Fr.). D.M. and L.E.K. thank Drs. A. Bax and M. Ikura (NIH) for their collaborations. L.E.K. would like to acknowledge the Medical Research Council of Canada for providing financial support. E.R.P.Z. acknowledges the collaboration with Drs. S.W. Fesik and E.T. Olejniczak in the 3D project at Abbott Laboratories.

REFERENCES

1. K. Wüthrich, "NMR of Proteins and Nucleic Acids", Wiley, New York (1986).
2. R. R. Ernst, G. Bodenhausen, and A. Wokaun, "Principles of Nuclear Magnetic Resonance in One and Two Dimensions" Clarendon Press, Oxford (1987).
3. C. Griesinger, O. W. Sørensen, and R. R. Ernst, *J.Am. Chem. Soc.* **109**, 7227 (1987).
4. [a] H. D. Plant, T. H. Mareci, M. D. Cockman, and W. S. Brey, "27th Experimental NMR Conference, Maryland, April 13-17, 1986".
   [b] G. W. Vuister and R. Boelens *J. Magn. Reson.* **73**, 328 (1987).
5. H. Oschkinat, C. Griesinger, P. J. Kraulis, O. W. Sørensen, R. R. Ernst, A. M. Gronenborn, and G. M. Clore, *Nature* **332**, 374 (1988).
6. G. W. Vuister, R. Boelens, and R. Kaptein, *J. Magn. Reson.* **80**, 176 (1988).
7. S. W. Fesik and E. R. P. Zuiderweg, *J. Magn. Reson.* **78**, 588 (1988).
8. D. Marion, L. E. Kay, S. W. Sparks, D. A. Torchia, and A. Bax, *J. Am. Chem. Soc.* **111**, 1515 (1989).
9. E. R. P. Zuiderweg, and S. W. Fesik, Biochemistry **28**, 2387 (1989).
10. R. Boelens, G. W. Vuister, T. M. G. Koning, and R. Kaptein, *J. Am. Chem. Soc.* **111**, 8525 (1989).
11. S. W. Fesik, H. L. Eaton, E. T. Olejniczak, E. R. P. Zuiderweg, L. P. McIntosh, and F. W. Dahlquist, *J.Am. Chem. Soc.* **112**, 886 (1990).
12. L. E. Kay, M. Ikura, and A. Bax, *J. Am. Chem. Soc.* **112**, 88 (1990).
13. M. Ikura, L. E. Kay, R. Tschudin, and A. Bax, *J. Magn. Reson.* **86**, 204 (1990).

14. E. R. P.Zuiderweg, L. P. McIntosh, F. W. Dahlquist, and S. W. Fesik, *J. Magn. Reson.* **86**, 210 (1990).
15. L. E. Kay, G. M. Clore, A. Bax, and A. Gronenborn, *Science* **249**, 411 (1990).
16. M. Ikura, L. E. Kay, and A. Bax, *Biochemistry* **29**, 4659 (1990).
17. L. E. Kay, M. Ikura R. Tschudin, and A. Bax, *J. Magn. Reson.* **89**, 496 (1990).
18. C. Griesinger, O. W. Sørensen, and R. R. Ernst, *J. Magn. Reson.* **84**, 14-63 (1989).
19. K. Nagayama, T. Yamazaki, M. Yoshida, S. Kanaya, H. Nakamura, Poster at NATO workshop, Il Ciocco, Italy (1990).
20. C. Griesinger, O. W. Sørensen, and R. R. Ernst, *J. Magn. Reson.* **73**, 574 (1987).
21. R. Boelens, G. W. Vuister, A. Padilla, G. J. Kleywegt, P. de Waard, T. M. G. Koning, and R. Kaptein, R. *in* "Biological Structure," Dynamics, Interactions and Expression." *Proceeding of the Sixth International Conversation in Biomolecular Stereodynamics*, R. H. Sarma and M. H. Sarma, eds., pp. 63-81, Adenine Press, New York, (1990).
22. G. W. Vuister, A. Padilla, R. Boelens, G. J. Kleywegt, and R. Kaptein, *Biochemistry* **29**, 1829 (1990).
23. A. Padilla, G. W. Vuister, R. Boelens, G. J. Kleywegt, and R. Kaptein, *J. Amer. Chem. Soc.* *112*, 5024 (1990).
24. C. Griesinger, G. Otting, K. Wüthrich, and R. R. Ernst, *J. Am. Chem. Soc.* **110**:7870 (1988).
25. J. N. Breg, R. Boelens, G. W. Vuister, and R. Kaptein, *J. Magn. Reson.* **87**, 646 (1990).
26. L. E. Kay, D. Marion, and A. Bax, *J. Magn. Reson.* **84**, 72 (1989).
27. D. Marion and K. Wüthrich *Biochem. Biophys. Res. Commun.* **113**, 967 (1983).
28. D. J. States, R. A. Haberkorn, and D. J. Ruben, *J. Magn. Reson.* **48**, 286 (1982).
29. P. C. Driscoll, G. M. Clore, D. Marion, P. T. Wingfield, and A. M. Gronenborn, *Biochemistry* **29**, 3542 (1990).
30. M. A. Delsuc and J. Y. Lallemand, *J. Magn. Reson.* **56**, 34 (1986).
31. D. Marion, P. C. Driscoll, L. E. Kay, P. T. Wingfield, A. Bax, A. M. Gronenborn, and G. M. Clore, *Biochemistry* **28**, 6150 (1989).

THE ROLE OF SELECTIVE TWO-DIMENSIONAL NMR CORRELATION METHODS IN SUPPLEMENTING COMPUTER-SUPPORTED MULTIPLET ANALYSIS BY MARCO POLO

Lyndon Emsley and Geoffrey Bodenhausen

Section de Chimie
Université de Lausanne
Rue de la Barre 2
CH-1005 Lausanne, Switzerland

ABSTRACT

Computer-supported analysis of two-dimensional correlation spectra encounters a number of obstacles which are due, in the simplest case, to insufficient digital resolution, and, in more severe cases, to fundamental ambiguities in multiplet patterns. In this paper, we discuss various strategies involving selective two-dimensional experiments that are designed to circumvent these limitations and lift ambiguities.

INTRODUCTION

Computer-supported analysis of multiplets in two-dimensional correlation spectra can yield a great deal of information about the networks of coupled spins under investigation[1-9]. The strategy used in our MARCO POLO program[7] (*multiplet analysis by reduction of cross peaks and ordering of patterns in an overdetermined library organization*) does not use any prior knowledge about the type of molecule under investigation, i.e., it can be applied even if we do not know a priori whether we are dealing with a peptide, a nucleic acid, an alkaloid, etc. This "open-mindedness" of our strategy has

This work has been supported by the Swiss National Science Foundation, by the Commission pour l'Encouragement de la Recherche Scientifique, and by Spectrospin AG, Fällanden, Switzerland.

Computational Aspects of the Study of Biological Macromolecules by Nuclear Magnetic Resonance Spectroscopy, Edited by J.C. Hoch *et al.,* Plenum Press, New York, 1991

151

a number of advantages. If we apply MARCO POLO to an unusual peptide such as Cyclosporin (which contains a few rare amino acids such as Sar, Abu and Bmt) we do not run into any particular problems, as we do not rely on a structural library with its corresponding restrictions. On the other hand, MARCO POLO suffers from some limitations that arise precisely from its lack of prejudice. For example, the terminal methyl groups of valine and leucine residues in Cyclosporin are often not identified as such because the corresponding quartet or heptaplet structures in the two-dimensional multiplets are not properly recognized[8,9]. In the area typical for chemical shifts of methyl groups, the program therefore often "sees" isolated protons and fails to "realize" that these signals are actually due to groups of three magnetically equivalent protons each. This type of misinterpretation could be avoided if the fragment library contained some information regarding the likelihood that signals between, say, 1 and 2 ppm are due to methyl protons. Although this proposal may appear quite simple, it is actually hard to integrate such knowledge into an "unprejudiced" approach.

As a result we favor an alternative strategy: whenever MARCO POLO runs into difficulties, one should carry out complementary experiments specifically designed to lift the ambiguities. Suppose the digital resolution is insufficient to permit a proper analysis of some of the multiplets. Instead of simply increasing the resolution of the entire two-dimensional spectra, one should focus on those multiplets which give rise to problems, by carrying out some selective experiments to "zoom in" on those multiplets[10-13]. Typically, while a non-selective COSY spectrum with a decent resolution may require an overnight run, a soft COSY spectrum of a selected multiplet might require 10 minutes, and yields much better resolution than could ever be expected from a non-selective experiment. Under suitable computer control, one could investigate some 72 multiplets in a 12-hour overnight series of experiments!

THE MARCO POLO APPROACH AND ITS LIMITATIONS

We have described the strategy of our MARCO POLO program in three earlier papers[7-9]. The basic idea is that each multiplet reflects a *fragment* of the coupling network. A pattern recognition algorithm simultaneously

---

Figure 1. Non-selective two-dimensional z-filtered COSY spectrum of *cyclo*-(L-Pro^1-L-Pro^2-D-Pro^3) with a spectral width of 1724 Hz in both dimensions, recorded with $1K \times 2K$ data points in t_1 and t_2 respectively, zero-filled to $2K \times 2K$ data points (0.84 Hz/point). Above: one-dimensional spectrum, where the shifts of the D-Pro^3 residue have been highlighted. Below: multiplets identified by MARCO POLO and fragments assigned to these multiplets by automated analysis. Upper right: global network obtained from automated assembly of these fragments, which corresponds to the D-proline^3 subunit. Adapted from Novič et al.[4] and Pfändler et al.[7]

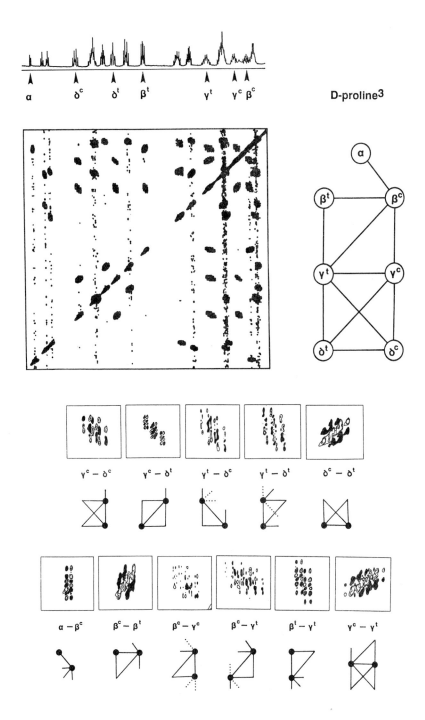

D-proline³

γ^c − δ^c   γ^c − δ^t   γ^t − δ^c   γ^t − δ^t   δ^c − δ^t

α − β^c   β^c − β^t   β^c − γ^c   β^c − γ^t   β^t − γ^t   γ^c − γ^t

scans a pair of non-selective two-dimensional spectra (z-COSY and anti-z-COSY), recorded following prescriptions given elsewhere[14]. The multiplets, which may be partly overlapping, are separated by symmetrization[4,7], and classified according to criteria that depend on the topology but not on the magnitudes of the scalar couplings. The characteristics of an experimental multiplet are compared with a library containing some ten thousand different fragment topologies. A suitable program then assembles fragments into global coupling networks[9]. In general, several fragments may appear to be compatible with a given experimental multiplet; misleading fragments are discarded at the assembly stage.

Fig. 1 gives an example where MARCO POLO has been quite successful. In the z-COSY spectrum of cyclo-(L-Pro^1-L-Pro^2-D-Pro^3), a set of multiplets were identified that are shown enlarged. In fact, many more multiplets were found, but they belong to other subsystems (i.e., other proline residues) and are not shown here, for the assembly program has been successful in separating subsystems. The multiplets shown were automatically assigned to the fragments that are represented symbolically. In fact, many more possible fragments were assigned to each experimental multiplet; the fragments shown here are those that were retained at the assembly stage. For each fragment, the filled dots represent active spins; solid lines represent couplings that were correctly identified (including their numerical values and, if possible, their relative signs), and dotted lines represent passive couplings that were "overlooked" in the course of automated analysis. These errors are not surprising if we consider the complexity of the multiplets in this demanding system, and the poor quality of some of the experimental multiplets, for example the $\beta^{cis} - \gamma^{cis}$ multiplet. In spite of these errors, the assembly of the fragments into the global network (top right in Fig. 1) was successful in this case. This network corresponds to the D-proline^3 subunit, in accordance with earlier work[15]. However, the networks of the L-Pro^1 and L-Pro^2 subunits were not identified completely by our program, due to poor digital resolution, strong coupling, and extensive overlaps of cross peaks.

We believe that these problems cannot be solved by improving the strategy of our analysis. The spectra used as input simply do not contain enough information. However, the fragments "suggested" by MARCO POLO may trigger the user into asking intelligent questions. In many cases, the spectroscopist easily realizes why MARCO POLO encounters a problem. Suitably tailored experiments can then be carried out to round off the picture.

SOFT COSY EXPERIMENTS

Non-selective two-dimensional experiments covering wide spectral ranges usually suffer from coarse digital resolution. Typically, the correlation spec-

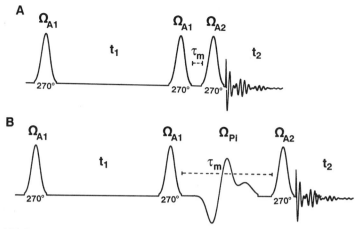

Figure 2. (A) Pulse sequence for selective correlation spectroscopy ("Soft COSY") using three self-refocusing pulses with 270° on-resonance flip angles and Gaussian envelopes, centered at frequencies corresponding to the chemical shifts Ω_{A1} and Ω_{A2} of the spins that are active in the t_1 and t_2 periods, respectively. (B) Pulse sequence for selectively inverted soft correlation spectroscopy ("SIS COSY"), where a G^3 Gaussian cascade at frequency Ω_{Pi} is inserted in the τ_m interval of the soft COSY sequence to invert the polarization of a passive spin P_i. Reproduced with permission from Emsley et al.[13]

tra that we have analyzed by MARCO POLO have a resolution of the order of 1 Hz per point after zero-filling and Fourier transformation. As a result, some ambiguity may arise in the assignment of multiplet patterns. In such cases, selective or "soft" two-dimensional experiments[10,11] can often help to lift ambiguities. A variety of selective experiments can be tailored to specific hypotheses regarding the topology of the coupling network[13,18]. The simplest approach consists in increasing the resolution by restricting the bandwidth. The selective or soft COSY experiment of Fig. 2A provides a simple means for achieving this purpose. Much has been written about this subject, and we shall merely give an eloquent example. Fig. 3B shows a high-resolution multiplet of N-Methyl-valine-11 in Cyclosporin A[16], while Fig. 3C shows the corresponding simulation. The multiplet structure gives a detailed view of the scalar couplings in the H^α-H^β-H_3^γ-$H_3^{\gamma'}$ system. In fact, all couplings except $J(H^\gamma, H^{\gamma'})$ can be measured from this multiplet, *including the long-range $J(H^\alpha, H^\gamma)$ and $J(H^\alpha, H^{\gamma'})$ couplings*, which are responsible for a slight "tilt" of the resonances within each heptet. Fig. 3A shows the corresponding multiplet extracted from a non-selective COSY experiment, where it is impossible to recognize any fine-structure. At first sight, the comparison might appear fallacious, as if the linewidth in Fig. 3A were deliberately increased to make the comparison more favorable to Fig. 3B. In fact, the digital resolution of Fig. 3A (3.7 Hz per point) is quite typical for non-selective COSY experiments, and it is often not practical to obtain better resolution, simply because the experiments would require several days of spectrometer time. To achieve the resolution of Fig. 3B (0.39 Hz per point) for the entire

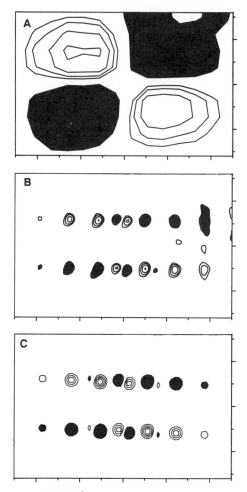

Figure 3. Fine-structure of the H^α-H^β cross-peak multiplet of the H^α-H^β-H_3^γ-$H_3^{\gamma'}$ system N-Methyl-valine-11 in Cyclosporin A. (A) Low-resolution multiplet extracted from an ordinary non-selective double-quantum filtered COSY spectrum with a spectral width of 3788 Hz in both dimensions, recorded with $1K \times 2K$ data points in t_1 and t_2 respectively, zero-filled to $2K \times 2K$ data points (1.85 Hz/point in both dimensions). Without zero-filling, the digital resolution would be only 3.7 Hz/point in the vertical ω_1 dimension. A Lorentz-Gauss lineshape transformation was used ($LB = -3.5$ in t_1, -5.0 in t_2, $GB = 0.15$ in t_1, 0.1 in t_2). Only the active coupling $J(H^\alpha$-$H^\beta)$ can be identified (although not measured!); all passive splittings, including the heptaplet structure due to the two methyl groups, are obscured by the linewidth. (B) High-resolution multiplet obtained under identical conditions with selective "soft" COSY, with a spectral width of 25 Hz in ω_1 and 250 Hz in ω_2, recorded with 64×512 data points in t_1 and t_2 respectively, zero-filled to 256×1024 data points (0.1 Hz/point in ω_1 and 0.24 Hz/point in ω_2). Without zero-filling, the digital resolution would be 0.39 Hz/point in ω_1 and 0.49 Hz/point in ω_2. Lorentz-Gauss lineshape transformation with $LB = -1.2$ in both dimensions and $GB = 0.15$ in ω_1, 0.2 in ω_2. (C) Simulation including all 8 protons in N-Methyl-valine-11, with $J(H^\alpha, H^\beta) = 11.0$ Hz, $J(H^\beta, H^\gamma) = J(H^\beta, H^{\gamma'}) = 6.5$ Hz, $J(H^\alpha, H^\gamma) = J(H^\alpha, H^{\gamma'}) = -0.3$ Hz. These latter two long-range couplings give rise to a slight slope of the interleaved heptaplets. Of the seven peaks of each heptaplet, which should have binomial intensities 1:6:15:20:15:6:1, only the inner five resonances can be observed. As usual, the ω_1 and ω_2 axes run from top to bottom and from right to left, respectively. The tick spacing is 5 Hz in both domains.

156

double-quantum filtered COSY spectrum, one would in principle need 9700 t_1 increments, which would require 54 hours. By comparison, the experiment of Fig. 1B took only 10 minutes! Note that soft COSY requires a mere phase alternation (two steps for each t_1 value), whereas double-quantum filtered COSY requires an 8-step phase-cycle. There are only 45 different cross-peak multiplets (since there are only 45 J couplings between the 111 protons of Cyclosporin), so it would require no more than 45×10 min $= 7\frac{1}{2}$ hrs to "zoom in" on all cross-peak multiplets in succession. Such a data base would be much more useful for automated analysis by MARCO POLO, or indeed could also greatly improve manual analysis.

SELECTIVELY-INVERTED SOFT COSY EXPERIMENTS

The soft COSY approach, using the sequence of Fig. 2A, may be sufficient to lift some ambiguities, but is limited in scope as it merely provides a "passive" solution. Fig. 2B shows a more sophisticated approach, which has come to be known as the SIS-COSY experiment (selectively inverted soft COSY)[13]. In addition to the three self-refocusing 270° pulses[12] that are used in soft-COSY, a G^3 Gaussian cascade[17] with a frequency Ω_{Pi} is inserted in the τ_m interval to invert the polarization of a passive spin P_i.

Fig. 4 shows the effect of this sequence on a cross-peak multiplet in a four-spin system. In both multiplets, there are four interleaved square patterns consisting of two positive and two negative peaks (the latter have been filled in for clarity). The breadth and height of the squares is given by the active coupling constant $J_{MX} = 8.2$ Hz. In Fig. 4A (normal soft COSY), the four quadratic patterns are displaced with respect to each other by vectors $\mathbf{J}_A = (J_{AM}, J_{AX}) = (4.1, 6.1$ Hz) and $\mathbf{J}_K = (J_{KM}, J_{KX}) = (5.1, -4.9$ Hz). In Fig. 4B (SIS-COSY), the four patterns are displaced by $\mathbf{J}'_A = (-J_{AM}, J_{AX}) = (-4.1, 6.1$ Hz) and $\mathbf{J}_K = (J_{KM}, J_{KX})$. In effect, the multiplet behaves as if the sign of the coupling J_{AM} (or, equivalently, of J_{AX}) has been changed. This leads to a dramatic change in the overall appearance of the multiplet. This modification confirms that the RF frequency of the G^3 Gaussian cascade in the sequence of Fig. 2B does indeed coincide with the chemical shift Ω_A of the passive proton A.

Fig. 5 shows another example of the manipulation of multiplets by SIS-COSY. In a normal soft COSY (Fig. 5A), the H^α-H^β-$H^{\beta'}$-H^γ-H_3^δ-$H_3^{\delta'}$ system of N-Methyl-leucine gives rise to an $\alpha - \beta$ cross peak comprising basic antiphase squares split by $J(H^\alpha$-$H^\beta) = 11.2$ Hz, separated by a sloping vector $\mathbf{J}_{\beta'} = [J(H^\alpha$-$H^{\beta'}), J(H^\beta$-$H^{\beta'})] = (4.5, -15.0$ Hz), and split in the horizontal direction by the vector $\mathbf{J}_\gamma = [J(H^\alpha$-$H^\gamma), J(H^\beta$-$H^\gamma)] = (0, 3.0$ Hz). The long-range couplings to the methyl protons H_3^δ and $H_3^{\delta'}$ are not resolved. In Fig. 5B, the slope of the displacement vector has been reversed,

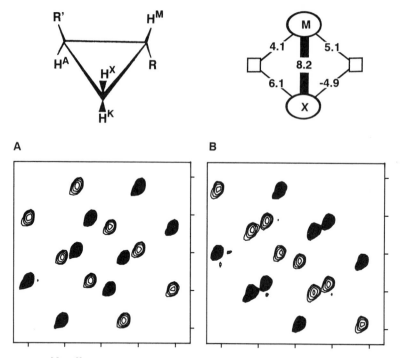

Figure 4. (A) H^M-H^X cross-peak multiplet of a cyclopropane derivative ($R = COOD_3$, $R' = C_6H_5$) recorded with selective "soft" COSY, with a spectral width of 30 Hz in ω_1 and 250 Hz in ω_2, 64 × 1024 data points in t_1 and t_2 respectively, zero-filled to 256 × 1024 data points (0.12 Hz/point in ω_1 and 0.24 Hz/point in ω_2). A Lorentz-Gauss lineshape transformation was used ($LB = -0.6$, $GB = 0.15$ in both dimensions). (B) Same H^M-H^X cross-peak multiplet recorded with selectively inverted soft COSY, where the passive proton A was inverted in the mixing interval. The tick spacing is 5 Hz.

i.e., $\mathbf{J}'_{\beta'} = \left[-J(H^{\alpha}\text{-}H^{\beta'}), J(H^{\beta}\text{-}H^{\beta'})\right] = (-4.5, -15.0 \text{ Hz})$, because the G^3 Gaussian cascade was applied at the chemical shift of the passive proton $H^{\beta'}$.

In general, these experiments are ideally suited to confirm or reject a hypothesis about a coupling network, i.e., about the existence of passive spins, their number and their chemical shifts. A yet more sophisticated way, which has been described elsewhere[18], encompasses a variety of experiments that combine two or more selective inversion pulses, inserted either in the mixing interval or in the middle of the evolution period of soft COSY experiments. These experiments allow one to select a particular network on the grounds of its topology, and are extremely effective to elucidate situations where MARCO POLO fails.

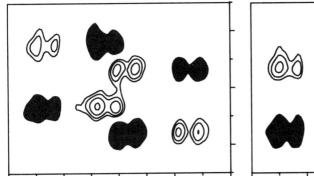
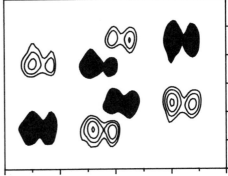

Figure 5. (A) H^α-H^β cross-peak multiplet of N-Methyl-leucine-9 in Cyclosporin A, recorded with soft COSY, with a digital resolution of 0.14 Hz/point in ω_1 and 0.19 Hz/point in ω_2 after zero filling. (B) Same H^α-H^β cross-peak multiplet recorded with the same resolution with SIS-COSY, where the proton $H^{\beta'}$ was inverted selectively. The tick spacing is 5 Hz.

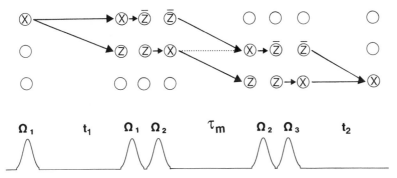

Figure 6. Pulse sequence and operator evolution graphs appropriate for a selective relayed magnetization transfer experiment. All RF pulses have the same phase (e.g., along the y-axis of the rotating frame), except for the first pulse which is alternated $\pm y$ in conjunction with addition and subtraction of the signals.

SELECTIVE HOMONUCLEAR RELAYED MAGNETIZATION TRANSFER

In some cases, the complexity of the spin system is so daunting, or the degree of overlap so severe, that it is worthwhile to carry out more sophisticated experiments that are specifically tailored to probe a particular coupling network. As a further example of the kind of information selective two-dimensional correlation experiments can provide in such cases, we have explored the possibility of implementing *selective relayed magnetization transfer* experiments.

Fig. 6 shows a pulse sequence designed for selective relayed magnetization transfer. One first excites transverse magnetization of a spin I_1, which is transferred to spin I_2 and, in a second step, to spin I_3. The first two 270° Gaussian pulses, separated by the evolution period t_1, are applied at the

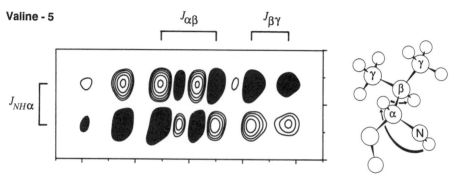

Valine - 5

$J_{\alpha\beta}$ $J_{\beta\gamma}$

$J_{NH\alpha}$

Figure 7. Cross-peak multiplet obtained by relayed magnetization transfer $H^N \to H^\alpha \to H^\beta$ in Valine-5 in Cyclosporin A, using the pulse sequence of Figure 6, with a spectral width of 30 Hz in ω_1 and 200 Hz in ω_2, 50 × 512 data points in t_1 and t_2 respectively, zero-filled to 256 × 1024 data points (0.12 points/Hz in ω_1 and 0.20 points/Hz in ω_2). A Lorentz-Gauss lineshape transformation was used ($LB = -1.5$ in both dimensions, $GB = 0.05$ in ω_1, 0.1 in ω_2). The tick spacing is 5 Hz.

chemical shift Ω_1 of the first spin I_1; the next two pulses, separated by a fixed interval $\tau_m \simeq (2J_{12})^{-1} \simeq (2J_{23})^{-1}$, are applied at the chemical shift Ω_2 of I_2; finally, the last pulse is applied at the chemical shift Ω_3 of I_3. All RF pulses have the same phase (e.g., along the y-axis of the rotating frame), except for the first which is alternated $\pm y$, in conjunction with addition and subtraction of the resulting signals.

The upper part of Fig. 6 shows a so-called operator evolution graph, following a set of graphical conventions discussed elsewhere[19]. The circles represent, from top to bottom, spins $\mathbf{I_1}$, $\mathbf{I_2}$ and $\mathbf{I_3}$; the symbols x, y and z indicate whether these spins are associated with I_x, I_y or I_z operators. An empty circle represents a unity operator. Horizontal solid arrows represent the effect of a chemical shift (e.g., a term $\sin \Omega_1 t_1$); the horizontal dashed arrow indicates that the precession under chemical shift should not occur (e.g., a term $\cos \Omega_2 \tau_m$). Sloping arrows indicate transformations under scalar couplings (e.g., from left to right, $\sin J_{12}t_1, \sin J_{12}\tau_m, \sin J_{23}\tau_m, \sin J_{23}t_2$.) The operator evolution graph shows how transverse magnetization of spin I_1 (or, more accurately, in-phase single-quantum coherence I_{1x}) is converted into transverse magnetization of spin I_3, passing through anti-phase coherences of spin I_2.

Fig. 7 shows a cross-peak multiplet obtained with the sequence of Fig. 6, by relayed magnetization transfer $H^N \to H^\alpha \to H^\beta$ in Valine-5 in Cyclosporin A. In the vertical ω_1 domain, one observes an antiphase doublet centered at $\Omega(H^N)$ with a splitting $J(H^N, H^\alpha)$. In the horizontal ω_2 domain, one observes a multiplet centered at $\Omega(H^\beta)$ with an antiphase splitting $J(H^\alpha, H^\beta)$ and an in-phase heptaplet structure with a splitting $J(H^\beta, H^\gamma) \simeq J(H^\beta, H^{\gamma'}) \simeq 7.2$ Hz. Only 4 of the seven lines (which should have binomial amplitudes 1:6:15:20:15:6:1) are visible, due to partial cancellation effects. This single multiplet gives all couplings in the entire va-line system.

This type of selective relayed experiment, while perhaps a bit more complicated to set up than a simple soft-COSY, can obviously yield a great deal of insight into the topology of the coupling network and the magnitudes of the coupling constants.

DISCUSSION

In the previous sections we have concentrated on some molecules of biological importance which have molecular weights below 2,000. Unfortunately, in larger systems (e.g., containing more than 50 amino acids), where the transverse relaxation times are short and the lines broad, current approaches based on a MARCO POLO type philosophy often lead to ambiguous answers. In such systems it is unlikely that sophisticated selective experiments will be of much use, since long selective pulses lead to poor sensitivity when T_2 is shorter than the pulse widths. In such cases, it will be necessary to provide some prior information about the nature of the spin systems, e.g., typical chemical shifts and coupling patterns of amino-acid residues together with information about the composition of the particular molecule under investigation[20]. The complexity of the spectra of macromolecules often compels one to use three-dimensional methods, which, while they resolve most of the problems arising from overlaps, tend to suffer from much poorer digital resolution than their two-dimensional counterparts. In non-selective three-dimensional spectra, it is unlikely that the fine-structure of multiplets would ever be sufficiently resolved for automated multiplet analysis to be successful, and we believe that selective three-dimensional experiments will be helpful in such circumstances.

In small or medium sized molecules, however, the objectives of pattern recognition can be very ambitious: we have shown that in some moderately difficult cases such as *cyclo*-triproline, it is possible to identify complete coupling networks, including all chemical shifts, degeneracies, magnitudes and signs of scalar coupling constants. Often, however, limited digital resolution leads to ambiguities. This paper describes a few recently-developed methods that allow one to focus on problematic multiplets, thereby allowing one to extract more information from those areas of the spectrum that are most resistive to analysis.

Thus, particularly in cases where the analysis cannot be made completely automatic, it is highly desirable that the computer enter into dialogue with the spectroscopist, to remove some of the burden of repetitive assignment, and to suggest alternative interpretations that might not come to one's mind because of prejudice. One should of course be extremely careful about misleading computer outputs and erroneous results that, although they may appear dressed in the objectivity traditionally ascribed to a machine, merely reflect limitations of the strategy.

We believe that the dialogue between user and computer should be able to produce better, more reliable results in cases of uncertainty, thus confirming the old wisdom that collaborative research is often more productive than working on one's own — even if one of the partners in the relationship is a mere computer.

REFERENCES

1. J. C. Hoch, S. Hengyi, M. Kjær, S. Ludvigsen, and P. M. Poulsen, *Carlsberg Res. Commun.* **52**, 111 (1987).
2. Z. L. Mádi, C. Griesinger, and R. R. Ernst, *J. Am. Chem. Soc.* **112**, 2908 (1990).
3. B. U. Meier, Z. L. Mádi, and R. R. Ernst, *J. Magn. Reson.* **74**, 565 (1987).
4. M. Novič, and G. Bodenhausen, *Anal. Chem.* **60**, 582 (1988).
5. M. Novič, U. Eggenberger, and G. Bodenhausen, *J. Magn. Reson.* **77**, 394 (1988).
6. P. Pfändler, G. Bodenhausen, B. U. Meier, and R. R. Ernst, *Anal. Chem.* **57**, 2510 (1985).
7. P. Pfändler and G. Bodenhausen, *J. Magn. Reson.* **79**, 99 (1988).
8. P. Pfändler and G. Bodenhausen, *Magn. Reson. Chem.* **26**, 888 (1988).
9. P. Pfändler and G. Bodenhausen, *J. Magn. Reson.* **87**, 26 (1990).
10. R. Brüschweiler, J. C. Madsen, C. Griesinger, O. W. Sørensen, and R. R. Ernst, *J. Magn. Reson.* **73**, 380 (1987).
11. J. Cavanagh, J. P. Waltho, and J. Keeler, *J. Magn. Reson.* **74**, 386 (1987).
12. L. Emsley, and G. Bodenhausen, *J. Magn. Reson.* **82**, 211 (1989).
13. L. Emsley, P. Huber, and G. Bodenhausen, *Angew. Chem.* **102**, 576 (1990); *Angew. Chem. Int. Ed. Engl.* **29**, 517 (1990).
14. H. Oschkinat, A. Pastore, P. Pfändler, and G. Bodenhausen, *J. Magn. Reson.* **69**, 559 (1986).
15. H. Kessler, W. Bermel, A. Friedrich, G. Krack, and W. E. Hull, *J. Am. Chem. Soc.* **104**, 6297 (1982).
16. H. Kessler, H. R. Loosli, and H. Oschkinat, *Helv. Chim. Acta* **68**, 661 (1985).
17. L. Emsley and G. Bodenhausen, *Chem. Phys. Lett.* **165**, 469 (1990).
18. L. Emsley and G. Bodenhausen, *J. Am. Chem. Soc.* **113**, 3309 (1991).
19. U. Eggenberger, and G. Bodenhausen, *Angew. Chem.* **102**, 392 (1990); *Angew. Chem. Int. Ed. Engl.* **29**, 374 (1990).
20. G. J. Kleywegt, R. M. J. N. Lamerichs, R. Boelens, and R. Kaptein, *J. Magn. Reson.* **85**, 186 (1989).

APPLICATION OF MAXIMUM ENTROPY METHODS TO NMR SPECTRA OF PROTEINS

M. A. Delsuc[†], M. Robin[†], C. Van Heijenoort[†],
C. B. Reisdorf[*], E. Guittet[†], and J. Y. Lallemand[†]

([†])Laboratoire de R.M.N.
ICSN-CNRS
91190 Gif-Sur-Yvette, France
([*])Département de Chimie de Synthèse Organique Ecole
Polytechnique
91128 Palaseau, France

INTRODUCTION

Maximum Entropy Processing (*MaxEnt*) has proven to be a serious alternative to Fourier transform for the processing of NMR data sets[1-4]. However, use of *MaxEnt* processing of protein NMR spectra presents several difficulties that this paper intends to address. Examples using the GIFA algorithm are shown which include lorentzian deconvolution of 1D and 2D spectra of proteins, J-deconvolution of 2D COSY experiments, lorentzian deconvolution, and truncation artifact reduction of homonuclear 3D spectra.

PREPROCESSING THE DATA-SETS

Typical NMR spectra of proteins are lage data-sets (2D and 3D experiments can be up to several hundred Megabytes). They present a large dynamic range due to the presence of an intense water signal in the middle of the spectral window. Artifacts are not uncommon, due for instance to slow variations of frequencies during the experiment (so called t_1-noise) or out of phase "signal-like" phantoms due to quadrature images. Signal features may be quite different from one line to another; for instance, amide protons in exchange with the solvent generally show broader lines than aliphatic or aromatic protons. These features imply that some care should be taken when processing these data-sets with *MaxEnt*.

On the other hand, the "classical" maximum entropy approach forces the reconstructed spectrum to be strictly real positive. This arises from

Computational Aspects of the Study of Biological Macromolecules by Nuclear Magnetic Resonance Spectroscopy, Edited by J.C. Hoch *et al.,* Plenum Press, New York, 1991

the fact that the spectrum is handled as a probability density function, and follows from the classical definition of the entropy of a spectrum f_i:

$$S = - \sum \frac{f_i}{\sum f_i} \log \left(\frac{f_i}{\sum f_i} \right)$$

It should be noted that P. Hore recently proposed an expression for the entropy of complex signals which permits a rigorous processing of complex out-of-phase NMR spectra[5]. However, we will not detail this proposition here.

The points presented above (phased spectra, artifacts, large dynamic range and large data-sets) imply that some kind of processing should be done on NMR data sets in order to be able to use *MaxEnt*.

A simple approach to this preprocessing is the following: the data-set is first Fourier transformed (with no apodization), then phase corrected and inverse Fourier transformed; the result of this operation being used as a pseudo-experimental data-set. Another typical preprocessing consists in "extracting" one small region of the spectrum after the phase correction, which is then transformed back to the time-domain. This last measure allows the reduction of the size of the giant data-sets found in today's NMR. The 2D or 3D NMR data sets are generally too large for the computer memory, but also for the numerical dynamics of the computer and the stability of the algorithms. This is the only practical way to process the practical data-sets found today in protein studies.

PRESERVING CAUSALITY

All these preprocessing examples have a major drawback: they correspond to modifications that cannot be done by a simple linear filter. In other words, no NMR spectrometers could ever generate such data. The main symptom of this, is the fact that the FID generated is no longer causal; i.e., the signal starts before the time t_o, before the very pulse that caused the FID!

This is exemplified in Fig. 1 where the presence of a large water signal causes artifacts when extracting the amide region of a ^{1}H spectrum of a protein. When this region is inverse Fourier transformed, the FID shows a strong non-causality. This is due to the fact that information from the water is still present in this window by its dispersive contribution found in the imaginary part of the spectrum.

What we would like to find is a preprocessing strategy that conserves the causal character of the processed data. The causality principle is simple to express in the time domain, but how does it translate in the frequency domain?

164

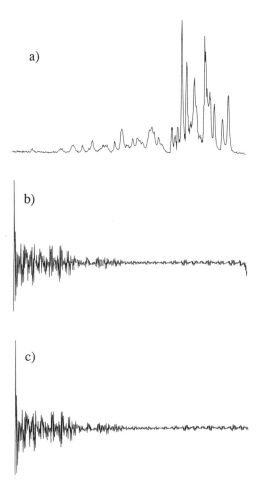

Figure 1. The amide/aromatic region of the ^{1}H spectrum of Cryptogein has been chosen as an example of non-causal behavior. (a) the spectral region as extracted from the whole spectrum. (b) the result of the inverse Fourier transform of the same region after extraction. Note the artifacts and the end of the FID. This is characteristic of a non-causal behavior (see text). (c) the result of the inverse Fourier transform of the same region applied as explained in the text, note the disappearance of the non-causal artifacts.

For a spectrum corresponding to the Fourier transform of a causal time domain data-set, the Bayard-Bode (or Kramers-Krönig) theorem states that the real ($\Re$) and the imaginary ($\Im$) parts are related by a Hilbert transform[6]:

$$\Re(\nu) = \frac{1}{\pi} \int_{-\infty}^{+\infty} \frac{\Im(\nu')}{\nu - \nu'} d\nu'$$

$$\Im(\nu) = -\frac{1}{\pi} \int_{-\infty}^{+\infty} \frac{\Re(\nu')}{\nu - \nu'} d\nu'$$

Stating that a time domain signal is zero before a time t_o (causal) or stating that the real and imaginary parts of its Fourier transform are Hilbert transform of each other, are two strictly equivalent propositions. These relations can be extended for multidimensional hypercomplex data-sets in straightforward manner[6].

The Bayard-Bode relation thus shows that the real and imaginary parts do not hold independent information. The Hilbert transform can be seen as the "phase shift" transform that permits us to go from one to the other.

Now we can design a preprocessing strategy that conserves the causality of the data set: let's process as usual, but before applying the inverse Fourier transform, throw away the imaginary part and reconstruct it from the Hilbert transform of the real part, thus leading to a causal time domain pseudo data-set.

THE DIGITAL FOURIER TRANSFORM

However, this procedure must be adapted to work on a computer. Here the mathematical Fourier transform cannot be used any longer ($\int_{-\infty}^{+\infty}$ does not make any sense for numerical processing), instead a digital Fourier transform (DFT) will be used which performs an equivalent of the Fourier transform on digitalized time and frequency domains. The DFT differs from the Fourier transform by several characteristics that are worth pointing out:

1. The DFT is an in-place operation, the number of data points before and after the DFT is the same. So if one has N experimental data-points, one will have $\frac{N}{2}$ points in both real and imaginary parts of the spectrum. Since no information is lost by the DFT (it is invertible), the real and imaginary part **do not** hold equivalent information any more. So when one looks at only the real channel of the spectrum, one is losing half of the information. To bring all the available information in the real channel of the spectrum, one has to double the number of points available in this channel. A way of doing this is by zero-filling the time domain data-set before DFT. This explains the well known rule of thumb: *"always zero-fill at least once"*.

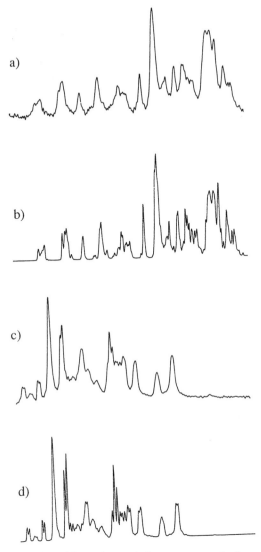

Figure 2. The aromatic and amide regions of the same protein have been processed by *MaxEnt*. The protein was 1.5 mM in H_2O, the spectrum was obtained at 600 MHz on a Bruker AM600 spectrometer. The same processing was performed with a lorentzian deconvolution of 1.6 Hz for the aromatic region, and of 2 Hz for the amide region. Iteration was applied until $\frac{\chi^2}{N} < 1.5$ and $\gamma < 0.05$, for a total of 30 iterations. (a) the amide region processed with classical treatment. (b) the amide region processed with *MaxEnt*. (c) the aromatic region processed with classical treatment. (d) the aromatic region processed with *MaxEnt*.

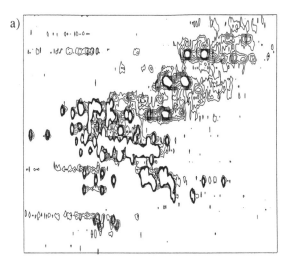

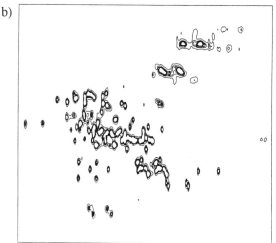

Figure 3. The ^{1}H-^{13}C HMQC of the same as Fig. 2 was performed at 400 MHz for ^{1}H on an AM400 Bruker spectrometer. Experimental time was 24 hours. (a) the aliphatic region processed with classical treatment. (b) the aliphatic region processed with *MaxEnt*. A Lorentzian deconvolution of 5 Hz was performed in both dimensions.

2. The time domain data-set is periodized before Fourier transform, i.e., a periodic function is built by tiling the entire time domain $[-\infty, +\infty]$ with the data-set. This means that the right border of the data-set $(t = t_{max})$ corresponds also to the region just **before** the t_o. So this is where noncausal behavior will show up (see Fig. 1).

3. The DFT is usually implemented with a Fast Fourier Transform algorithm (FFT). The FFT algorithm is much faster than the simple DFT algorithm. FFT algorithms exist for transform from a complex data-set to a complex data-set and for transform from a real data-set to a complex one (in which case there is no need to store zeros for the imaginary part of the real data-set), it is the so-called Real FFT (RFT). The Fourier transform of a real function is a hermitian function, the real part is even $(f(x) = f(-x))$ and the imaginary part is odd $(f(x) = -f(-x))$. So the RFT produces only one half of the data (on $[0, +\infty[$, the other part holding redundant information. This operation is done at N constant, and thus with no loss of information. The Hilbert transform of a real data-set can be easily implemented on the computer by means of this RFT operation: *apply RFT on the real data-set, thus generating a complex data-set; *multiply this complex data-set by the complex number $e^{i\frac{\pi}{2}}$; *applying then the inverse RFT will generate the Hilbert transform of the starting data-set.

MaxEnt PROCESSING

We are now ready to apply this technique to process protein NMR data-sets for *MaxEnt*. In the examples presented below, we have applied the following procedure: *The acquired data-sets (either 1D or 2D) have been zero-filled once and Fourier transformed with no apodization. *The resulting spectra have been phase corrected, imaginary parts discarded and regions of interest extracted. *The sub-spectra thus generated have been inverse Fourier transform by the mean of an RFT algorithm, thus generating complex causal FIDs, that will be used for *MaxEnt* processing.

The data-sets thus generated have then been processed by a generic *MaxEnt* algorithm. In a nut-shell the principle is as follows. The purpose is to find a spectrum which fits the data as close as possible, and at the same time maximizes the entropy. So a function Q is constructed and maximized:

$$Q = S - \lambda \chi^2$$

where λ is a Lagrange multiplier,

$$\chi^2 = \sum \frac{(D_i - R_i)^2}{\sigma_i}$$

is the classical evaluator of the fit quality, D_i is the experimental data-set, σ_i is the amount of noise in the ith channel, and R_i is the time domain function reconstructed by applying the Transfer function Tf of the measure to the current spectrum,

$$R_i = \sum_j Tf_{ij}(f_j).$$

This inverse approach easily permits one to deconvolute blurred spectral features, by inserting the blurring function in the expression of Tf. The optimization is done by using the GIFA (*Generalized Iterative Fixed-point Algorithm*) algorithm which is an enhancement of the Gull and Daniell[7] algorithm based on fixed point solution of the problem. The Gull and Daniell algorithm tends to be very unstable, and as such is barely usable. The GIFA fixes this problem as well as it introduces a sophisticated control of the λ parameter[8]. The convergence is stopped when the χ^2 criterion reaches N (the number of data points). The quality of the reconstructed spectrum is measured by the number $\gamma = \cos(\theta)$ where θ is the angle between the two gradients ∇S and $\nabla \chi^2$. The smaller the γ, the better the reconstruction. The time required for the processing with the GIFA algorithm is typically of 6 computations of the transfer function Tf per iteration. The main processing burden in the computation of Tf is the Fourier transform, implemented as a FFT, and as such is very fast. Overall, the time involved is of the order of a hundred FFT computations of the data-set.

RESULTS

This procedure has first been applied to the ^{1}H 1D spectrum of the Cryptogein, a protein of 98 amino acids, implicated in the attack of plants by a fungus. Fig. 2 presents the aromatic and the amide regions which have been processed by *MaxEnt*. Different line-width deconvolutions have been applied in each subspectra. Fig. 3 shows the result of a *MaxEnt* computation on a small extract of the ^{1}H-^{13}C HMQC spectrum of the same protein. The region shown is the aliphatic region. Here the line-widths have been deconvoluted, but the gain in resolution comes also from the fact that the data-set is badly truncated, most notably in the F1 direction, because of the limited acquisition time. In Fig. 4 the result of the J-deconvolution of the finger-print region of the same protein is shown. The J-deconvolution consists in deconvoluting the coupling patterns by expressing in the transfer function Tf, the sine modulation (in the case of anti-phase coupling)

170

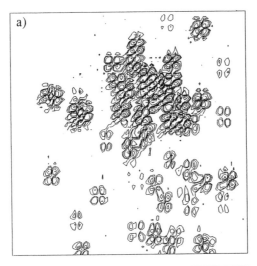

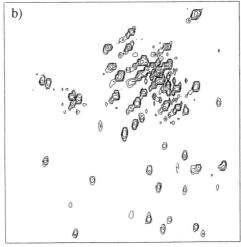

Figure 4. The DQF-COSY of the same sample as Fig. 2 was performed at 400 MHZ for ^{1}H on an AM400 Bruker spectrometer. Experimental time was 24 hours. The finger print region (NH-Hα region) was processed both by classical and *MaxEnt* methods. (a) the finger print region of the DQF-COSY processed with classical treatment. (b)the same region processed with *MaxEnt*. A deconvolution of 3 Hz was performed, as well as an anti-phase deconvolution of 7 Hz. The region is 128×256 points, reconstructed on 256×256 spectral points. 30 iterations were needed for a final $\frac{\chi^2}{N}$ of 1.5 and $\gamma < 0.1$.

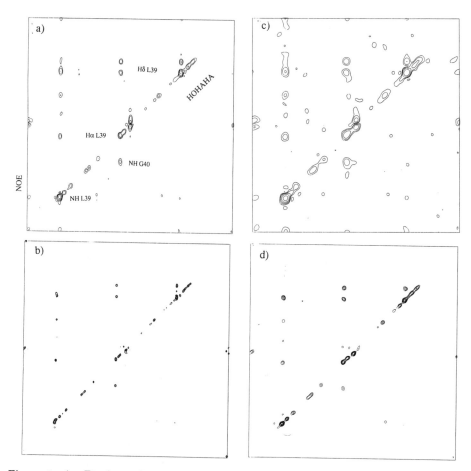

Figure 5. An F3-plane plane of the HOHAHA-NOESY 3D spectrum performed on the α-cobratoxin protein. The sample was 2 mM in H_2O. The experiment was run in 4 days at 400 MHz on an AM400 Bruker spectrometer. 120 experiments in F1, 128 experiments in F2 and 8 scans were performed. The F1 dimension was first processed classically, then the F3-plane was extracted and processed as a 2D experiment. The classical processing was: cosine-bell apodization, zerofilling to 512×512 and hypercomplex Fourier transform. The *MaxEnt* processing included the preprocessing as explained in the text, needed because of the 180° first order phase correction. (a) The F3 plane processed classically. (b) The same plane processed with *MaxEnt*. The reconstruction took place on 256×256 points. 20 iterations were performed, with a lorentzian deconvolution of 5 Hz.for a final $\frac{x^2}{N}$ of 1.5 and $\gamma < 0.1$. (c) The same plane was truncated to 64×64 points and processed classically. (d) The same plane was truncated to 64×64 points and processed with *MaxEnt*. Same processing as in (b).

introduced by this coupling. This has the effect of reducing all the active coupling patterns to a single line centered on the multiplet[4,8].

As a final example, the result on *MaxEnt* on a 3D data-set is shown. The approach here is not to process an extract of the whole 3D spectrum, but rather to process as a 2D data-set a plane extracted from the 3D. The experiment is a HOHAHA-NOESY homonuclear 3D experiment[9] from the α-cobratoxin protein, a 71 amino acids protein extracted from the venom of *Naja Naja Siamensis*. A linear (first order) phase correction of 180° was necessary in the F1 dimension, because of limitations of the spectrometer. A plane perpendicular to the F3 dimension was extracted in the amide region. The whole plane was then processed, since the number of points and the number of signals is small enough. A first example with all the data points (120×128) was first realized, showing a good enhancement of the resolution and of the signal to noise ratio. The same data-set was then truncated to 64×64 points and processed by *MaxEnt*. The effects of the truncation are not as drastic in the *MaxEnt* spectrum as they are in the classical one. Most of the features of the complete data-set can be seen. It should be noted that such a 3D data-set would have required on the order of a day to acquire.

CONCLUSION

We have shown that *MaxEnt* is a useful technique for the treatment of protein NMR spectra. It can produce higher quality spectra than the classical approaches. Resolution enhancement, noise separation, and spectral deblurring are the main benefits that one can expect from such a method.

All the processing shown in this paper has been realized with the GIFA (1D, 2D, *MaxEnt*), and GIFB (3D) programs. This set of programs is available from the authors. The programs are today available in VAX/VMS and UNIX/X_Window versions. The UNIX version has been ported to the SUN4, Silicon Graphics and Alliant machines. Ports to SUN3/Sun View and to IBM Risc6000 are currently in progress.

REFERENCES

1. S. Sibisi, J. Skilling, R. G. Brereton, E. D. Laue, and J. Staunton, *Nature* **311**, 446 (1984).
2. E. D. Laue, J. Skilling, J. Staunton, S. Sibisi, and R. G. Brereton, *J. Magn. Reson.* **62**, 437(1985).
3. P. Hore, *J. Magn. Reson.* **62**, 561 (1985).
4. M. A. Delsuc and G. C. Levy, *J.Mag.Reson.* **76**, 306 (1988).
5. G. J. Daniell and P. J. Hore, *J. Magn. Reson.* **84**, 515 (1989).
6. R. R. Ernst, G. Bodenhausen, and A. Wokaun, "Principles of Nuclear Magnetic Resonance in One and Two Dimensions," p. 94-96 and p. 306-308, Oxford University Press (1987).

7. S. F. Gull and G. J. Daniell, *Nature* **272**, 686 (1978).
8. M. A. Delsuc, "Maximum Entropy and Bayesian Methods, Cambridge, England, 1988," J. Skilling, ed., Kluwer Academic Publishers, Dordrecht, The Netherlands (1989).
9. G. W. Vuister, R. Boelens, and R. Kaptein, *J. Magn. Reson.* **80**, 176 (1988).

PATTERN RECOGNITION IN TWO-DIMENSIONAL NMR SPECTRA OF PROTEINS

Hans Robert Kalbitzer, Klaus-Peter Neidig,
Matthias Geyer, Rainer Saffrich, and Michael Lorenz

Max-Planck-Institut für Medizinische Forschung
Abt. Biophysik
Jahnstrasse 29
D-6900 Heidelberg, Germany

GENERAL CONSIDERATIONS

Today, two-dimensional NMR spectroscopy has become a well-accepted method for the determination of the three-dimensional structures of biological
macromolecules such as proteins in solution. The primary evaluation of the
spectroscopic data is very time-consuming and complicated, therefore computer assistance grows more and more important. It is not surprising that a
number of groups work in the field of pattern recognition in n-dimensional
NMR spectra of macromolecules ($n \geq 2$). Correspondingly, a number of different methods were described that can solve at least some special aspects
of this problem[1-29].

From a more general point of view pattern recognition in two-dimensional
NMR spectra can be considered as a special application of image analysis.
Image analysis usually comprises 4 different layers of operations, improvement of image quality and feature enhancement, separation of relevant objects from the background, classification of objects, and finally interpretation
of objects and classes of objects.

IMPROVEMENT OF IMAGE QUALITY

Improvement of the image quality is routinely done in NMR spectroscopy by
time domain filtering since the early days of FT NMR. In addition, methods
for removal of strong solvent signals in two-dimensional NMR spectra[30,31]

Computational Aspects of the Study of Biological Macromolecules by Nuclear Magnetic Resonance Spectroscopy, Edited by J.C. Hoch *et al.,* Plenum Press, New York, 1991

or for the suppression of t_1- and t_2-ridges by processing of the time domain data[32] are available. In the frequency domain, various methods for the suppression of experimental artifacts[33-40] and for enhancing spectral features by making use of local[4,10,13,16,41,42] or global symmetries[41,43-46,48] were proposed.

In the following we will shortly mention a generalization of the method for removal of t_1-ridges proposed earlier by Klevit[33] as it is implemented in the program AURELIA (*AUtomatic REsonance LInes Assignment*).

In the original method for suppression of t_1-ridges[33] a 'mean row' is subtracted from all rows of the data set. The mean row is simply obtained as an average of rows of an area defined by the user that contains only random noise and t_1-ridges but no resonance lines from the sample. The averaging of rows suppresses the influence of statistical noise on the mean row and gives only the characteristic profile from base plane offsets and t_1-ridges. This method works well for absolute-value spectra, but has only small influence on phase-sensitive spectra. Here, it only corrects for base plane offsets that are constant in the ω_1-direction, but has almost no effects on the t_1-ridges, because these show an oscillatory pattern which is averaged out by constructing the mean row. Phase-sensitive spectra are better corrected by different methods. Zolnai et al. propose a two step procedure[37,38]: after subtracting a mean row as defined before they test if the data points are inside or outside the range of $\pm D E(j)$ with D a constant usually set between 3 and 5 and $E(j)$ the r.m.s. noise calculated from the mean row area for each column j. Data points inside this range are zeroed, data points outside are left intact. The method implemented in AURELIA uses a separate processing of negative and positive parts of the 2D-spectrum. First two 'mean rows' $M^+(j)$ and $M^-(j)$ are calculated as averages of the positive or negative parts of the rows defined by the user. The points $R'(i,j)$ of the corrected row i are obtained from

$$
\begin{aligned}
R'(i,j) &= \max((R(i,j) - M^+(j), 0) \text{ for } R(i,j) \geq 0 \\
R'(i,j) &= \min((R(i,j) + M^-(j), 0) \text{ for } R(i,j) < 0
\end{aligned} \tag{1}
$$

If a stronger suppression of t_1-ridges appears to be useful one can subtract the 'sky line projections' $S^+(j)$ and $S^-(j)$ instead of $M^+(j)$ and $M^-(j)$. $S^+(j)$ and $S^-(j)$ contain not the averages but the maxima and minima of the user defined rows.

One of the earliest methods for improving the quality of two-dimensional NMR spectra were the spectrum symmetrizations proposed by Baumann et al.[43,44]. In the beginning of biological applications of two-dimensional NMR these symmetrization procedures were often necessary for extracting any information out of the noisy spectra, with the technical progress and the improvement of the genuine spectral quality they became less important.

Nowadays, the application of symmetrization procedures is often rejected with the argument that they may produce artifacts or suppress important information. But this argument is not limited to procedures that make use of the global symmetry in 2D-spectra, but it is an inherent property of any feature enhancement procedure. For example, the frequently used Gaussian filtering[49]. leads to the suppression of broad peaks and to the appearance of negative side lobes, that is to a loss of information and to the production of artifact peaks. Therefore, it is not the question, to use such procedures or not, but to find the procedure optimally tuned to the enhancement of the information just wanted.

In principle, there are many different ways to use the symmetry information in two-dimensional NMR spectra for enhancing symmetric features in the spectrum[41,43−46]. It appears worthwhile to develop a general formalism that shows the characteristic properties of the symmetrization procedures already published. As we shall see, such a formalism leads also to a natural generalization of the known methods[48].

All symmetry enhancement procedures published so far can be described by the following procedure: the intensities I_0 in the symmetry related positions (i,j) and (j,i) in the spectrum are replaced by the new intensities I. The relation between the new intensities $I(i,j)$ and $I(j,i)$ and the old intensities $I^0(i,j)$ and $I^0(j,i)$ is given by the functions $f_{\max}$ and $f_{\min}$ of a match factor m and the intensities $I^0(i,j)$ and $I^0(j,i)$

$$
\begin{aligned}
I(i,j) &= f_{\max}(m, I^0(i,j), I^0(j,i)) \text{ for } |I^0(i,j)| \geq |I^0(j,i)| \\
I(j,i) &= f_{\min}(m, I^0(i,j), I^0(j,i)) \\
I(i,j) &= f_{\min}(m, I^0(i,j), I^0(j,i)) \text{ for } |I^0(i,j)| < |I^0(j,i)| \\
I(j,i) &= f_{\max}(m, I^0(i,j), I^0(j,i)).
\end{aligned}
\tag{2}
$$

The normalized match factor $m(0 \leq m \leq 1)$ is a measure of the symmetry between the two locations, ideally, it equals 1 for complete symmetry and zero if no positive correlation exists. In the most simple cases m uses only the limited information in positions (i,j) and (j,i). More elaborate procedures use also morphological features of an environment E_{ij} and E_{ji} of the data points (i,j) and (j,i) considered.

We propose three different sets of transformations $f_{\max}$ and $f_{\min}$, the match proportional symmetry enhancement (A), the additive symmetry enhancement (B), and the difference proportional symmetry enhancement (C). Without restriction of the generality we assume in the following definitions that the absolute value of $I^0(i,j)$ in position (i,j) is greater/equal the intensity $I^0(j,i)$ in position (j,i). In the match proportional symmetry enhancement the function $f_{\max}$ is defined by

$$
f_{\max}^A = m I^0(i,j)
\tag{3}
$$

$$f_{\min}^A = m I^0(j, i).$$

The match proportional symmetrizations are rather effective in the suppression of noise and artifacts, but peaks that are not present simultaneously in the two symmetry related positions can be lost. In contrast, the additive symmetry enhancement does not lead to the disappearance of any peak, since they are still observable at low contour levels, but the gain in spectral quality can only be moderate. This method is defined by

$$
\begin{aligned}
f_{\max}^B &= (m + B) \; I^0(i, j) \quad (B > 0) \\
f_{\min}^B &= (m + B) \; I^0(j, i).
\end{aligned}
\tag{4}
$$

Usually, we set B to 1, but smaller values of B can be used to increase the artifact suppression, the limiting case B=0 results in the match proportional procedure.

The difference proportional symmetry enhancement has properties somewhat intermediate between the first two methods. In addition, the symmetry of the resulting spectrum can be tuned by a suitable choice of the constants D_1 and D_2. The procedure is described by

$$
\begin{aligned}
f_{\max}^C &= I^0(i, j) - D_1(1 - m)\Delta I \quad (0 \leq D_1 \leq 1) \\
f_{\min}^C &= I^0(j, i) + D_2 m \Delta I \quad (0 \leq D_2 \leq 1) \\
with & \\
\Delta I &= I^0(i, j) - I^0(j, i).
\end{aligned}
\tag{5}
$$

D_1 and D_2 are free parameters that influence the remaining spectral assymmetry, selecting e.g., $D_1 = D_2 = 1$ leads to completely symmetric spectra. The match factor m can be defined in different ways, the selection of a certain match factor depends on experimental details, on the information wanted and on the computing time accepted.

All symmetry enhancement procedures published so far can be expressed in the above formalism, more important, the application of the corresponding match factors in conjunction with the different definitions of $f_{\max}$ and $f_{\min}$ leads to new classes of symmetrization procedures. Also, the generalization of methods only defined for absolute value spectra to phase-sensitive spectra is a natural outcome of the proposed formalism.

Earlier we defined a match factor[46] m by

$$
\begin{aligned}
m = \cos \alpha \; &= \underline{r}_{ij} \underline{r}_{ji} / |r_{ij}||r_{ji}| \quad \text{for } 0 \leq \cos \alpha \leq 1 \\
m = 0 \; & \qquad\qquad\qquad\quad \text{otherwise}
\end{aligned}
\tag{6}
$$

with α the angle between the two vectors $\underline{r}_{ij}$ and $\underline{r}_{ji}$ that contain the intensities from rectangular environments of points (i, j) and (j, i) of the data matrix. This match factor has the advantage that it equals 1 for two identical but differently scaled cross peaks and that it is near zero if two data windows containing random noise are compared. However, it is rather sensitive to base plane variations as they usually occur·in absolute value spectra. This effect can be diminished by introducing an offset corrected match factor m' obtained by comparison of the modified vectors $\underline{r}'_{ij}$ and $\underline{r}'_{ji}$

$$
\begin{aligned}
\underline{r}'_{ij} &= \underline{r}_{ij} - r^0_{ij}\underline{1} \\
\underline{r}'_{ji} &= \underline{r}_{ji} - r^0_{ji}\underline{1}
\end{aligned}
\qquad (7)
$$

with $\underline{1}$ a vector with all components equal to 1 and r^0_{ij} and r^0_{ji} the average of all components of $\underline{r}_{ij}$ and $\underline{r}_{ji}$. Fig. 1 shows the application of the offset corrected match factor in conjunction with the three symmetry enhancement procedures defined by (3), (4), and (5) to a phase-sensitive double-quantum filtered COSY spectrum of the HPr protein.

All spectra were normalized to the intensity of the cross peak at (1.47, -0.18) ppm that has a well-defined symmetry related partner. As to be expected the match proportional (Fig. 1c) and the difference proportional (Fig. 1c) symmetry enhancement lead to a very effective noise and artifact suppression with a gradually better suppression by the match proportional method. The additive symmetry enhancement (Fig. 1d) still provides a reasonable spectral improvement with the advantage that one can be sure that no relevant cross peaks have been lost, that is this method can be applied in almost any case.

In summary, symmetry enhancement is a useful and very flexible method for enhancing the spectral quality, but it should be applied, similar to time-domain filtering according to the properties of the spectrum under investigation and according to the type of information wanted. Match factors calculated from data windows with dimensions greater 1 should be used if the local symmetries are rather well conserved since they use the symmetry information more effectively. The inherent local symmetries can be strongly affected by very different degrees of zero-filling and/or by the use of very different filter functions in t_1 and t_2 direction. In this case single point comparisons are superior. Here the generalizations of the procedures described by Bolton[41,45] appear to be best-suited.

Usually, 2D NMR spectra of macromolecules, especially NOESY and RCT spectra, have a very large dynamic range. Weak cross peaks are found next to very intense cross peaks. In this case, it is impossible to find plot levels that represent all these peaks equally well. Too low plot levels show the broad tails of the intense peaks, too high plot levels do not display weak, but significant peaks. Here segmentation combined with local rescaling of the data[47] can be helpful. (Peak recognition and segmentation procedures

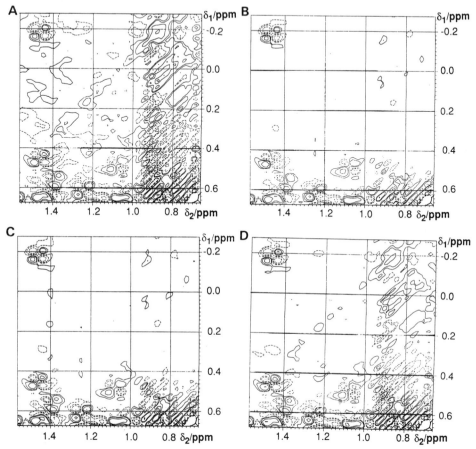

Figure 1. Symmetry enhancement in 2D NMR spectra. Part of a 500 MHz phase-sensitive DQF-COSY spectrum of HPr protein dissolved in 90% H_2O and 10% D_2O. Original spectrum (a) and symmetry enhanced spectra (b), (c), (d) using the offset corrected match factor calculated for a data window of 7×7 data points. Spectrum (b) match proportional symmetry enhancement, spectrum (c) difference proportional symmetry enhancement with $D_1 = 1$ and $D_2 = 0$, spectrum (d) additive symmetry enhancement with $B = 1$. Contour lines depicted correspond to multiples of 2 of the lowest contour level.

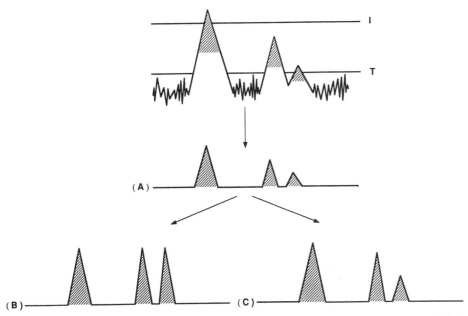

Figure 2. Schematic representation of the local rescaling. The three variants, method A, B, and C, are depicted schematically. (Top) Peak recognition and segmentation.

are also integral parts of the procedures necessary for pattern recognition; they are described in the corresponding section.)

The basic idea can be understood from Fig. 2: After identifying and segmenting all peaks i with intensities I_i above a threshold $T(T > 0; |I_i| > T)$ all peaks are cut at their half heights and the lower parts are removed (A).

Now it is possible to plot all peaks at their individual half heights using only a single (the lowest) contour level. Since linewidths at half heights are smaller than near the base plane, this leads to a considerable apparent resolution enhancement. The intensity information is still preserved. For describing the peak shape of the individual peaks sufficiently, it is still necessary to use a large number of contour levels. This can be prevented by giving all peaks the same intensity (B). Now only a limited number of contour levels is necessary for the description of the peak shape. The disadvantage is that the intensity information is now completely lost. Alternatively, one can project all peaks inside a user selected range of intensities between $-I$ and I into a fixed range from $-C$ and C. Peaks with intensities outside $(-I, I)$ are set to $-C$ and C, respectively (C). Fig. 3 shows an example how this method works.

In Fig. 3a a part of the original spectrum, an absolute value relayed coherence transfer spectrum from the HPr protein, is depicted. The lowest contour level has been chosen in such a way that all significant cross peaks are visible. Apparently, the strong cross peaks in the upper part of the figure cannot be resolved using these contour levels. Local rescaling using method C (Fig. 3b) leads to an improved resolution in the plot, the peak shapes can be represented with only a few contour levels, nevertheless qualitative intensity information is still preserved.

In addition, the segmentation procedure allows to remove spikes or broad back ground peaks since the line widths of all peaks are determined during this procedure. A favorable side effect is that the automated estimation of peak volumes is possible and very simple because one has only to add up the voxels belonging to a given peak.

Since the peak shapes at half heights are usually defined only by a few data points (k, d_k), it is advisable to interpolate the polygons of contour lines. In our experience, the interpolation by a Bezier polynomial $P(u)$ (see e.g., ref. 50) is best-suited. $P(u)$ is defined as

$$P(u) = \sum_{i=0}^{3} b_i B_i(u) \tag{8}$$

with

$$B_i = \binom{3}{i} (1 - u)^{3-i} u^i$$

Interpolated values for the pair of data (k, dk) and $(k + 1, dk + 1)$ are obtained from the condition $u - k = 0$. The Bezier coefficients b_i are not defined completely by (8), a definition of b_i known from the literature as 'smooth' Bezier interpolation gives the best results in our applications. The effect of this interpolation can be best seen from Fig. 3c that shows the same data as Fig. 3b but after application of the interpolation routine.

SEPARATION OF RELEVANT OBJECTS

Resonance peaks are the basic objects of an 2D NMR spectrum. The basic peak recognition is done in AURELIA by peak picking: the spectrum is searched for relative maxima and minima defined as data points (i,j) having intensities $I(i, j) > I(k, l)$ and $I(i, j) < I(k, l)$ $(k = i - 1, ..., i + 1$ and $l = j - 1, ..., j + 1)$; $(i, j) \neq (k, l))$, respectively. To reduce the number of noise and artifact peaks extrema $I(i, j)$ are discarded if $|I(i, j)| < |T|$

Figure 3. Contour plots of locally rescaled 2D NMR spectra.
Part of an absolute value RCT spectrum of HPr protein from *Staphylococcus aureus* dissolved in H_2O (10% D_2O). Frequency 500 MHz, digital resolution 6.8 Hz/point. Part of the original spectrum (a) and of the same spectrum processed with method C without (b) and with Bezier interpolation of data (c).

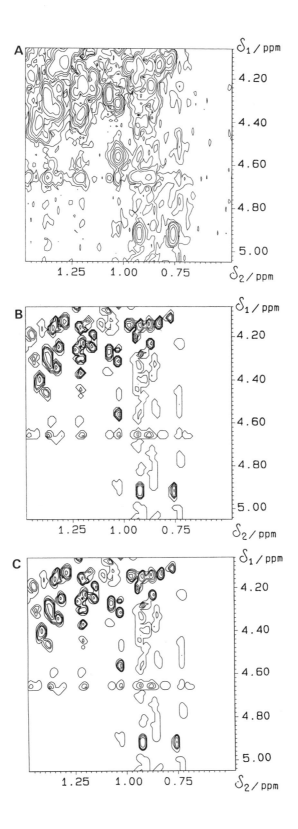

183

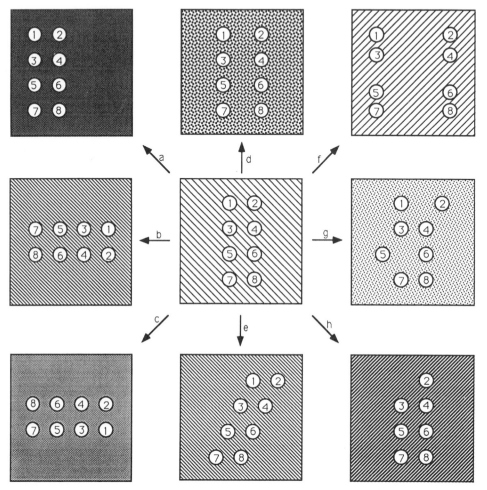

Figure 4. Definitions of multiplets and their transformation. The center picture symbolizes a theoretical multiplet pattern defined in the data base which could be shifted (a), rotated (b), reflected (c), uniformly scaled (d), and sheared (e). Additional degrees of freedom are variation of the individual distances between the subpeaks (without changing the overall symmetry) (f), small shifts of individual peaks (g), and missing peaks (h).

184

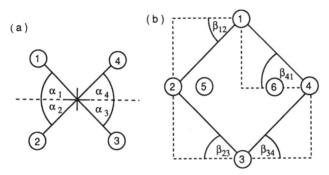

Figure 5. Definition of match factors. The match factor m_a is defined by the angle between the vectors a_e and a_t containing the angles α_i in the experimental and theoretical patterns (a), the matching factor m_b is defined by the angle between the vectors b_e and b_t containing the angles β_{ij} in the experimental and theoretical patterns (b).

with T a user defined threshold. Moreover, the number of recognized peaks that have to be considered in subsequent steps can be reduced by defining interactively areas where no valid peaks are to be expected. These areas are excluded from the search. An additional criterion for the validity of peaks can be their line widths. They can be obtained by segmentation of data by an iterative region growing algorithm[9] that allows definition of the data points belonging to one given peak.

CLASSIFICATION AND INTERPRETATION OF OBJECTS

In our context interpretation of objects and classes of objects corresponds to recognition of multiplets and spin systems.

In the past, many different methods have been proposed for multiplet recognition in two-dimensional NMR spectra. An important class of methods makes use of the internal symmetries of the multiplets[4,9,10,13,15-17]. Within their limitations these methods work rather effectively; in the following a more general approach is presented which allows recognition of any arbitrarily defined multiplet pattern[2]. At this point, it is necessary to define what the program should recognize as a pattern. In the simplest case, the multiplet pattern is defined by the exact coordinates of its subpeaks. The variation of one distance between a pair of subpeaks caused by a change of J-couplings, for example, would mean a different pattern. Such a definition is in most cases impractical due to variations of J-couplings with the dihedral angles. Fig. 4 shows which transformations of a given multiplet are of importance.

We start for example with a multiplet consisting of 8 subpeaks with a given sign and intensity distribution. It is clear that the multiplet should be recognized if its position is changed in the spectrum. The same is true

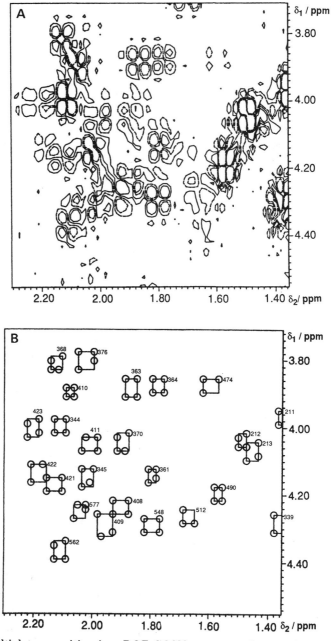

Figure 6. Multiplet recognition in a DQF-COSY spectrum of a protein. Contour plot of $H_\alpha - H_\beta$ region of double-quantum filtered COSY spectrum of HPr protein from *S. aureus* (a) and recognized multiplets (b).

for uniform scaling caused by changes in digital resolution. The variation of J-couplings leads to changes in the relative distances of the subpeaks. The limited digital resolution and superposition with other multiplets often leads to a displacement of some of the peaks or even to their disappearance. All these cases should be comprised within our recognition procedure. Other possible transformations like shearing, inversion or reflection are excluded in our present implementation; by definition they produce new multiplet patterns. As we will see, if one really would like to recognize them in the spectrum, they could easily be taken into consideration by using different definitions of match factors.

Before starting the procedure one has to decide which multiplet pattern(s) should be recognized in the spectrum. The multiplet patterns are contained in a problem-related data base that can be modified interactively by the user. Before the true multiplet recognition is started, the resonance peaks found after peak search are clustered using an unsupervized single-pass cluster algorithm[51]. By this cluster analysis all peaks that possibly can be part of a common multiplet are grouped together. Every cluster is searched, if it contains a given multiplet pattern.

The pattern recognition procedure consists of 5 steps: (1) Theoretical and experimental patterns are placed in a common coordinate system, (2) the theoretical pattern is shifted in such a way that the centers of gravity of both patterns coincide, (3) the theoretical pattern is homogeneously rescaled, so that the radii of gyration of experimental and theoretical pattern are identical, (4) the subpeaks of the theoretical and the experimental pattern are pairwise correlated by minimizing the sum of the pairwise distances of subpeaks having the same sign, (5) a match factor m between theoretical and experimental pattern is calculated. The match factor m is a measure for the similarity between theoretical and experimental pattern, only patterns with a match factor high enough are retained in the final multiplet list. Ideally, the match factor m should be invariant for the allowed transformations of the multiplet pattern. We define m as the weighted sum of the match factors m_a, m_b and m_c (Fig. 5):

$$m = q_a m_a + q_b m_b + q_c m_c \quad . \tag{9}$$

In our experience, a value of 0.25 for the weights q_a and q_b and a value of 0.5 for the weight q_c is best suited. The match factors themselves are designed according to the same scheme: they are defined as the cosine of the angle between two vectors, one for the theoretical pattern, one for the experimental pattern. The components of the vectors are certain quantities that characterize the patterns. The match factor m_a is calculated from the vectors containing the angles α_i relative to the center of gravity. The match factor m_b compares the overall shape of the multiplet, the corresponding vectors contain the angles β_{ij} between two pairs of subpeaks located at the border on the multiplet pattern. Finally, the match factor m_c compares the relative intensities of corresponding subpeaks. The match factor m defined in this way is invariant for translation and uniform scaling and it is rather insensitive for the remaining allowed transformations.

As a typical example Fig. 6 shows the multiplet recognition in a double-quantum filtered COSY spectrum of HPr protein. In the part of the spectrum depicted very strong, well defined cross peaks as well as very weak ones can be observed. Some of the cross peaks are strongly superposed, near intense cross peaks filtering artifacts have almost the same intensity as the weak cross peaks. The result of the multiplet search appears rather satisfactory: Most of the multiplets are recognized correctly, even if one of the subpeaks is missing or shifted considerably.

How good are the results one obtains in general? Usually, more than 90% of all multiplets are recognized in 2D-spectra. Of course, the results strongly depend on the quality of the spectrum and the degree of mutual superposition. However, besides the correctly recognized multiplets, the program usually recognizes a number of meaningless multiplets originating for example from t_1-noise. Therefore, before using the found multiplets for the search of spin systems, the multiplet list has to be cleaned up. This is done interactively in the graphical environment of AURELIA. Here multiplets can be removed from or added to the final list.

Most probably computer aided data evaluation will get more and more important in future. There are two limiting cases demanding different strategies: For small molecules that are soluble enough that spectra with an excellent signal/noise ratio can be obtained one should try to define a basic set of 2D-experiments that allow a complete automated analysis. For large molecules with a limited signal-to-noise ratio a completely automated analysis is in our opinion impossible in the near future, simply because the molecules one would like to study become larger and larger. Here, one should develop a computer assisted, interactive strategy that is as flexible as possible.

REFERENCES

1. K.-P. Neidig, H. Bodenmüller, and H. R. Kalbitzer, *Biochem. Biophys. Res. Comm.* **125**, 1143-1150 (1984).
2. K.-P. Neidig, R. Saffrich, M. Lorenz, and H. R. Kalbitzer, *J. Magn. Reson.*, **89**, 543-552 (1990).
3. B. U. Meier, G. Bodenhausen, and R. R. Ernst, *J. Magn. Reson.* **60**, 161-163 (1984).
4. B. U. Meier, Z. L. Mádi, and R. R. Ernst, *J. Magn. Reson.* **74**, 565-573 (1987).
5. P. Pfändler, G. Bodenhausen,, B. U. Meier, and R. R. Ernst, *Anal. Chem.* **57**, 2510-2516 (1985).
6. P. Pfändler and G. Bodenhausen,, *J. Magn. Reson.* **70**, 71-78 (1986).

7. P. Pfändler and G. Bodenhausen,, *Magn. Reson. Chem.* **26**, 888-894 (1988).
8. P. Pfändler and G. Bodenhausen,, *J. Magn. Reson.* **79**, 99-123 (1988).
9. S. Glaser and H. R. Kalbitzer, *J. Magn. Reson.* **74**, 450-463 (1987).
10. J. C. Hoch, S. Hengyi, M. Kjær, S. Ludvigsen, and F. M. Poulsen, *Carlberg Res. Comm.* **52**, 111-122 (1987).
11. Z. L. Mádi, B. U. Meier, and R. R. Ernst, *J. Magn. Reson.* **72**, 584-590 (1987).
12. M. Novic, H. Oschkinat, P. Pfändler, and G. Bodenhausen, *J. Magn. Reson.* **73**, 493-511 (1987).
13. M. Novic, U. Eggenberger, and G. Bodenhausen, *J. Magn. Reson.* **77**, 394-400 (1988).
14. H. Egli, *Magn. Reson. Chem.* **26**, 876-860 (1988).
15. M. Novic, and G. Bodenhausen,, *Anal. Chem.* **60**, 582-591 (1988).
16. B. U. Meier and R. R. Ernst, *J. Magn. Reson.* **79**, 540-546 (1988).
17. Z. L. Mádi, and R. R. Ernst, *J. Magn. Reson.* **79**, 513-527 (1988).
18. U. Eggenberger, P. Pfändler, and G. Bodenhausen, *J. Magn. Reson.* **77**, 192-196 (1988).
19. H. Grahn, F. Delaglio, M. A. Delsuc, and G. C. Levy, *J. Magn. Reson.* **77**, 294-307 (1988).
20. M. Billeter, V. J. Basus, and I. D. Kuntz, *J. Magn. Reson.* **76**, 400-415 (1988).
21. C. Ciesler, G. M. Clore, and A. M. Gronenborn, *J. Magn. Reson.* **80**, 119-127 (1988).
22. C. Ciesler, T. A. Holak, and H. Oschkinat, *J. Magn. Reson.* **87**, 400-407 (1990).
23. C. D. Eads and I. D. Kuntz, *J. Magn. Reson.* **82**, 467-482 (1989).
24. W. J. Goux, *J. Magn. Reson.* **85**, 457-469 (1989).
25. G. J. Kleywegt, R. M. J. N. Lamerichs, R. Boelens, and R. Kaptein, *J. Magn. Reson.* **85**, 186-197 (1989).
26. G. J. Kleywegt, R. Boelens, and R. Kaptein, *J. Magn. Reson.* **88**, 601-608 (1990).
27. P. J. Kraulis, *J. Magn. Reson.* **84**, 627-633 (1989).
28. V. Stoven, A. Mikou, D. Piveteau, E. Guittet, and J.-Y. Lallemand, *J. Magn. Reson.* **82**, 163-168 (1989).
29. H. R. Kalbitzer, K.-P. Neidig, and W. Hengstenberg, *Physica B* **164**, 180-192 (1990).
30. D. Marion, M. Ikura, and A. Bax, *J. Magn. Reson.* **84**, 425-430 (1989).
31. Y. Kuroda, A. Wada, T. Yamazaki, and K. Nagayama, *J. Magn. Reson.* **84**, 604-610 (1989).
32. G. Otting, H. Widmer, G. Wagner, and K. Wüthrich, *J. Magn. Reson.* **66**, 187-193 (1986).
33. R. E. Klevit, *J. Magn. Reson.* **62**, 551-555 (1985).
34. P. H. Bolton, *J. Magn. Reson.* **64**, 352-355 (1985).
35. S. Glaser and H. R. Kalbitzer, *J. Magn. Reson.* **68**, 350-345 (1986).
36. Z. Zolnai, S. Macura, and J. L. Markley, *Comput. Enhanced Spect.* **3**, 141-145 (1986).
37. Z. Zolnai, S. Macura, and J. L. Markley, *J. Magn. Reson.* **80**, 60-70 (1988).
38. Z. Zolnai, S. Macura, and J. L. Markley, *J. Magn. Reson.* **82**, 496-504 (1989).
39. I. L. Barsukov and A. S. Arseniev, *J. Magn. Reson.* **73**, 148-149 (1987).
40. L. Mitschang, K.-P. Neidig, and H. R. Kalbitzer, *J. Magn. Reson.*, **90**, 359-362 (1990).
41. P. H. Bolton, *J. Magn. Reson.* **68**, 180-184 (1986).
42. P. H. Bolton, *J. Magn. Reson.* **70**, 344-349 (1986).
43. R. Baumann, A. Kumar, R. R. Ernst, and K. Wüthrich, *J. Magn. Reson.* **44**, 76-83 (1981).
44. R. Baumann, G. Wider, R. R. Ernst, and K. Wüthrich, *J. Magn. Reson.* **44**, 402-406 (1981).
45. P. H. Bolton, *J. Magn. Reson.* **67**, 391-395 (1986).
46. K.-P. Neidig and H. R. Kalbitzer, *Magn. Reson. Chem.* **26**, 848-851 (1988).
47. K.-P. Neidig and H. R. Kalbitzer, *J. Magn. Reson.* **88**, 155-160 (1990).
48. K.-P. Neidig and H. R. Kalbitzer, *J. Magn. Reson.*, **91**, 155-164 (1991).

189

49. A. G. Ferrige, and J. C. Lindon, *J. Magn. Reson.* **31**, 337-340 (1978).
50. J. D. Foley, and A. van Dam, "Fundamentals of Interactive Computergraphics," Addison-Wesley Publishing Company, London (1984).
51. J. A. Richards, "Remote Sensing Digital Image Analysis," Springer, Heidelberg (1987).

THE APPLICATION AND DEVELOPMENT OF SOFTWARE TOOLS FOR THE PROCESSING AND ANALYSIS OF HETERONUCLEAR MULTI-DIMENSIONAL NMR DATA

E. T. Olejniczak, H. L. Eaton, E. R. P. Zuiderweg, and S. W. Fesik

Pharmaceutical Discovery
Abbott Laboratories
Abbott Park, Il 60064

ABSTRACT

Software for the processing and analysis of multi-dimensional data will be described. A variety of tools has been implemented to improve the processing of multi-dimensional data including the ability to extrapolate the time domain data with linear prediction. Predicting several additional time domain points can significantly improve the data of multi-dimensional NMR experiments where one of the dimensions is undersampled due to accumulation time constraints. This data extrapolation allows a more favorable windowing of the data before Fourier transformation. It will be shown that when used conservatively the augmented data are very reliable and that improved resolution and sensitivity are obtained in the frequency domain data.

For analyzing multi-dimensional data, sorting through and keeping track of data are important, but mundane, problems that are ideally suited for a computer. Peak-picking is one important aspect of the bookkeeping that is needed. Our implementation of a peak-picking algorithm that uses volume integrals instead of heights will be described.

The analysis of multi-dimensional data is greatly facilitated if the computer has the ability to present the user with information about slices in the data set that fulfill criteria defined by the spectroscopist. For example, a useful criterion for finding adjacent or nearby amides in alpha helical sections of proteins, is to find different amide protons that have a large number of NOEs to protons at common frequencies. The computer can sort through the 3D $^{15}N - ^1H - ^1H$ NOE data and find the amide protons that best match this criterion, and then present the user with the top choices. This simple aid greatly facilitates the assignment of these amides in the spectra. Examples of this software aid will be demonstrated using simulated BPTI data.

INTRODUCTION

Increasing the resolution of ^1H NMR experiments by spreading the signals over the chemical shift range of an attached heteronucleus has great potential

Computational Aspects of the Study of Biological Macromolecules by Nuclear Magnetic Resonance Spectroscopy, Edited by J.C. Hoch *et al.,* Plenum Press, New York, 1991

191

for significantly expanding the size range of macromolecules that can be studied by high resolution NMR techniques[1-3]. The improved resolution greatly simplifies the spectra by removing many ambiguities due to overlap that would occur in the corresponding two dimensional spectra. An obvious consequence of the improved resolution is that the number of data points increases rapidly with each new dimension. Thus, even though the data are simplified, software is still needed to aid the spectroscopist to sort through the data and to keep track of the information obtained. In this paper we will give brief descriptions and examples of several software tools that we have developed to improve the processing and analysis of multi-dimensional data.

USING LINEAR PREDICTION TO IMPROVE WINDOWING OF DATA

In multi-dimensional NMR experiments, the acquisition of time points in the higher dimensions may take hours for 3D and even days for 4D[2,3]. Because of limitations in experimental time, it is generally not possible to fully sample these dimensions. In order to properly process such truncated data using Fourier transformation, stringent apodization functions need to be applied, which causes the last points to make almost no contribution to the frequency domain signal. To alleviate this problem we propose a hybrid technique which uses Linear Prediction to extend the time domain data followed by normal Fourier transform techniques.

In Linear Prediction (LP)[5-7], future points are predicted based on a time series of previous points (i.e., X_n) of the form:

$$X_n = C_1 X_{n-1} + C_2 X_{n-2} + C_3 X_{n-3} + \cdots + C_k X_{n-k}. \tag{1}$$

The minimum number of LP coefficients (i.e., C_k) necessary is twice the number of frequency components present.

Several different approaches can be used to obtain the Linear Prediction coefficients[5-7]. In our implementation we found that satisfactory results for noisy data can be obtained by using more poles than twice the number of real frequencies. We have chosen to use a high number of poles, equal to three quarters of the data points. For example, for a thirty-six point complex time domain data set, twenty four poles were used. For complex time domain data, an independent LP calculation was done for the real and imaginary data. LP coefficients are often corrected by adjusting the magnitude of the roots of their characteristic equation so that none of the signals increase with time. This is the most time consuming part of the computational procedure and in our hands did not improve the predictions significantly and therefore has not been used in our data processing. Errors

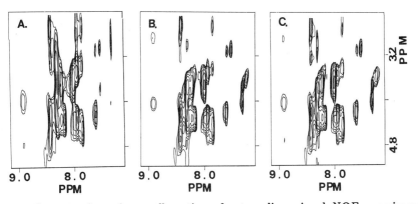

Figure 1. Contour plots of a small section of a two dimensional NOE experiment on the small peptide Pro(10)AAP (7-23) (8). Data obtained from 48 t_1 data points (B) is compared to a 36 t_1 data point set (A) and a 48 point data set obtained by using linear prediction to predict the last twelve data points (C). All of the data was processed using a cosine window function in t_1 and are plotted at the same contour levels.

in the predicted time domain points are further mitigated by the windowing of the data prior to Fourier transformation[9].

In Fig. 1 the hybrid method is compared to standard data processing using a small section of a two dimensional NOE experiment. Time domain data augmented with twelve LP predicted points in t_1 (Fig. 1C) result in a spectrum that is nearly as good in both sensitivity and resolution as the data with an equal number of experimental points (Fig. 1B). The gain in sensitivity and resolution of the LP augmented data can be seen by comparing it to spectra obtained without using the twelve additional LP data points (Fig. 1A).

The benefits of this procedure increase as the number of dimensions and therefore the accumulation time per data point increases. The ability to reliably extend time domain data can lead to a significant reduction in the accumulation time of multi-dimensional NMR experiments. Shorter accumulation times will reduce long term spectrometer stability problems. Alternatively, the time savings could be applied to increasing the number of scans, the phase cycling or to extensions to additional time domain dimensions.

PEAK-PICKING USING VOLUMES

The first step in any computer assisted assignment procedure is the identification of peaks in the spectra. A variety of methods has been used for peak-picks[10]. The most common method is to check for a maximum height in all directions. Other methods have also been proposed which use integrals instead of peak heights to find peak maxima[10]. The criteria for a good

193

peak-picking routine are that it should identify as many peaks as possible and only identify a small number of incorrect peaks. Unfortunately, as the criteria used in the peak-pick are modified to find more peaks the number of incorrect peak-picks also increases.

We have been investigating the use of integrals of small regions of the multi-dimensional data set to find peaks. The advantage of using integrals is that broader, lower intensity peaks can in principle be identified[10]. For three dimensional data our algorithm can be summarized as follows:

(1) Baseline correct the data.

(2) Select direction of search. This defines the longest dimension of each small "rectangular box" that will be searched. The size in the other 2 dimensions is 10 points. Each box is searched independently.

(3) Using a random number generator select a small number of points in box and determine the standard deviation of the noise.

(4) Replace each point in box by the sum of the amplitudes of the 3 by 3 square of points surrounding it.

(5) Scan through data to find points where the value of the point as determined in step (4) is greater than $X * (\sigma)$ and the sum of the points in the 3 by 3 by 3 cube is greater than $X * (\sigma)$. If this criteria fails then a 6 by 6 square and 6 by 6 by 6 cube is tested to determine if a broad peak is present. The value of X is inputted by the user. In most cases X is chosen to be greater than 3.0.

(6) Do a normal peak-pick on the points that pass step (5). Note that because of step (4), the program will be comparing volumes rather than heights.

(7) Improve on the position of each peak maximum by using parabolic interpolation.

In step (2) the user should select the dimension of data that has the most streaks in it. Since each box is treated separately this greatly reduces the number of false maxima caused by artifacts in the data. A good estimate of the threshold to be used for each box can be obtained from a small fraction of the actual number of points to be tested. This is done in three loops, in each of which points are included in the estimate if the difference between the amplitude of the point x and the average of the points in the previous loop quantity squared is less than ten times σ^2 of the previous loop.

$$(x - X(avg))^2 > 10 * \sigma^2. \tag{2}$$

In step (5) peak regions are identified using volume criteria based on two different size cubes and squares. This gives both broad and sharp peaks a reasonable chance to pass the criteria for a peak. Points that pass the criteria described in step (5) are then examined using criteria very similar to what would be used if the peak-pick were based on heights. For three dimensional data twenty six points surrounding the point being tested are used. These twenty six points define thirteen different directions that are tested to see if the value at point X is a local maximum.

Fairly restrictive criteria are used to test for peaks that are shoulders on other peaks. In our experience it is very difficult to find badly overlapping peaks without including an unacceptably large number of erroneous peak-picks. Thus we limit our tests to points that are local maxima in all but one direction. In the one direction that they are not local maxima, the volume at the point must be greater than the sum of the point before and after divided by 1.95. Passing this criteria indicates that the value of this point is significantly different than what would be expected for a point on the side of a gaussian or Lorentzian line except for points near a peak maximum. Therefore a check is then made to make sure this is not a point close to the top of another local maximum.

The location of peak maxima can be improved using inverse parabolic interpolation. For well-resolved peaks parabolic interpolation gives an accurate position for the peak maxima. The interpolation can in some cases greatly improve the position of the peak maximum for broad peaks. In all cases tested the procedure improved the position of the peak maximum; however, in many cases the improvement was small. The improvement in peak position that is obtained by this procedure is obtained at almost no CPU cost. For a maximum at point N, reverse interpolation using point N-1 and N-2 uses the equations

$$del = -.50 \left(X(N+1) - X(N-1) \right) / \left(2 * X(N) - X(N+1) - X(N-1) \right) \tag{3}$$

$$N_{corrected} = N + del. \tag{4}$$

The most noticeable advantage of the peak-pick routine described here is that it can find broader peaks without adding many incorrect peaks. By determining its own threshold for each box, the routine limits the number of incorrect peaks in streaks. It still does a reasonable job on peaks that are in the noise from these streaks.

IMPROVING INTERACTIVE COMPUTER-AIDED DATA ANALYSIS

The analysis of multi-dimensional NMR data is greatly facilitated by having the computer provide the user with information about slices in the data set that fulfill criteria defined by the spectroscopist. The computer can sort through the data and find the resonances that best match the criterion and then prompt the user with the top choices. The user can then quickly review and test each choice.

An example of how a computer can facilitate an assignment strategy was tested using simulated 3D $^{15}N - {}^1H - {}^1H$ data for BPTI. Chemical shifts used in the simulation were obtained from the literature when known or placed in the correct range of ppm values based on atom type[11,12]. NOE intensities were obtained from simulations based on distances obtained from a crystal structure of BPTI[13].

The first step in the assignment procedure consisted of identifying resonances on adjacent residues. For α-helical portions of the protein adjacent amides can be identified by searching for amides with a strong and symmetrical NH-NH NOE and a large number of NOEs to resonances at common proton frequencies. The adjacent amide is generally one of the top three choices found by this criteria. In contrast, β-sheets are identified by the presence of large NH-H$^\alpha$ coupling constants and weak NOEs to this alpha proton. Once an amide has been identified which satisfies this criteria candidates for the preceding residue are found by using the characteristic strong NOE between the amide and the alpha proton on the preceding residue. Using these criteria the computer can present the user with the best amides in the data set which fulfill the given criterion. The software itself does not make any assignments but it makes it possible for the user to quickly evaluate and test all of the choices which satisfy the user defined criteria.

Once several adjacent residues have been identified these residues can be placed in the sequence by using ^{15}N correlation data assigned using proteins selectively labeled with ^{15}N. For example, sections which contain two selective labels can generally be uniquely placed in the sequence. Additional assignments can be made by repeating this procedure looking for unique sequences which fit the data.

In our trial case we found that the exact amino acid type for the selective labels is not important but it is very important to have them spread throughout the sequence. The need for selective labels is not a major drawback for the procedure because the 2D $^{15}N - {}^1H$ correlation experiments are very sensitive and thus only small amounts of the selectively enriched samples are necessary.

Alanines can often be used as landmarks even when a selective label is not available. Candidates for Alanine residues can often be identified by the strong NOE between the NH and its α and β. The candidates found in this way can be checked with a long range HMQC experiment to look for the correlation between the HN and β methyl group. This strategy generally finds a few but not all of the alanines in the sequence.

A computer can aid a user to test the choices defined by the procedure outlined here. All assignments are still made by the user but the drudgery of scanning through the data is left to the computer. The procedure outlined here is viable for the sequential assignment of large proteins since it relies on NOE data and on a small amount of correlated data between protons with large coupling constants. In this test case the amides of this small protein were correctly assigned in less than a day. Other strategies would be necessary to assign the sidechain protons.

CONCLUDING REMARKS

Resolving NMR data in multiple dimensions removes many ambiguities in the interpretation of NMR data due to overlap. Improvements in the data can be made using different processing techniques such as extrapolation of the time domain data with LP. The larger size of multi-dimensional data sets makes sorting without software aids much more arduous. A important step in sorting through the data is to identify important regions of the spectra by identifying peaks using some sort of peak-picking routine. Automated or computer aided assignment procedures will rely heavily on the results generated by such routines. Simple software tools as those described here can aid the user to efficiently interpret the much larger data sets involved and to quickly move to sections of the data set that fulfill criteria defined by the user.

ACKNOWLEDGEMENTS

We thank R.T. Gampe and J. Curtis for discussions and technical assistance.

REFERENCES

1. S. W. Fesik and E. R. P. Zuiderweg, *J. Magn. Reson.* **78**, 588 (1988).
2. S. W. Fesik and E. R. P. Zuiderweg, *Quarterly Rev. Biophysics* **23**, 97 (1990).
3. L. E. Kay, D. Marion, and A. Bax, *J. Magn. Reson.* **84**, 72 (1989).
4. R. R. Ernst, G. Bodenhausen, and A. Wokaun, "Principles of Nuclear Magnetic Resonance in One and Two Dimensions", p. 106, Clarendon Press, Oxford (1987).
5. W. H. Press, B. P. Flannery, S. A. Teukolsky, and W. T. Vetterling, "Numerical Recipes: The Art of Scientific Computing", Chap. 12, Cambridge Univ. Press, Cambridge. (1986).
6. H. Barkuijsen, R. De Beer, W. M. M. J. Bovee, and D. Van Ormondt, *J. Magn. Reson.* **61**, 465 (1985).
7. R. Kumaresan and R. J. Tufts, *IEEE Trans.* **ASSP-30**, 833 (1982).
8. R. T. Gampe Jr., P. J. Connolly, T. Rockway, and S. W. Fesik, *Biopolymers*, **27**, 313 (1988).
9. E. T. Olejniczak and H. L. Eaton, *J. Magn. Reson.* **87**, 628 (1990).
10. S. J. Nelson and T. R. Brown, *J. Magn. Reson.* **75**, 229 (1987).
11. ^{15}N shifts for BPTI were kindly supplied to us from D. Cowburn, personal communication.

12. G. Wagner and K. Wüthrich, *J. Mol. Biol.* **155**, 347 (1982).
13. Coordinate set obtained from the Brookhaven data base.

DISTANCE GEOMETRY IN TORSION ANGLE SPACE: NEW DEVELOPMENTS AND APPLICATIONS

Werner Braun

Institute for Molecular Biology and Biophysics
ETH Zürich
CH-8093 Zürich, Switzerland

ABSTRACT

Distance geometry aims at the determination of all macromolecular conformations compatible with distance and dihedral angle constraints. One particular method, the variable target function method in torsion angle space, has been frequently used in the determination of polypeptide and protein structures. This method can be efficiently vectorized on a supercomputer. With the improved program, DIANA, sampling and convergence properties of this method can be studied in detail. Another feature concerns the flexibility in the type of covalent structures which the method can handle. A new graphics tool, GEOM, was developed to deal with linear or cyclic structures that frequently arise in designing new drugs.

INTRODUCTION

In recent years, the determination of three-dimensional structures of macromolecules in solution from n.m.r. data has evolved more and more towards routine work where established calculational methods are used[1-5]. Distance geometry calculation is currently the most extensively used method when no approximate structure is available. Restrained molecular dynamics[6,7] is most useful as a refinement tool.

One particular distance geometry approach is based on the metric matrix method[8-11]. Distances are converted to three-dimensional cartesian coordinates by a partial diagonalization of a certain matrix, the metric matrix. The basic equations used in this approach are directly applicable in cases when all distances between all atoms in a protein are exactly known. For experimental data sets arising in practice the basic equations represent an approximation. In the first applications of this method for the calculation of molecular structures from NMR data[12-15] new programs were written to interface the embedding procedure with a standard library of amino acid residues (e.g., ECEPP[16]) and additional heuristic data processing lead to improved convergence. Treatment of complete covalent polypeptide structures required an efficient way of handling large distance matrices. In practice, this method showed good convergence properties, and it is in widespread use in the software packages DISGEO from T. Havel and DSPACE from Hare Research. However, it has been recognized that the metric matrix approach has a certain bias in regions of a protein with no or only a few experimental distance constraints[2,17−19]. It is not clear if the source of this behavior is due to problems in certain implementations or if it is a more fundamental problem. I will sketch some mathematical ideas, which might be the reason for the poor sampling property of the metric matrix method.

A second method, the variable target function method in torsion angle space, as implemented in the program DISMAN[20] has been successfully applied to determine the tertiary structure of several polypeptides[21−26] and proteins[27−31,17,32−36]. New developments include the program DIANA[37], which is an improved implementation of the variable target function method in torsion angle space, and a new graphics tool GEOM[38], which is used to extend the capabilities of the distance geometry approach in torsion angle space to other covalent structures besides polypeptides and proteins. This tool is especially useful in the field of drug design.

DISTANCE GEOMETRY VERSUS RESTRAINED MOLECULAR MECHANICS/DYNAMICS

Distance geometry means the characterization of all macromolecular conformations compatible with distance and dihedral angle constraints. It is not known, what the best algorithm is for this problem and what the performance of an ideal algorithm would be. In practice, heuristic reasoning prevails: as bound smoothing[10], checking of triangle inequalities for the distances within bounds[13], sometimes also called metrization[39], or different trials for the initial distance distribution in the metric matrix approach. Similarly, the best choice of the different levels for including the experimental distance and dihedral angle constraints during the optimization in the DISMAN approach is still under debate. However, distance geometry has several advantages compared to other calculational methods often applied in the computation of macromolecular structures, restrained energy minimization or restrained molecular dynamics. The answer to the problem does not depend on specific parametrization of the force fields and the global minimum target function value for a consistent data set is a priori known $(T = 0)$. A further important advantage for this approach, in practice, is the short cut-off distance which could be used for the nonbonded interactions.

This makes the list of nonbonded interactions quite small as compared to corresponding lists in energy minimization. This is also the reason why a number of molecular dynamics programs have been modified to this geometric scheme, as in the distance driven molecular dynamics[2,18] or in some simulated annealing schedules[4].

THE VARIABLE TARGET FUNCTION METHOD

The distance geometry problem is formulated in this method as a pure nonlinear optimization problem. Variable target function means that during the optimization, the function to be minimized is changed at several levels. The first implementation of this idea in the calculation of protein structures from NMR data used torsion angles as independent variables and the distance constraints were gradually taken into account. First, only the short range distance information was included in the optimization, and then other constraints with higher levels, i.e., with a greater difference in the sequence numbers, were gradually included. This method has been implemented in the program DISMAN[20].

This method has some resemblance to the build-up procedure which has been used in energy minimization problems[40,41]. Torsion angles are used as independent variables to keep the number of independent variables as small as possible. This approach is also quite natural for the type of the NMR data. Local distance constraints can be measured with higher accuracy. Recently, several new experiments have been proposed for measuring more accurate coupling constants[42-44].

The program DISMAN minimizes a series of functions, which are approximate to a final target function. This final target function adopts the value of zero for a structure which fulfills all constraints. The target function is a measure of how good the distance constraints are fulfilled. There are many ways to construct such target functions.

A typical form of the target function is given by

$$T = \sum_{i<j} [\theta(D_{ij} - U_{ij})(D_{ij} - U_{ij}) + \theta(L_{ij} - D_{ij})(L_{ij} - D_{ij})]. \quad (1)$$

The function $\theta(x)$, which is 0 for $x \leq 0$ and 1 for $x > 0$, is used to sum up all distance violations. The summation over the atom pairs i and j is only over those pairs where there are constraints. The function T is 0 for a solution, positive for all conformations not satisfying the constraints perfectly, and increases as the distance constraints violations get worse. Usually the target functions are variations of Eq. 1, using only the square of distances to allow efficient computation, and such that they are also continuously differentiable at the boundaries $D_{ij} = L_{ij}$ and $D_{ij} = U_{ij}$. This can be done by taking some powers of the distance violations.

Explicit restrictions on torsional angles from spin-spin coupling constants[45] can be implemented easily. This is done in DISMAN following the same philosophy used in constructing the target function from distance constraints. For each restricted torsion angle to an allowed region, i.e., to a region compatible with the NMR data, the target function is defined as zero within the allowed region, has continuous first derivatives at both region boundaries and increases smoothly with the amount of deviation from the allowed region.

The method of variable target functions means that one does not try to minimize T at once but rather to minimize gradually a series of functions which approximate T. More specifically, for a polypeptide chain of n residues the target functions $T_{k,l}$ $k = 1, 2, \ldots, n$ and $l = 1, 2, \ldots, n$ only include those terms of the form as in (1) for atom pairs belonging to residues with the difference of their sequence numbers less than k if the upper or lower limits are from NMR data or less than l if the lower limits are the sum of repulsive core radii. The strategy consists in first minimizing $T_{k,l}$ with small values of k and l and then gradually increasing k and l up to n. The final solution of the problem consists of one or several conformations having zero values for $T_{n,n} = T$. When a problem is overdetermined, the best conformation consistent with the input distance information and stereo-chemical criteria is the one which gives the global minimum of the target function.

This strategy was shown to be effective if good distance information of a short range nature is available. Exact characterization of good distance information which can guide the conformation from correct short range to medium or long range conformations is missing. More extensive numerical experience is certainly needed. Some heuristic ideas of describing the success of the method are as follows. Short and long range distance constraints impose different type of restrictions on the polypeptide conformation. Once short range distance constraints are fulfilled, the polypeptide chain keeps a large amount of "flexibility" for those conformational changes maintaining the short range distance information. Small local changes can give rise to drastic global changes. So these small changes can be used to satisfy the long range distance constraints.

PRACTICAL APPLICATIONS AND STRATEGIES FOR IMPROVING THE CONVERGENCE

In Table 1, polypeptide and protein structures determined so far with the variable target function method in torsion angle space (VTF) are listed. Most of these structures were calculated with the program DISMAN, some with a modified version, DADAS. I expect that in future an increasing number of structures will be calculated by the improved implementation, DIANA[37]. The list ranges from small cyclic polypeptides to proteins of

Table 1. Polypeptide and Protein structures calculated from NMR data by the VTF method

Protein (Number of Residues)	PDB Code	Reference
cyclic bouvardin analog (6)		25
oxytocin (9)		56
conotoxin G1 (*Conus geographus*) (13)		26
heat-stable enterotoxin (*E. coli*) (19)		57
S4 segment of the sodium channel protein (21) protein (21)		55
endothelin (human) (21)		21
melittin (bee venom) in methanol (26)		22,23
micelle-bound melittin (26)		24
neurotoxin ATX (sea anemone) (46)	1ATX	28
anthopleurin A (sea anemone) (49)		48
epidermal growth factor (mouse) (53)		29,30
transforming growth factor (human) (53)		31
trypsin inhibitor (bovine pancreas) (58)		17
cardiotoxin CTX (60) (*Naja mossambica mossambica*) (60)		58
Cd_7-metallothionein (rat) (61)	1MRT,2MRT	50
Cd_7-metallothionein (human) (61)	1MHU,2MHU	51
Cd_7-metallothionein (rabbit) (62)	1MRB,2MRB	27,49
α-thrombin inhibitor (*Hirud medicinalis*) (65)		59
homeodomain (*Antennapedia*) (68)		32
α-amylase inhibitor (74)	2AIT, 3AIT	60,33
anaphylatoxin C5a (human) (74)		61
acylphosphatase (horse muscle) (98)		35,36

about 100 amino acids. This range covers about the size of proteins where high resolution NMR data set can be expected with currently available NMR techniques. The list represents a significant fraction of all polypeptide and protein structures determined so far. Some of the protein structures and the nuclear magnetic resonance data have been deposited to the Brookhaven Protein Data Bank. These structures are given with their PDB code.

In the structure determination of the cyclic bouvardin analogue 212[25], distance geometry calculations in torsion angle space with the program DISMAN were extended to cyclic structures. Previous methods in calculation of cyclic peptide structures from NMR data were based on a grid search[46,47] which are still limited to rather small ring systems. Two different strategies were compared in their efficiency to give a high number of convergent structures. In the method A, first only the distance constraints for the ring closure condition were included in the calculation, starting from 500 random conformations. The best 100 cyclic structures were selected and then minimized in a second step with all NMR distance constraints. In the second method B, all constraints were simultaneously applied. The quality of the best structures, calculated with both methods, were roughly similar, but the second approach yielded a higher number of good structures.

In case of the protein structures of the α-amylase inhibitor Tendamistat[33] and ATX[28], the DISMAN program was first used to calculate small segments of the protein. By statistical analysis of the occurring dihedral angles in the converged structures of the segments, dihedral angle constraints could be extracted which were then used in further calculations in addition to the experimental dihedral angle constraints.

For trypsin inhibitor[17] and anthopleurin[48], both distance geometry methods, the metric matrix approach and the variable target function method, were compared. For both proteins, the best structures calculated by both programs, and the quality of these structures in terms of residual violations were about the same. A different behavior was observed in the sampling property for the trypsin inhibitor[17]. In regions with low numbers of distance constraints, the structures calculated by the program DISGEO did not represent a realistic picture of all possible solutions. Below, I will discuss some possible explanations for this observation.

Metallothionein turned out to be a quite interesting case. The NMR structure determinations were done on proteins from three different species, rabbit liver[27,49], rat liver[50], and human liver[51]. All three proteins showed the same global fold and Cd-cysteine connectivities. However, a significant difference between the NMR result and an independent X-ray study[52] was observed. Quite recently, a new X-ray structure analysis has been performed. In this new X-ray structure, all significant differences to the NMR structure vanished. The backbone rmsd values between the best NMR structure of rat liver MT-2 and the new X-ray structure (C. D. Stout personal communication) are 2.1 Å for the β-domain and 1.8 Å for the α-domain. These values are similar to the average of the pairwise rmsd values between the 10 best rat liver NMR structures[50]. A detailed comparison will be published elsewhere.

In case of acylphosphatase, the first enzyme structure determined by NMR[35,36], the program DISMAN was used as a consistent check for the secondary structure assignment. The secondary structure elements were calculated first and then combined in a final round of distance geometry calculations to get the global fold of the protein.

RECENT DEVELOPMENTS

Recent developments for distance distance geometry calculations in torsion angle space include two new programs DIANA[37] and GEOM[38].

The program DIANA is a new implementation of the variable target function method in torsion angle space. Full use has been made of the short list of nonbonded interactions which is periodically updated. The time consuming parts have been vectorized. The cpu time to calculate a complete protein structure for a typical protein (e.g., BPTI) through all levels is about 1 minute on a Cray X-MP and about 20 minutes on a Sun 4. This compares favorably to the times quoted for the vectorized version of the metric matrix distance geometry as implemented in the program VEMBED[5]. Details on the times required are given in Guentert et al.[37] A new feature included in DIANA is the treatment of distance constraints for diastereotopic pairs of protons. It involves new distance constraints for the pseudo atom representing the pair of protons. These distance constraints are tighter than using the conventional pseudo atom corrections[53]. Test calculations showed that this procedure gives most pronounced advantages for data sets with no stereospecific assignments.

The program GEOM is a new graphics tool for use of distance geometry programs in torsion angle space with linear or cyclic structures. In designing new drugs, it is important to be flexible in the choice of building blocks so that one is not restricted to a library of standard amino acids or nucleic acids. With GEOM, monomeric organic entities can be sketched by the user on a graphics screen. These are regularized by the program package to a standard geometry. These building blocks can then be deposited into a library which is used by the distance geometry program DISMAN. After the calculation the user can analyze the structures on the basis of residual distance and dihedral angle constraints. The similarity of a set of best structures can be checked by graphical superpositions and RMSD values. This package has been used in the structure determination of a cyclic bouvardin analogue[25] and in a model study of cyclosporin A[38].

SAMPLING PROPERTIES OF THE METRIC MATRIX METHOD APPROACH

Comparing the different distance geometry approaches, notably the DIS-GEO and the DISMAN approach, both programs agree quite well in their calculated structures. In regions where distance constraints are quite sparse, however, the rmsd values of the metric matrix approach severely underestimate the true variation of the set of compatible structures[2,17-19]. This bias might be an intrinsic property of the metric matrix approach. I suggest the following explanation. After the bound smoothing, the next step in the EMBED procedure consists in calculation the metric matrix from a set of estimated distances within the given bounds.

$$G_{ii} = \frac{1}{N} \sum_{j}^{N} D_{ij}^2 - \frac{1}{2N^2} \sum_{j,k}^{N} D_{jk}^2. \tag{2}$$

Usually D_{ij} are chosen randomly between the bounds or they are metrized, i.e., treated so that they satisfy the triangle inequality and are still within the given bounds[11,13]. In either case, the dominant part of the summation in Eq. (2) comes from distances, which were derived from insignificant upper and lower bounds in case of a sparse data set. The law of large numbers then suggests that all diagonal elements get similar values for large values of N, if the distances D_{ij} are chosen independently.

$$G_{ii} \rightarrow \frac{1}{2}\langle d^2 \rangle + \epsilon. \tag{3}$$

This is valid for any initial distance distribution. As the distance of the atom i from the centroid is given by $\sqrt{G_{ii}}$, this asymptotic behavior forces all atoms to be near a surface of a sphere. This explains the similarity of the metric matrix distance geometry structures in unconstrained loop regions of a protein. The observed bias is then a statistical effect by choosing the distances as independent variables. If one introduces in the first phase of the embedding procedure correlations through the triangle inequalities, one must carefully choose the parameters to avoid introducing an additional bias.

These remarks should show that the reason for this strange behavior might be deeper than a simple implementation problem. Quite recently, a more detailed mathematical and numerical study[54] confirmed the proposed hypothesis and came to similar conclusions.

REFERENCES

1. W. Braun, *Quart. Rev. Biophys.* **19**, 115-157 (1987).
2. R. Kaptein, R. Boelens, R. M. Scheek, and W. F. van Gunsteren, *Biochemistry* **27**, 5389-5395 (1988).
3. A. Bax, *Annu. Rev. Biochem.* **58**, 223-256 (1989).
4. G. M. Clore, and A. M. Gronenborn, *Crit. Rev. Biochem. Mol. Biol.* **24**, 479-564 (1989).
5. I. D. Kuntz, J. F. Thomason, and C. M.Oshiro, in "Methods in Enzymology," Oppenheimer and James, eds. Vol. 177, p. 159-203 (1989).
6. R. Kaptein, E. R. P. Zuiderweg, R. M. Scheek, R. Boelens, and W. F. van Gunsteren, *J. Mol. Biol.* **182**, 179-182 (1985).
7. A. T. Brünger, G. M. Clore, A. M. Gronenborn, and M. Karplus, *Proc. Natl. Acad. Sci.* **83**, 3801-3805 (1986).
8. G. M. Crippen, *J. Comp. Phys.* **26**, 449-452 (1977).
9. G. M. Crippen "Distance Geometry and Conformational Calculations", in "Chemometrics Research Studies Series," Vol. 1, D. Bawden, ed., New York: Research Studies Press (1981).
10. G. M. Crippen, and T. F. Havel, *Acta Cryst.* A, **34**, 282-284 (1978).
11. T. F. Havel, I. D. Kuntz, and G. M. Crippen, *Bull. Math. Biol.* **45**, 665-720 (1983).
12. W. Braun, G. Wider, K. H. Lee, and K. Wüthrich, *J. Mol. Biol.* **169**, 921-948 (1983).

13. W. Braun, C. Bösch, L. R. Brown, N. Gō, and K. Wüthrich, *Biochim. Biophys. Acta* **667**, 377-396 (1981).
14. G. M. Crippen, N. Oppenheimer, and M. Conolly, *Int. J. Pept. Prot. Res.* **17**, 156-169 (1981).
15. T. F. Havel and K. Wüthrich, *J. Mol. Biol.* **182**, 281-294 (1985).
16. F. A. Momany, R. F. McGuire, A. W. Burgess, and H. A. Scheraga, *J. Phys. Chem.* **79**, 2361-2381 (1975).
17. G. Wagner, W. Braun, T. F. Havel, T. Schaumann, M. Gō, and K. Wüthrich, *J. Mol. Biol.* **196**, 611-641 (1987).
18. M. Nilges, G. M. Clore, and A. M. Gronenborn, *FEBS Letters* **229**, 317-324 (1988).
19. W. J. Metzler, D. R. Hare, and A. Pardi, *Biochemistry* **28**, 7045-7052 (1989).
20. W. Braun and N. Gō, *J. Mol. Biol.* **186**, 611-626 (1985).
21. S. Endo, H. Inooka, Y. Ishibashi, C. Kitada, E. Mizuta, and M. Fujiino, *FEBS Letters* **257**, 149-154 (1989).
22. R. Bazzo, M. J. Tappin, A. Pastore, T. S. Harvey, J. A. Carver, and I. D. Campbell, *Eur. J. Biochem.* **173**, 139-146 (1988).
23. A. Pastore, T. S. Harvey, C. E. Dempsey, and I. D. Campbell, *European Biophysical Journal* **16**, 363-367 (1989).
24. F. Inagaki, I. Shimada, K. Kawaguchi, M. Hirano, I. Terasawa, T. Ikura, and N. Gō, *Biochemistry* **28**, 5985-5991 (1989).
25. H. Senn, H. R. Loosli, M. Sanner, and W. Braun, *Biopolymers* **29**, 1387-1400 (1990).
26. Y. Kobayashi, T. Ohkubo, Y. Kyogoku, Y. Nishiuchi, S. Sakakibara, W. Braun, and N. Gō, *Biochemistry* **28**, 4853-4861 (1989).
27. W. Braun, G. Wagner, E. Wörgötter, M. Vašák, J. H. R. Kägi, and K. Wüthrich, *J. Mol. Biol.* **187**, 125-129 (1986).
28. H. Widmer, M. Billeter, and K. Wüthrich, *Proteins* **6**, 357-371 (1989).
29. G. T. Montelione, K. Wüthrich, E. C. Nice, A. W. Burgess, and H. A. Scheraga, *Proc. Natl. Acad. Sci. USA*, **84**, 5226-5230 (1987).
30. D. Kohda, N. Gō, N., K. Hayashi, and F. Inagaki, *J. Biochem.* **103**, 741-743 (1988).
31. D. Kohda, I. Shimada, T. Miyake, T. Fuwa, and F. Inagaki, *Biochemistry* **28**, 953-958 (1989).
32. Y. Q. Qian, M. Billeter, G. Otting, M. Müller, W. J. Gehring, and K. Wüthrich, *Cell* **59**, 573-580 (1989).
33. A. D. Kline, W. Braun, and K. Wüthrich, *J. Mol. Biol.* **204**, 675-724 (1988).
34. E. R. P. Zuiderweg, D. G. Nettesheim, K. W. Mollison, and G. W. Carter, *Biochemistry* **28**, 172-185 (1989).
35. V. Saudek, R. A. Atkinson, R. J. P. Williams, and G. Ramponi, *J. Mol. Biol.* **205**, 229-239 (1989).
36. V. Saudek, V., Wormald, M. R., Williams, R. J. P., Boyd, J., Stefani, M., and Ramponi, G., *J. Mol. Biol.* **207**, 229-239 (1989).
37. P. Güntert, W. Braun, K. Wüthrich, *J. Mol. Biol.*, **217**, 517-530 (1991).
38. M. Sanner, A. Widmer, H. Senn, and W. Braun, *J. Comp. Aided Molecular Design* **3**, 195-210 (1989).
39. T. H. Havel, and K. Wüthrich, *Bull. Math. Biol.* **46**, 673-698 (1984).
40. M. Vasquez, and H. A. Scheraga, *J. Biomol. Struct. Dynam.* **5**, 705-755 (1988).
41. M. Vasquez, and H. A. Scheraga, *J. Biomol. Struct. Dynam.* **5**, 757-784 (1988).
42. G. T. Montelione, M. E. Winkler, P. Rauenbuehler, and G. Wagner, *J. Magn. Res.* **82**, 198-204 (1989).
43. G. Wider, D. Neri, G. Otting, and K. Wüthrich, *J. Magn. Res.* **85**, 426-431 (1989).
44. L. E. Kay, and A. Bax, *J. Magn. Res.* **86**, 110-126 (1990).
45. M. Karplus, *J. Amer. Chem. Soc.* **85**, 2870-2871 (1963).

46. J. S. Taylor, D. S. Garret, and M. J. Wang, *Biopolymers* **27**, 1571-1593 (1988).
47. G. M. Smith, and D. F. Veber, *Biochim. Biophys. Res. Commun.* **134**, 907-914 (1986).
48. A. W. Torda, B. C. Mabbutt, W. F. van Gunsteren, and R. S. Norton, *FEBS Letters* **239**, 266-270 (1988).
49. A. S. Arseniev, P. Schultze, E. Wörgötter, W. Braun, G. Wagner, M. Vašák, J. H. R. Kägi, and K. Wüthrich, K., *J. Mol. Biol.* **201**, 637-657 (1988).
50. P. Schultze, E. Wörgötter, W. Braun, G. Wagner, M. Vašák, J. H. R. Kägi, and K. Wüthrich, *J. Mol. Biol.* **203**, 251-268 (1988).
51. B. Messerle, A. Schäffer, M. Vasak, J. H. R. Kägi, and K. Wüthrich, *J. Mol. Biol.*, in press (1990).
52. W. F. Furey, A. H. Robbins, L. L. Clancy, D. R. Winge, B. C. Wang, and C. D. Stout, *Science* **231**, 704-710 (1986).
53. K. Wüthrich, M. Billeter, and W. Braun, *J. Mol. Biol.* **6**, 357-371 (1989).
54. T. F. Havel, *Biopolymers* **29**, 1565-1585 (1990).
55. D. Mulvey, G. F. King, R. M. Cooke, D. G. Doak, T. S. Harvey, and I. D. Campbell, *FEBS Letters* **257**, 113-117 (1989).
56. Y. Kuroda, S. Endo, A. Wada, and K. Nagayama, in "Spectroscopy of Biological Molecules — New Advances," D. E. Schmidt, F. W. Schneider, and F. Siebert, John Wiley: New York (1988).
57. T. Ohkubo, Y. Kobayashi, Y. Shimonishi, Y. Kyuogoku, W. Braun, and N. Gō, *Biopoymers* **25**, S123-S134, (1986).
58. W. E. Steinmetz, P. Bougis, H. Rochat, O. D. Redwine, W. Braun, and K. Wüthrich, *Eur. J. Biochem.* **172**, 101-116 (1988).
59. H. Haruyama, and K. Wüthrich, *Biochemistry* **28**, 4301-4312 (1989).
60. A. D. Kline, W. Braun, and K. Wüthrich, *J. Mol. Biol.* **189**, 377-382 (1986).
61. E. R. P. Zuiderweg, J. Henkin, K. W. Mollison, G. W. Carter, and J. Greer, *Proteins* **3**, 139-145 (1988).

STRUCTURE DETERMINATION BY NMR: THE MODELING OF NMR PARAMETERS AS ENSEMBLE AVERAGES

R. M. Scheek, A. E. Torda, J. Kemmink, and
W. F. van Gunsteren

Physical Chemistry Department
University of Groningen
The Netherlands

INTRODUCTION

High-resolution NMR has become a well established technique for determining three-dimensional structures of small proteins in solution[1]. Procedures for assigning resonances to individual spins are being automated in various places. There is general agreement how NMR parameters like initial NOE build-up rates and J couplings must be translated into interatomic distances and dihedral angles. However, there is no consensus yet about the way such parameters are best modeled in three-dimensional structures. Most of the established procedures for interpreting a set of interproton distances result in a family of static structures. A structure is accepted as a member of such a family if it is consistent with all the measured NMR parameters simultaneously. This approach denies the fact that such NMR parameters as NOE's and J couplings are properties of ensembles of molecules. Recently we presented a procedure, based on molecular dynamics simulation techniques, which generates such an ensemble of molecules[2,3]. While individual members of the ensemble may violate the experimental constraints, the ensemble taken as a whole must reproduce the data. Using this approach, the resulting set of structures roughly resembles the proper Boltzmann distribution over the conformational states that are accessible at the temperature of the NMR experiments, although the always limited simulation time hinders a complete sampling of such states. There are good reasons to have doubts about the correctness of the various force fields used in MD simulations, but there can be little doubt that the model is physically more realistic

Computational Aspects of the Study of Biological Macromolecules by Nuclear Magnetic Resonance Spectroscopy, Edited by J.C. Hoch *et al.*, Plenum Press, New York, 1991

than the usual set of static structures resulting from distance geometry and even restrained MD techniques. In an application to a high-quality NMR data set measured on the protein tendamistat it was shown that the experimental data is better reproduced by the MD ensemble than by a set of distance-geometry structures[3].

In this paper we present an alternative approach to obtain an ensemble of molecules that accounts for measured NMR parameters. Instead of starting with one copy of a molecule and generating an ensemble while simulating its dynamic behaviour through time, we start with an ensemble of molecules and try to modify it as a whole, using restrained MD techniques, until the experimental data is reproduced.

METHODS

Although for the actual calculations presented here we used an artificial testcase, we shall describe how the calculations would proceed in a real-life case, starting with the interpretation of the NMR spectra, via classical distance geometry and restrained MD techniques towards the final stages where ensemble averaging will be introduced.

INTERPRETATION OF NMR SPECTRA

NOE buildup rates are measured and translated into distances using a fixed distance for calibration. These are interpreted as the target distances to be imposed on the ensemble of molecules during the last stage of the modeling procedure. The experimental uncertainty in the measurement of an average distance is reflected by the difference between the upper and lower bounds taken for that distance. In addition the absence of an NOE is interpreted as a lower bound on the corresponding interproton distance, except in cases where the protons involved show no NOE's with other protons either. NOE's involving magnetically equivalent protons are referred to pseudo positions, as described before. In cases where no stereospecific assignments are available for protons or groups of equivalent protons in prochiral centers, the pairs are labeled arbitrarily and the absolute chirality is not enforced on the structures. This procedure was introduced as the floating chirality approach by Paul Weber et al.[4].

DISTANCE GEOMETRY CALCULATIONS

In the initial modeling step a set of structures is calculated using a distance geometry software package that was developed from the original EMBED program[5]. Inspired by some ideas of Tim Havel we implemented a routine for randomized metrization to improve the sampling properties of the embedding step. In short, the routine performs a random picking of distances from within the given bounds while adjusting the bounds on the other distances after each choice to obey the triangle inequality relationships. This assures that the resulting trial distances also obey the triangle inequality relationships. This feature, which was named metrization, was already realised in the program DISGEO, written by Havel in 1985[6]. Randomized metrization implies that the order of the atom pairs for which distances are chosen is random, thus avoiding an undesired bias in the resulting cluster of configurations that was noted by Havel[7]. The sampling properties of the resulting algorithm are significantly better than those of earlier clones of the EMBED program. In addition the number of successful embeddings is significantly larger in those difficult cases where only short- or medium-range distance information is available from the NMR experiments, as is usually the case for small linear molecules like peptides and nucleic acid fragments.

In our standard protocol we first embed each of the distance matrices into 4D space (a trick inspired by Jeff Blaney[8], which helps to avoid the conflict between distance and chirality constraints in cases where a chiral center needs to be inverted during the next optimization stage) and improve the resulting structures by minimizing an error function consisting of two parts. One part reflects the violations of the upper and lower bounds by the distances, defined in 4D space; the other part reflects the deviations of the signed volumes of those quadruples of atoms for which the relative atomic positions are fixed, from their target values, which were calculated from a reference structure with ideal geometry. This part of the error function uses only the 3D coordinates corresponding to the three largest eigenvalues of the metric matrix that can be calculated from the trial distance matrix. Thus:

$$
E = K^{dc} * \left[\sum_{d_{i,j} > u_{i,j}} \left(d_{i,j}^2 - u_{i,j}^2 \right)^2 + \sum_{d_{i,j} < l_{i,j}} \left(l_{i,j}^2 - d_{i,j}^2 \right)^2 \right]
$$
$$
+ K^{ch} * \sum_{quadruples\ i} (V_i - V_i^t)^2 \tag{1}
$$

where u and l stand for the upper and lower bounds on an interatomic (4D) distance d, measured in Å and the sums are over all pairs for which d violates a bound, V is the (3D) signed volume of a quadruple of atoms with V^t the target value for that quadruple, measured in Å^3. Typically, with coordinates in Å, $K^{dc} = K^{ch} = 1$.

In the next step the 4D structures are embedded into 3D space by calculating 3D coordinates from the metric matrices of the structures, using only the three largest eigenvalues and the corresponding eigenvectors. Then another optimization follows using the same error function (1), now defined completely in 3D space.

DISTANCE-BOUNDS DRIVEN DYNAMICS

Finally distance-bounds driven dynamics[9,10] is started in two stages, one at an elevated temperature with a rather tight coupling to a bath of that temperature (typically a few hundred steps of 0.002 ps at 500 or 1000 K, with a coupling time constant of 10 steps)[11], in the next stage the temperature is brought down slowly to 1 K (usually in 1000 steps with a temperature coupling of 50 steps). Holonomic constraints (calculated from the reference structure) and experimental constraints (calculated from the NOE build-up rates) are given the same weights. The last annealing step was found to increase the number of acceptable solutions significantly, even if the previous optimization was not very successful. Normally no structures are discarded up to this point. In cases where significant violations remain in the final structures a last optimization step is included in which only the holonomic constraints and the target chiral volumes are imposed on the structures. Violations remaining after this regularization step only involve experimental constraints, which may be due to errors in the list of constraints, or to errors in the underlying assumption that all constraints are to be realized in one static structure.

THE PROBLEM

Once the remaining violations in the individual structures are considered real and not caused by mistakes, there are some ways to proceed. One way would be to use restrained MD with time-dependent restraints taking the distance-geometry structures as starting structures. This legitimate approach will be discussed by Andrew Torda elsewhere in this volume. Another way would be to divide the experimental constraints in subsets which can be imposed simultaneously on static structures. This approach was followed in the case of a somatostatin analogue by Pepermans et al.[12] and will not be described here either. A third, widely used approach would be to relax the experimental constraints by using somewhat higher values for the upper bounds and lower values for the lower bounds, with the argument that in reality the structures are flexible. There is a serious risk that this will result in unrealistic structures that could be seen as transition states between two conformations. A better approach would be to *include this flexibility* in the model, as was done by Torda et al.[2,3] or as will be described below. Only then one can use all the information that is available from the experiments and possibly arrive at a realistic description of the molecule's behaviour in solution.

ENSEMBLE DYNAMICS

In ensemble dynamics at least two copies of the molecule are taken as starting points for a parallel MD simulation. We used the simplified MD algorithm referred to above as distance-bounds driven dynamics, to test the procedure. The atoms are given a unit mass and kinetic energies corresponding to the desired temperature, which is controlled by scaling the velocities of all atoms after each step, following Berendsen et al.[11] Only the distance-part of the error function (Eq. 1) is taken as the potential and gradients derived from it are treated as forces. Now we discriminate between the holonomic constraints, which must be imposed on each member of the ensemble, and the experimental constraints derived from the NOE's, which must be imposed on the ensemble as a whole. Thus the holonomic components of the restraining forces are calculated directly as the derivatives of the holonomic part of the above error function with respect to the atomic coordinates, as in normal DDD:

$$\mathbf{F}_i = -K^{holo} * \left[\sum_{d_{i,j}>u_{i,j}} 4*\left(d_{i,j}^2 - u_{i,j}^2\right)*\mathbf{r}_{i,j} - \sum_{d_{i,j}<l_{i,j}} 4*(l_{i,j}^2 - d_{i,j})^2 \mathbf{r}_{i,j} \right] \tag{2}$$

where the sums run over all atom pairs for which d violates a holonomic constraint. With all coordinates in Å, typically $K^{holo} = 1$. Following Torda et al., we define an average distance d as follows:

$$\bar{d}_{i,j} = \left\langle r_{i,j}^{-3} \right\rangle^{-1/3} \tag{3}$$

where the square brackets denote an average over the ensemble.

For the atom pairs involved in experimental constraints we use the following form of the restraining force:

$$\mathbf{F}_i = -K^{exp} * \left[\sum_{\bar{d}_{i,j}>u_{i,j}} 4*\left(\bar{d}_{i,j}^2 - u_{ij}^2\right)*\mathbf{r}_{i,j}/|r_{i,j}| \right.$$

$$\left. - \sum_{\bar{d}_{i,j}<l_{i,j}} 4*\left(l_{i,j}^2 - \bar{d}_{i,j}^2\right)*\mathbf{r}_{i,j}/|r_{i,j}| \right] \tag{4}$$

replacing the interatomic distance d in individual structures by the averaged distance $\bar{d}$, defined above (Eq. 3). The sums then run over all pairs of atoms for which this ensemble-averaged distance exceeds an experimentally determined bound. Typically we choose $K^{exp} = K^{holo}/N$, where N is the number of molecules per ensemble. We realize that this form of the force is no longer a proper derivative of the potential with respect to the atomic coordinates, but it has all the desired properties, as will be demonstrated below.

RESULTS

We examined the new protocol in an artificial testcase, consisting of 22 atoms taken from the backbone of octo-alanine: $(N - C\alpha - C)_7 - N$. The distance between the first and last nitrogen atoms was restrained between 4 and 5 Å. The holonomic constraints were the same as those used for proteins, with the exception that the ω backbone dihedral was treated the same way as the ϕ and ψ dihedrals, whereas in proteins it is usually fixed to 180°. 64 structures were calculated in 4 stages: embedding, optimization by a conjugate gradient algorithm, distance bounds driven dynamics on the separate structures, and finally a parallel distance bounds driven dynamics step with 2, 4, 8, 16, 32, or 64 of the 64 structures treated as ensembles. The restrained N1 to N8 distance was calculated for all 64 structures at the end of each stage of the calculations and plotted as histograms in Fig. 1. As was noted before the raw structures after the embedding stage show a range of N1 to N8 distances well outside 4 to 5 Å(Fig. 1a), which necessitates the optimization stage. After the optimization stage no violations remain, but the distribution shows undesired peaks at the boundaries (Fig. 1b). These result from stage-1 structures with N1 to N8 distances outside the allowed range, which move along the gradient towards the allowed range until the gradient disappears exactly at the boundaries. Distance bounds driven dynamics results in a more even distribution of the 64 structures within the allowed range (Fig. 1c).

A fundamentally different distribution results when we change the criteria for acceptance: once structures are taken together and treated as ensembles for which the N1 to N8 distance restraint must be satisfied only after averaging over the ensemble, much wider distributions result (Fig. 1d-1f). This is the desired property of the protocol we were hoping to find: individual structures are allowed to violate the restraint as long as the ensemble on average does not. An important new parameter introduced is the number of structures treated as an ensemble. In this test case the peak in the distribution shifts to higher values if more structures are grouped together as ensembles. Thus if we divide the 64 structures over 32 ensembles of 2 structures each, the N1 to N8 distance distribution peaks around 4-5 Å, whereas we find the peak around 7 Å if we create 2 ensembles of 32 structures each. In all cases the final 64 structures satisfy the holonomic constraints and the ensemble-averaged N1 to N8 distance falls within the allowed range of 4 to 5 Å. For comparison, an unrestrained DDD run on 64 molecules results in an ensemble-averaged N to N distance around 9.5 Å.

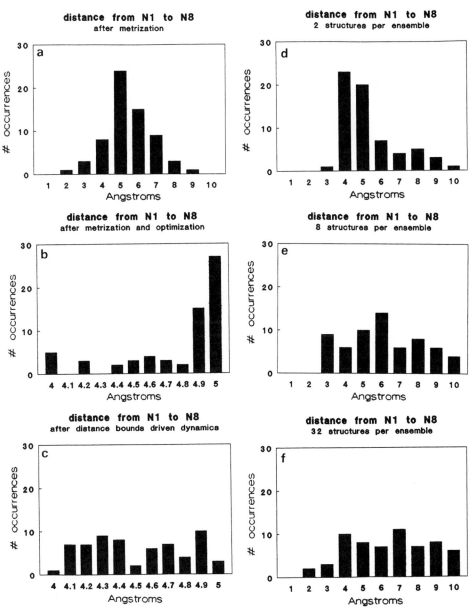

Figure 1. Histograms showing the distribution of the restrained N1 to N8 distances in the 64 test 'molecules' (see text) after different stages of the modeling protocol. 1a: after the embedding stage (using a randomized metrization technique) (7). 1b: after embedding and a conjugate-gradients optimization stage to minimize violations of all restraints simultaneously. 1c: after 2 ps of distance bounds driven dynamics at 300 K and cooling to 1 K. 1d-1e: after another 2 ps of distance bounds driven dynamics at 300 K and subsequent cooling to 1 K, grouping together 2, 8, or 32 of the 64 molecules for ensemble-averaging the N1 to N8 distances before comparison with the upper and lower bounds.

DISCUSSION

The above calculations demonstrate that with the modified restrained MD technique introduced here it is possible to adjust an ensemble of molecules such that ensemble-averaged atom-to-atom distances are made to agree with experimentally obtained values. The advantage of this approach is most obvious in cases where no single static conformation exists that satisfies all the experimental and holonomic restraints simultaneously and where agreement with experimental restraints can be obtained only at the expense of violating the holonomic constraints and hence increasing the internal energy of the conformation. Other test calculations (not shown here) demonstrate that the ensemble approach results in a set of structures that are much more relaxed in the force field applied while on average the experimental restraints are still satisfied. However, also in cases where a single, static solution can be found, the ensemble approach will often lead to different conclusions about the dynamic behaviour of the molecule and about the accuracy of the structure determination.

It can be argued that there may be many different solutions given a set of experimentally obtained distance restraints. Some of these solutions may be physically unrealistic because the individual molecules in the ensemble can never be in a fast-exchange conformational equilibrium. This is a problem, since to our knowledge there is no way to decide at which rate two different conformations interconvert. We therefore take a conservative approach, where we start the ensemble calculations with a number of molecules in the same initial conformation and let them diverge during the restrained MD calculations. Ideally, with a correct force field, the ensemble will approach a proper Boltzmann distribution over conformations that are accessible at the temperature used in the simulation. The remaining problem using this conservative approach is that large parts of conformation space are not sampled given the limits on the simulation time. If the agreement with the experimental restraints remains unsatisfactory, this can be used to test a model in which a new conformation is introduced in the existing ensemble, e.g., by flipping a few dihedral angles by hand or by heating the ensemble to overcome barriers that had excluded possibly important conformations.

Presently we are incorporating these ideas in the Gromos molecular dynamics simulation programs[11].

REFERENCES

1. K. Wüthrich, "NMR of Proteins and Nucleic Acids," Wiley, New York (1986).
2. A. Torda, R. M. Scheek and W. F. van Gunsteren, *Chem. Phys. Letters* **157**, 289 (1989).
3. A. Torda, R. M. Scheek and W. F. van Gunsteren, *J. Mol. Biol.* **214**, 223 (1990).

4. P. Weber and J. Blaney, personal communication (1987).

5. T. F. Havel, I. D. Kuntz, and G. M. Crippen, *Bull. Math. Biol.* **4S**, 665 (1983).

6. T. F. Havel and K. Wüthrich, *Bull. Math. Biol.* **46**, 673 (1984).

7. T. F. Havel, *Biopolymers* **29**, 1565 (1990).

8. J. Blaney, personal communication (1987).

9. R. Kaptein, R. Boelens, R. M. Scheek, and W. F. van Gunsteren, *Biochemistry* **27**, 5389 (1988).

10. R. M. Scheek, W. F. van Gunsteren, and R. Kaptein, *in* "Methods in Enzymology", **177**, Nuclear Magnetic Resonance, Part B, Structure and Mechanism, N. J. Oppenheimer and T. L. James, eds., pp. 204-218, Academic Press, 1989.

11. H. J. C. Berendsen, J. P. M. Postma, W. F. van Gunsteren, A. DiNola, and J. R. Haak, *J. Chem. Phys.* **81**, 3684 (1984).

12. H. Pepermans, D. Tourwe, G. van Binst, R. Boelens, R. M. Scheek, W. F. van Gunsteren, and R. Kaptein, *Biopolymers*, **27**, 323 (1988).

13. W. F. van Gunsteren, and H. J. C. Berendsen, "Groningen Molecular Simulation (GROMOS) Library Manual", Biomos b.v., Groningen, The Netherlands (1987).

TIME AVERAGED DISTANCE RESTRAINTS IN NMR BASED STRUCTURAL REFINEMENT

Andrew E. Torda, Ruud M. Scheek* and
Wilfred F. van Gunsteren

Physical Chemistry Department
ETH Zentrum
CH-8092 Zürich, Switzerland
and
*Laboratory of Physical Chemistry
University of Groningen
Nijenborgh 16
9747 AG Groningen
The Netherlands

ABSTRACT

We discuss a method for the refinement of NMR based solution structures which attempts to account for the effect of mobility on measured NMR data. Rather than regard the NOE as a measure of a static distance, we attempt to account for the fact that the measured quantity actually reflects a time averaged property of the molecule.

A penalty function, suitable for incorporation into molecular dynamics simulations, is described which perturbs a system only to the extent necessary to ensure that a trajectory average is consistent with measured interproton distances.

Finally, we present some results which suggest wide applicability for the method, while also noting the limitations and scope for improvement.

INTRODUCTION

Since the earliest determinations of protein structures by NMR, several trends have become apparent in the literature. Firstly, studies of ever-larger molecules have been undertaken. Secondly, the apparent accuracy of the structures has increased to the point where the method may rival X-ray crystallography.

Computational Aspects of the Study of Biological Macromolecules by Nuclear Magnetic Resonance Spectroscopy, Edited by J.C. Hoch *et al.*, Plenum Press, New York, 1991

219

At the same time, the increased accuracy of structural determination has led a questioning of some assumptions implicit in earlier approaches to structural determination. For example, the assumption that an NOE can be regarded as a pairwise interaction has been challenged by methods incorporating the full relaxation matrix[1-3]. In a similar vein, we wish to approach the assumption that measured NOE's can be used as static distance bounds and the problem of structure generation viewed in purely geometrical terms.

In the case of small peptides, it is well accepted that measured NMR parameters often reflect an average over many conformations accessible on the NMR time scale. In these cases, it will often be found that techniques such as distance geometry or distance restrained molecular dynamics can not produce any individual structures which are consistent with all of the NMR data. Alternatively, structures may be produced, but it is not clear whether they exist to any significant extent in solution or are simply an artificial average which happens to satisfy the enforced distance restraints.

In the case of larger molecules, the situation is still less clear. Ensembles of closely grouped structures are often produced which show no evidence of conformational averaging, yet evidence such as averaged homonuclear J-coupling constants suggests that there must be fast motion on the NMR time scale. Given the non-linear dependence of the NOE on distance, it would not appear adequate to try to satisfy all NMR restraints simultaneously and refer to the result as an average solution structure[4].

To this end, we have introduced a pseudo-energy term, suitable for incorporation into MD simulations which enforces NOE's as an ensemble average property, where the ensemble consists of the structures generated throughout the course of a MD simulation. By using appropriate averaging of distances, we more closely model the physical NOE and attempt to better estimate the range of conformational space consistent with the experimental data.

Finally, we give an example of the method applied to the previously solved, high resolution solution structure of tendamistat[5].

THEORY

To enforce a static distance restraint by means of a pseudo-potential, one must construct a term so that the energy rises as the interproton distance violates the restraint. One such potential, quadratic with respect to violations is[6,7]

$$
\begin{aligned}
V_{dc}(r) &= 0 && \text{if } r \leq r^0 \\
&= \tfrac{1}{2} K_{dc}(r - r^0)^2 && \text{if } r^0 < r < r^0 + \Delta r \\
&= K_{dc}(r - r^0 \tfrac{\Delta r}{2})\Delta r && \text{if } r^0 + \Delta r \leq r
\end{aligned}
\tag{1}
$$

where $V_{dc}(r)$ is the potential due to the distance restraint term for a given pair of atoms, r is the instantaneous distance between the atoms and r^0

the distance restraint calculated from the measured NOE data. The force constant, K_{dc} controls the relative weight of the pseudo-energy term with respect to the physical terms in the force field.

Eq. 1 actually describes three regimes. Firstly, if the distance restraint is satisfied, V_{dc} is zero. If the violation of the restraint is less than an arbitrary distance Δr, the potential is quadratic with respect to the size of the violation. Finally, if the distance is greater than $(r^0 + \Delta r)$, the potential becomes linear so as to put an upper bound on the artificial force applied to the system.

This potential could be made to include time or ensemble averaging by replacing r with $\bar{r}(t)$, defined to be some kind of time average. For example one could define the time average

$$\bar{r}(t) = \left(\int_0^t [r(t')]^{-6} dt' \right)^{-\frac{1}{6}} \tag{2}$$

or the ensemble average

$$\bar{r}(t) = \left\langle r^{-6} \right\rangle^{-\frac{1}{6}} \tag{3}$$

where the angle brackets denote an average over time. On the short time scale of a MD simulation however, one should neglect the influence of angular fluctuations[8], so a more correct form of averaging is

$$\bar{r}(t) = \left(\int_0^t [r(t')]^{-3} dt' \right)^{-\frac{1}{3}}. \tag{4}$$

Using Eq. (4), one could calculate $\bar{r}(t)$ by a simple summation over the course of a trajectory and this is, indeed, the correct way to calculate $\bar{r}(t)$ for analysis of trajectories. Over the course of a MD simulation however, this kind of calculation would not lead to a suitable force. As the time of the simulation increased, the averaging would be conducted over an ever increasing period, meaning the average would become progressively less sensitive to instantaneous changes in r. To avoid this problem, one can incorporate a memory function into the averaging so at any point in time, recent history is more heavily weighted than past history. If this weighting is given an exponential form, one can define

$$\bar{r}(t) = \left(\frac{1}{\tau} \int_0^t e^{-\frac{t'}{\tau}} [r(t - t')]^{-3} dt' \right)^{-\frac{1}{3}}. \tag{5}$$

Simulations using Eq. 5 to define the average r showed that distance restraints could be enforced as a time average in the case of a simple model two dimensional system[9]. Unfortunately, a potential of this form gave rise to occasional large forces due to a fourth power term in the force with respect to $\frac{\vec{r}(t)}{r(t)}$. This led to the definition

$$
\begin{aligned}
\vec{F} &= \quad\quad 0 \quad\quad\quad\quad\quad\quad \text{if } \bar{r}(t) \leq r^0 \\
&= -K_{dc}\left[\bar{r}_{ij}(t) - r^0\right]\frac{\vec{r}_{ij}(t)}{r_{ij}(t)} \quad \text{if } \bar{r}(t) > r^0
\end{aligned}
\tag{6}
$$

where $\vec{f}_i(t)$ is the force on atom i due to atom j and $\vec{r}_{ij} = \vec{r}_i - \vec{r}_j$. Using this approach, there was no formal definition of the potential, but only the force given by Eq. 6.

From Eq. 5, it can be seen that at zero time, $\bar{r}(t)$ is not formally defined. That is, at the start of the simulation, one is free to choose an initial value for the running average to be used in calculating forces. In practice, we have found it advantageous to set $\bar{r}(t)$ for each restraint to be slightly less than the corresponding r^0. This means that at the start of a simulation, no artificial forces are applied and the artificial force will only gradually build up if the value of r remains greater than r^0 and $\bar{r}(t)$ rises correspondingly.

AN EXAMPLE SIMULATION

An example simulation using time averaged distance restraints was performed on the 74 residue polypeptide, tendamistat[10]. Previously, Kline et al.[5], had published structures of this molecule after very thorough refinement using the variable target function method[11]. In the work of Kline et al., there had been no reason to invoke molecular motions to explain the data and the work resulted in one of the most precisely determined solution structures up to that time[12].

For this test of time averaged distance restraints, two of the Zürich group's final structures were subjected to comparison runs, with static and with time averaged restraints. The restraints used in the simulations were identical to those used in the distance geometry calculations with only the exception of the hydrogen bond restraints. These were not put in explicitly since the GROMOS force field should reproduce these as a result of electrostatic and Lennard-Jones interactions.

All the structures were initially energy minimized to relax them in the GROMOS force field[13], followed by 2.5 ps MD with static distance restraints. 10 ps of MD were then performed for each structure with parallel runs (with static or time averaged distance restraints). Finally, each of the four simulations was continued for 20 ps where structures were saved and analysis performed.

As would be expected, all structures showed an improvement in potential energy after MD refinement. In contrast however, the structures showed a marked difference in their agreement with the experimental data

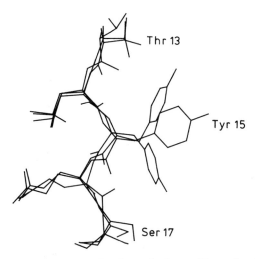

Figure 1. Mobility of a Tyr 15 of tendamistat during a 20 ps simulation. The peptide segment from residues 13 to 17 is shown with the aromatic ring in three positions, (TOP) 9 ps into trajectory, (MIDDLE) Distance geometry starting structure (BOTTOM) 16.2 ps into trajectory. Both MD structures were least squares fitted to the distance geometry structure on the basis of all backbone heavy atoms. Taken from Torda et al.[10]

and their internal mobility. Assessing the distance violations as trajectory averages using Eq. 3, runs using time averaged restraints had violations typically 70-80% of those using static restraints. Judging mobility by root mean square positional fluctuations of backbone α-carbon atoms, fluctuations were consistently larger and, at some sites, double those using conventional refinement[10].

Aside from the overall increase in mobility, the simulations using time averaged restraints provided some unexpected results for specific sites in the molecule. For example the sidechain of Tyr-15 had appeared well defined in the distance geometry structures, being restrained by 26 NOE's. In the MD simulations however, with the use of time averaged distance restraints, the sidechain moved continuously through a large region of space. Fig. 1 shows two snapshots from different times in a simulation trajectory superimposed, along with the starting structure. This interpretation of the data actually appeared consistent with other available information. For example, the sidechain was a surface residue and would not be expected to be rigidly positioned. The homonuclear J-coupling constant about χ_1 was consistent with rapid motion and the sidechain had vanishing electron density in the high resolution crystal structure[14].

Further examination of the motions of tendamistat showed that the motions of Tyr 15 were distinct, but not unique. The sidechain of Ile 61 showed similarly large motions as did many other parts of the molecule.

DISCUSSION

Aside from better reproducing experimental data, the use of time averaged distance restraints offers some additional advantages. Since the simulated system is not obliged to satisfy all restraints simultaneously, it is not forced to reside near a geometrical average forced by the distance restraints. This means it can spend more time in energetic minima which presumably correlate better with realistic solution behaviour. Furthermore, smaller force constants for the artificial terms will usually be sufficient to make the system conform to the experimental data. This in turn should lead to structures less distorted by artificial forces. Furthermore, the ability to span a wider area of conformational space leads to improved searching behaviour and a better estimation of the true size of the conformational space allowed to the molecule within the confines of the experimental data.

While these results are encouraging, work is continuing to extend the method. Some trial calculations have shown that conformations may be in rapid exchange on the NMR time scale, leading to averaged information, but that the corresponding energetic barriers may be insurmountable over the much shorter time course of a MD simulation. To this end we are investigating the feasibility of running simulations of multiple copies of molecules with averaging conducted over the simulated ensemble. In this case, it may not be necessary for molecules to cross barriers if different copies are already on different sides of these barriers.

Ultimately, one would want a penalty function that models the physical NOE as closely as possible with the inclusion of time averaging, multi-spin cross relaxation and the influence of different motions in different parts of molecules. As NMR methods improve in accuracy and structures such as DNA are undertaken requiring better data, it is appropriate that the interpretation of the data progresses in a similar manner.

REFERENCES

1. B. A. Borgias and T. L. James, *J. Magn. Reson.* **79**, 493-512 (1988).
2. R. Boelens, T. M. G. Koning, and R. Kaptein, *J. Mol. Struct.* **173**, 299-311 (1988).
3. P. Yip, and D. A. Case, *J. Mag. Reson.* **83**, 643-648 (1989).
4. O. Jardetzky, *Biochim. Biophys. Acta.* **621**, 227-232 (1980).
5. A. D. Kline, W. Braun, and K. Wüthrich, *J. Mol. Biol.* **204**, 675-724 (1988).
6. W. F. van Gunsteren, R. Kaptein, and E. R. P. Zuiderweg, in "Proceedings of a NATO/CECAM Workshop on Nucleic Acid Conformation and Dynamics," W. K. Olsen, ed., CECAM, Orsay, pp. 79-82, CECAM, Orsay (1984).
7. R. Kaptein, E. R. P. Zuiderweg, R. M. Scheek, R. Boelens, and W. F. van Gunsteren, *J. Mol. Biol.* **182**, 179-182 (1985).
8. J. Tropp, *J. Chem. Phys.* **72**, 6035-6043 (1980).
9. A. E. Torda, R. M. Scheek, and W. F. van Gunsteren, *Chem. Phys. Lett.* **157**, 289-294 (1989).

10. A. E. Torda, R. M. Scheek, and W. F. van Gunsteren, *J. Mol. Biol.* **214**, 223-235 (1990).
11. W. Braun, and N. Gō *J. Mol. Biol.* **186**, 611-626 (1985).
12. K. Wüthrich, *Science* **243**, 45-50 (1989).
13. W. F. van Gunsteren, and H. J. C. Berendsen, "Groningen Molecular Simulation (GROMOS) Library Manual", Biomos, Groningen, (1987).
14. M. Billeter, A. D. Kline, W. Braun, R. Huber, and K. Wüthrich, *J. Mol. Biol.* **206**, 677-687 (1989).

ANALYSIS OF BACKBONE DYNAMICS OF INTERLEUKIN-1β

Angela M. Gronenborn and G. Marius Clore

Laboratory of Chemical Physics
National Institute of Diabetes and Digestive and Kidney
Diseases
National Institutes of Health
Bethesda, MD 20892, U.S.A.

ABSTRACT

The ^{15}N T_1, T_2 and NOE data for interleukin-1β has been studied using inverse detected heteronuclear nuclear spectroscopy. It is shown that the ^{15}N relaxation data for a number of residues cannot be accounted for by the simple two parameter model free approach of Lipari and Szabo, and require the introduction of two distinct internal motions, one with a time scale much less than 100 ps, the other with an effective correlation time in the range 0.5-4 ns, slightly less than the overall rotational correlation time (8.3 ns) of the protein.

INTRODUCTION

^{13}C and ^{15}N nuclear magnetic relaxation data provide a wealth of information on the nature of internal motions of macromolecules in solution. In general, the fast internal motions can be described by two model independent quantities: a generalized order parameter S which provides a measure of the magnitude of the motion, and an effective correlation time τ_e[1,2]. This simple formalism has proved remarkably successful in accounting for relaxation data on small molecules and simple polymers, as well as for fragmentary data obtained from one-dimensional NMR measurements on peptides and proteins[3-5]. With the advent of new two-dimensional techniques for measuring heteronuclear relaxation[6,7] it has now become possible to obtain a detailed and comprehensive picture of these fast motions in proteins. We therefore undertook an in depth analysis[8] of the backbone dynamics of IL-1β using inverse detected ^{1}H-^{15}N 2D NMR methods. ^{15}N T_1, T_2 and NOE data were obtained for 90% of the backbone amide groups (128 out of a total of 144). In the course of this study we noticed that the ^{15}N T_1, T_2

Computational Aspects of the Study of Biological Macromolecules by Nuclear Magnetic Resonance Spectroscopy, Edited by J.C. Hoch *et al.*, Plenum Press, New York, 1991

and NOE data for the backbone amide groups of certain residues cannot be accounted for by the simple two parameter model-free approach and require the introduction of two distinct internal motions, one with a time scale less than 100 ps, the other with an effective correlation time in the range 1-3 ns.

We were able to show that all measurable residues exhibit very fast motions on a time scale < 20-50 ps. In addition, 32 residues display a second motion of significant amplitude on a time scale of 0.5-4 ns which is less than an order of magnitude smaller than the overall rotational correlation time (8.3 ns), while another 42 residues are characterized by an additional motion on the 30 ns to 10 ms time scale which leads to ^{15}N T_2 exchange line broadening.

The T_1 and T_2 relaxation times and the NOE enhancement of an amide ^{15}N spin relaxed by dipolar coupling to a directly bonded proton and by chemical shift anisotropy are given by Abragham[9]:

$$1/T_1 = d^2\{J(\omega_H - \omega_N) + 3J(\omega_N) + 6(J(\omega_H + \omega_N))\} + c^2 J(\omega_N) \quad (1)$$

$$1/T_2 = 0.5d^2\{4J(0) + J(\omega_H - \omega_N) + 3J(\omega_N) + 6J(\omega_H)$$
$$+ 6J(\omega_H + \omega_N)\} + c^2\{3J(\omega_N) + 4J(0)\}/6 \quad (2)$$

$$NOE = 1 + \{T_1(\gamma_H/\gamma_N)d^2[6J(\omega_H + \omega_N) - J(\omega_H - \omega_N)]\} \quad (3)$$

where

$$d^2 = 0.1\gamma_H^2\gamma_N^2 h^2 < 1/r_{HN}^3 >^2$$

and

$$c^2 = (2/15)\omega_N^2(\sigma\| - \sigma_\perp).$$

γ_H and γ_N are the gyromagnetic ratios of ^{1}H and ^{15}N, respectively, r_{HN} is the NH bond length (1.02 Å from neutron diffraction), H_o is the magnetic field strength, $\sigma_\|$ and $\sigma_\perp$ are the parallel and perpendicular components of the axially symmetric ^{15}N chemical shift tensor which differ by -160 ppm, and $J(\omega_i)$ is the spectral density function.

In the model-free formalism of Lipari and Szabo[1,2], the spectral density function for a molecule undergoing isotropic tumbling is given by

$$J(\omega_i) = S^2\tau_R/(1 + \omega_i^2\tau_R^2) + (1 - S^2)\tau/(1 + \omega_i^2 t^2) \quad (4)$$

which corresponds to an internal correlation function of $C_I(t) = S^2 + (1 - S^2)e^{-t/\tau_e}$, where S is the generalized order parameter, τ_R the overall isotropic rotational correlation time of the molecule, and $\tau = \tau_R\tau_e/(\tau_R + \tau_e)$ where τ_e is a single effective correlation time describing the internal motions. [Note that from the X-ray structure of IL-1β^{11}, the three principal components of the inertia tensor are calculated to be in the ratio of 1.00:0.77:0.93, indicating that IL-1β is a globular, almost spherical, protein which should behave isotropically in solution.]

For 53 of the 153 residues of IL-1β, τ_e is sufficiently small ($\ll 100$ ps) that the T_1 and T_2 data can be accounted for by the simplified spectral density function $J(\omega_i) = S^2 \tau_R / (1 + \omega_i^2 \tau_R^2)$ and the data for further 42 residues which had larger than average ^{15}N T_1/T_2 ratios with concomitantly larger linewidths could be fitted using the simplified spectral density function in conjunction with an additional chemical exchange term. For those residues, however, where the relaxation data cannot be fitted with this simplified spectral density function, we find that Eq. (4) can account for the T_1 and T_2 data at several spectrometer frequencies, but fails to account for the NOE data. This behaviour was found for the remaining 32 residues. In particular, the calculated values for the NOE are either too small or negative, whereas the observed ones are positive. To illustrate this point, the data for several amino acids are listed in Table 1 and compared to the best fit calculated values using Eq. (4).

The solution to this discrepancy between theory and experiment is obtained in a simple model-free manner by expanding the internal correlation function to a two exponential decay given by $C_I(t) = S_f^2 S_s^2 + (1 - S_f^2)e^{-t/\tau_f} + S_f^2(1 - S_s^2)e^{-t/\tau_s}$.[8,9,10] This corresponds to a spectral density function of the form

$$J(\omega_i) = S_f^2 S_s^2 \tau_R / (1 + \omega_i^2 \tau_R^2) + S_f^2(1 - S_s^2)\tau / (1 + \omega_i^2 \tau^2) \qquad (5)$$

where S_f^2 is the amplitude of the very fast motion whose effective correlation time τ_f can be neglected (i.e., it is $\ll 100$ ps), and S_s^2 is the amplitude of the slower motion with an effective correlation time $\tau_s = \tau_R \tau / (\tau_R - \tau)$ which is still faster than the overall correlation time τ_R. The best fit values of S_f^2, S_s^2 and τ_s together with the calculated values of the T_1, T_2 and NOE data are given in Table 1. It will be noted that the value of the overall order parameter, $S_f^2 S_s^2$, is the same as that of the generalized order parameter S^2 obtained using the spectral density function (4).

The errors in the experimental NOEs are ± 0.1. The errors for the calculated values of the various parameters are obtained from analysis of the variance-covariance matrix obtained from the least squares Powell optimization procedure. In all the fits, a value of 8.3 ns is used for τ_R which was obtained by a best fit of the T_1/T_2 ratios of all residues *simultaneously* (excluding 16 which had T_1/T_2 ratios outside $\pm$ 1SD from the mean) using Eqs. 1 and 2. In the fits obtained with Eq. 4, S^2 and τ_e were varied; and in the fits obtained with Eq. 5, S_f^2, S_s^2 and τ_s were varied (and S^2 is given by the product $S_f^2 S_s^2$. In the best fits using Eqs. 4, only the T_1 and T_2 data were used. In the best fits using Eq. 5 the T_1, T_2 and NOE data were used.

Table 1. Comparison of calculated and observed ^{15}N relaxation data for some backbone amide groups of Il-1β.

Residue	$T_1(600)$ (ms)	$T_2(600)$ (ms)	NOE(600)	S^2	τ_e (ps)	S^2	S_f^2	S_s^2	τ_s (μs)
S17									
expt	653± 38	107± 8.5	0.78						
calc Eq. 4	643± 23	107± 5	0.28± 0.03	0.73± 0.04	574± 18				
calc Eq. 5	659± 7	107± 1	0.76± 0.02			0.72± 0.01	0.86± 0.01	0.84± 0.0	2.6± 0.8
G22									
expt	784± 7	210± 11	0.32						
calc Eq. 4	784± 4	210± 6	-1.2± 0.04	0.33± 0.014	310± 8				
calc Eq. 5	782± 2	210± 2	0.33± 0.02			0.31± 0.005	0.68± 0.006	0.46± 0.007	1.2± 0.04
L26									
expt	745± 40	106± 2.4	0.79						
calc Eq. 4	745± 19	105± 1.1	0.42± 0.1	0.77± 0.01	107± 43				
calc Eq. 5	747± 7	106± 0.5	0.78± 0.02			0.76± 0.005	0.81± 0.008	0.93± 0.009	2.0± 1.0
D35									
expt	739± 6	128± 1.6	0.68						
calc Eq. 4	739± 3	128± 0.7	-0.24± 0.01	0.61± 0.004	250± 0.7				
calc Eq. 5	736± 2	128± 0.5	0.70± 0.02			0.59± 0.004	0.76± 0.008	0.78± 0.005	2.0± 0.3
D54									
expt	899± 35	142± 3.3	0.58						
calc Eq. 4	892± 16	142± 1.5	-0.008± 0.07	0.56± 0.007	104± 11				
calc Eq. 5	896± 9	142± 0.7	0.58± 0.02			0.55± 0.004	0.68± 0.006	0.81± 0.009	0.95± 0.08

The errors in the experimental NOEs are ±0.1. The errors for the calculated values of the various parameters are obtained from analysis of the variance-covariance matrix obtained from the least squares Powell optimization procedure. In all the fits, a value of 8.3 ns is used for τ_R which was obtained by a best fit of the T_1/T_2 ratios of all residues *simultaneously* (excluding 16 which had T_1/T_2 ratios outside ±1SD from the mean) using Eqs. 1 and 2. In the fits obtained with Eq. 4, S^2 and τ_e were varied; and in the fits obtained with Eq. 5, S_f^2, S_s^2 and τ_s were varied (and S^2 is given by the product $S_f^2 S_s^2$. In the best fits using Eq. 4, only the T_1 and T_2 data were used. In the best fits using Eq. 5 the T_1, T_2 and NOE data were used.

The effect of using the spectral density function (5) compared to (4) is easily understood by noting that the calculated values of τ_e (0.2-0.3 ns) are an order of magnitude smaller than those of τ_s (0.5-4 ns). For $\tau_R \sim 8$ ns and $\omega_N = 2\pi \times 60.8$ MHz, the NOE reaches a minimum value at an internal correlation time of ~ 0.25 ns. Thus the shift in internal correlation time to larger values which accompanies the introduction of the two internal motion formulation results in larger values of the NOE, whilst leaving the values of T_1 and T_2 unaffected.

These results clearly indicate that the ^{15}N relaxation data are sufficient to separate out two internal motions with time scales differing by more than one order of magnitude. Hence the correlation function for the internal motions of these residues can no longer be approximated to a single exponential of the Lipari and Szabo treatment, but rather requires two distinct exponentials. The slower motion, of the order of 0.5-4 ns, is slightly slower than the overall rotational correlation time (8-9 ns) for IL-1β, and can have a significant magnitude. The very fast motion, which must have a time scale of $\ll 100$ ps, reflects the fast random thermal motions that are manifested in molecular dynamics calculations.

ACKNOWLEDGEMENTS

This work was supported by the Intramural AIDS Targeted Antiviral Program of the Office of the Director of the National Institutes of Health. We thank Attila Szabo for enlightening discussions.

REFERENCES

1. G. Lipari and A. Szabo, *J. Am. Chem. Soc.* **104**, 4546-4559 (1982).
2. G. Lipari and A. Szabo, *J. Am. Chem. Soc.* **104**, 4559-4570 (1982).
3. D. C. McCain and J. L. Markley, *J. Am. Chem. Soc.* **108**, 4259 (1986).
4. A. J. Weaver, M. D. Kemple, and F. G. Predergast, *Biophys. J.* **54**, 1 (1988).
5. M. J. Dellwo and A. J. Wand, *J. Am. Chem. Soc.* **111**, 4571-4578 (1989).
6. L. E. Kay, D. A. Torchia, and A. Bax, *Biochemistry* **28**, 8972-8979 (1989).
7. N. R. Nirmala and G. Wagner, *J. Magn. Reson.* **82**, 659-661 (1989).
8. G. M. Clore, P. C. Driscoll, P. T. Wingfield, and A. M. Gronenborn, *Biochemistry*, **29**, 7387-7401 (1990).
9. A. Abragham, "The Principles of Nuclear Magnetism," Clarendon Press, Oxford (1961).
10. G. M. Clore, A. Szabo, A. Bax, L. E. Kay, P. C. Driscoll, and A. M. Gronenborn, *J. Am. Chem. Soc.* **112**, 4989-4991 (1990).
11. B. C. Finzel, L. L. Clancy, D. R. Holland, S. W. Muchmore, K. D. Watenpaugh, and H. M. Einspahr, *J. Mol. Biol.* **209**, 779-791 (1989).

A NEW VERSION OF DADAS (DISTANCE ANALYSIS IN DIHEDRAL ANGLE SPACE) AND ITS PERFORMANCE

Shigeru Endo[a], Hiroshi Wako[b], Kuniaki Nagayama[a], and Nobuhiro Gō[c]

[a]Biometrology Lab, JEOL Ltd., Musashino
Akishima Tokyo 196,
[b]Faculty of Social Science, Waseda University, Nishiwaseda
Shinjuku-ku, Tokyo 169
[c]Department of Chemistry
Faculty of Science
Kyoto University
Kitashirakawa, Sakyo-ku
Kyoto 606, Japan

ABSTRACT

Among the computational algorithms to determine solution structures of proteins, the minimization of a variable target function in the dihedral angle space with the first derivative has been widely applied with its success at generating unbiased structures. The program called DADAS (Distance Analysis in Dihedral Angle Space, sometimes called DISMAN) has been extended to include additional tools, besides the rapid first derivative calculation and the variable target function: a rapid second derivative calculation, effective Metropolis Monte Carlo sampling and simulated annealing based on the Monte Carlo simulation. The new version of DADAS, called DADAS90, has some advantages, as compared with the older one, of flexibility to design new hybrid algorithms consisting of the mentioned tools and the possibility to manipulate the target function minimization from short range to long range information. The performance of each of the tools and their combination has been tested with respect to the convergence to the exact structure with the use of simulated data sets of distance constraints. The points on how to apply the program to an actual system with a limited amount of NMR data on distance and angle constraints have been also studied.

INTRODUCTION

Various kinds of computational methods to determine the solution conformation of proteins have been hitherto proposed. Distance geometry using

Computational Aspects of the Study of Biological Macromolecules by Nuclear Magnetic Resonance Spectroscopy, Edited by J.C. Hoch *et al.*, Plenum Press, New York, 1991

the metric matrix formulation was first applied to generate structures[1]. This has been recently extended to linearized embedding, which can explicitly include the holonomic information of chemical structures such as bond angles and bond lengths[2]. Another early algorithm is a variable target function approach which employs first derivative minimization in dihedral angle space[3]. This algorithm is an optimization method extended from the conventional computational method used in energy minimization[4,5]. On the same line of approach restrained molecular dynamic calculation also provided a method for carrying out the optimization process[6,7]. The method together with its modification called dynamical simulated annealing[8] was applied to refine the initial structures obtained from metric matrix embedding[9]. The third approach recently proposed, NOESY simulation by back calculation, further refines obtained structures, to be faithful to the NMR spectral data[10]. These historical developments are closely interrelated and hence could be summarized with a coherent view on the information flow from the NMR spectra to the solution structures as:

(1) Metric Matrix Embedding (Level I)

$$
\begin{array}{cc}
\text{distance} & \text{structure} \\
\{d_{ij}\}_{\text{exp}} & \rightarrow \quad \{r_i\}_{\text{exp}}
\end{array}
$$

(2) Optimization (Level II)

$$
\begin{array}{ccc}
\text{distance} & \text{distance} & \text{structure} \\
\{d_{ij}\}_{exp} \quad \leftrightarrow & \{d_{ij}\}_{test} \leftarrow & \{r_i\}_{test} \\
\text{compare} & | \underline{\qquad\qquad} \uparrow & \\
& \text{optimization loop} &
\end{array}
$$

(3) NOESY Simulation (Level III)

$$
\begin{array}{ccccc}
\text{NOESY} & \text{NOESY} & \text{relaxation} & \text{distance} & \text{structure} \\
& & \text{matrix} & & \\
\{I_{ij}\}_{exp} \quad \leftrightarrow & \{I_{ij}\}_{test} \leftarrow & \{\sigma_{ij}\}_{test} \leftarrow & \{d_{ij}\}_{test} \leftarrow & \{r_i\}_{test} \\
\text{compare} & | \underline{\qquad\qquad\qquad\qquad\qquad\qquad} & & & \uparrow \\
& & \text{optimization loop} & &
\end{array}
$$

Nowadays these three levels of approaches are separately or hierarchically utilized depending upon the quality of available NMR data.

When we look at the problem from another angle, we can recognize many ideas which have originally been proposed in computational physics. They are not necessarily mutually exclusive and become the building blocks in a comprehensive approach to the structure calculation. The analytic optimization with the first derivative can be extended to include the second derivative method. The molecular dynamic approach can be replaced by the Metropolis Monte Carlo method. To make these extensions effective and to make the algorithms acceptably fast, however, the choice of independent variables to describe conformations is crucial. Dihedral angles around rotatable chemical bonds have been adopted[3] instead of Cartesian coordinates of constituent atoms for the following reasons. (1) With bond lengths and bond angles fixed to standard values the number of independent variables can be reduced to about one eighth of the number of variables in the Cartesian coordinates. The reduction of the number of variables is an important factor in computational analysis of large molecules in order to save memory and time. (2) The distance constraints and angle constraints for the determination of the solution structures of proteins are naturally built in the algorithm as compared in the Cartesian coordinate approach in which they have to be treated as an additional set of constraints such as a priori constraints for bond lengths, bond angles and chiralities. (3) A recurrent algorithm[4,5] and can be adopted in the dihedral angle space to make the second derivative calculation very fast and the sampling of Monte Carlo simulation in the conformational space effective[11].

On such consideration we have recently developed a new version of DADAS (Distance Analysis in Dihedral Angle Space, sometimes called DISMAN)[3]. In the older version (DISMAN) only the first derivative is calculated for the conjugate gradient minimization (MIN1) of the target function. In the new version, which we call DADAS90 from now on, the second derivative calculation for the Newton minimization (MIN2), effective sampling Monte Carlo method (MC) and simulated annealing based on the Monte Carlo sampling (SA) have been added as tools of the program. DADAS90, working as a command interpreter, can invoke these tools in an arbitrary stage of the calculation of the variable target function minimization. The details of the new version of program, DADAS90, and its performance is described.

ALGORITHM AND PROGRAM

Description of a Three-Dimensional Structure

In the space using dihedral angles as independent variables describing molecular conformations, a molecule can be represented by a tree structure composed of rigid bodies (referred to as units) connected by rotatable bonds which define the variables. When a chemical structure of a molecule is given, a standard conformation of the molecule is uniquely constructed by adopting and connecting units relevant to the molecule. With the standard conformation any three-dimensional structure of the molecule can be easily generated by the corresponding set of dihedral angles in the course of the structure calculation. We have prepared a library of units which have as standard bond lengths and bond angles of ECEPP/2 for the naturally occurring twenty amino acids[12,13] and developed a program, MKMOL, to produce a standard conformation for an arbitrary amino acid sequence.

MIN1, MIN2 and MC as Tools of Structure Calculation

In the tree-structural topology of a protein molecule a fast algorithm was developed, which enabled the calculation of first and second derivatives of conformational energy functions and/or objective functions with respect to dihedral angles in computing time proportional to n^2 (n, the number of variable dihedral angles)[4]. Afterwards the algorithm was revised in the form of recurrent equations[5]; one which needs a memory space only proportional to n, and another well adapted to a supercomputer[14]. In minimization of conformational energy and/or objective functions, the method using both first and second derivatives, the Newton method, is known to be more effective than a minimization method using only first derivatives, such as the conjugate gradient method with respect to the convergence to the exact minimum point and number of iterations in the vicinity of minimum points. However, the computational speed of first derivatives is much faster than that of second derivative matrix. We have, therefore, prepared two minimizers, MIN1 and MIN2. MIN1 is the conjugate gradient minimizer optimized by introducing a cut-off distance in the calculation of van der Waals contacts. It can calculate the first derivative of objective functions in computing time proportional to n^3. MIN2 is the minimizer of the Newton method based upon the vectorized calculation of the second derivatives of energy and/or objective functions[14]. In Monte Carlo simulation (MC) the second derivative matrix is used for effective sampling of the acceptable conformations[11]. Furthermore, simulated annealing (SA) was implemented by adopting the Metropolis Monte Carlo simulation with temperature manipulation.

DADAS90 as a Toolkit Driver

We have developed a new DADAS program, DADAS90, on UNIX and VAX/VMS operating systems. DADAS90 is composed of (1) the main routine as a command interpreter and (2) various tools such as MIN1, MIN2 and MC, and works as a toolkit driver to execute a command sequence input from a keyboard or a macro command file (cmdfile). Major commands of DADAS90 are listed in Table 1. Fig. 1 shows the schematics of DADAS90. DADAS90 is able to handle target functions made of structural constraints from NMR experiments and/or empirical energy functions in an arbitrary

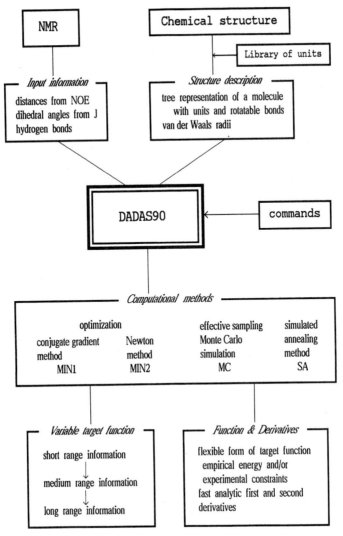

Figure 1. Schematics of DADAS90.

combination. The term of energy functions was not involved in the calculation reported here.

In DADAS90 various kinds of strategies for calculation can be employed to determine a protein structure. DADAS90, working as a command interpreter, can invoke various tools in the arbitrary stage of the calculation of the variable target function minimization. If the calculation is trapped in the early stage of the variable target function minimization, we may call MC to escape the trap. The mechanical stress arising from distance constraints and van der Waals contacts could be easily released. As an example, a flowchart of the Monte Carlo simulated annealing to relax severe violations in a random initial structure followed by the Newton minimization of a

Table 1. Major Commands in DADAS90

[Preparation]

title	;	sets a title header of the current job.
rmol	;	reads a molecular structure file in the dihedral angle space.
rrdt	;	prepares upper and/or lower distance constraints.
rrag	;	prepares angle constraints and/or fixes variable dihedral angles.

setpar │ ofunc   ; sets an objective function form:
       │   energy and/or experimental constraints.
       │ temp   ; sets temperature in Monte Carlo simulation.
       │ mcstep   ; sets an output interval of dihedral angle
       │   data in Monte Carlo simulation.

dfiang ; defines dihedral angle data of initial structures
         as an extended β-structure, random structures
         within angle constraints, or specified dihedral
         angle data.

stiang ; reinitializes all dihedral angle values to the next structure.

outmc ; opens an angle data file for structures
         generated during Monte Carlo simulation.

doconf │ nconf   ; begin of loop. nconf is the number
       │   of initial conformations.

enddo   ; end of doconf loop.

[Calculation]

min1[a] krngdc, krngvw, wrrr, wrvw, wrrt, rsize, limit, eps10p
       ; invokes minimization by conjugate gradient method.

krngdc	;	range of distance constraints in unit.
krngvw	;	range of van der Waals contact check in unit.
wrrr	;	weight for distance constraints.
wrvw	;	weight for van der Waals repulsion.
wrrt	;	weight for angle constraints.
rsize	;	size parameter of van der Waals radii.
limit	;	maximum iteration number.
eps10p	;	exponent of minimization convergent condition.

min2[a] krngdc, krngvw, wrrr, wrvw, wrrt, rsize, limit, eps10p
       ; invokes minimization by Newton method.

mc[a] krngdc, krngvw, wrrr, wrvw, wrrt, rsize, limit, tc
    ; invokes Monte Carlo simulation (tc = 0) or simulated
      annealing (tc < 0).

      tc ; exponent of temperature scaling factor as
           Temperature=T_0 * exp(tc*naccpt)
           where naccpt is number of accepted conformations
           and T_0 initial temperature.

Table 1. Major Commands in DADAS90 (continued)

[Evaluation]

wpdb	;	outputs coordinate data in PDB format.
aang	;	outputs dihedral angle data at append mode.
dvdw	;	displays van der Waals violations larger than the cutoff value.
drdt	;	displays distance constraint violations larger than the cutoff value.
drag	;	displays angle constraint violations larger than the cutoff value.
dss	;	displays geometrical configuration of S-S bonds.

[a]Execution of these commands can be interactively terminated.

variable target function (SA+VMIN2 procedure) is shown in Fig. 2 and its command procedure in Table 2. This flow of calculation is characterized by many parameters in Table 1 such as the unit range within which atom-pairwise interactions are included in the objective function, weight values for various constraint conditions, and van der Waals radii. One can use default values for these parameters or assign different values if necessary. By the calculation step of MIN2 in Table 2 the unit range is increased in such a way that the number of distance constraints newly added to the objective function in each of the step of range increment is approximately constant.

Distance Constraints

For the purpose of the strict assessment of the obtained structures against the "real structure", a *simulated* distance data set is produced from the three-dimensional atomic coordinates of bovine pancreatic trypsin inhibitor (BPTI) obtained by regularizing coordinates in the file 5PTI from the Protein Data Bank of Brookhaven National Laboratory in such a way that they have standard bond lengths and bond angles of ECEPP/2 with omega angles fixed to 180 degrees. All methyl protons and all methylene protons beyond beta position are replaced by pseudoatoms located at their mean positions[16], which emulates the typical NMR experimental situation. In this reference structure of BPTI, all proton pairs within 4 Å distance are identified. Pseudoatom-proton and pseudoatom-pseudoatom pairs within

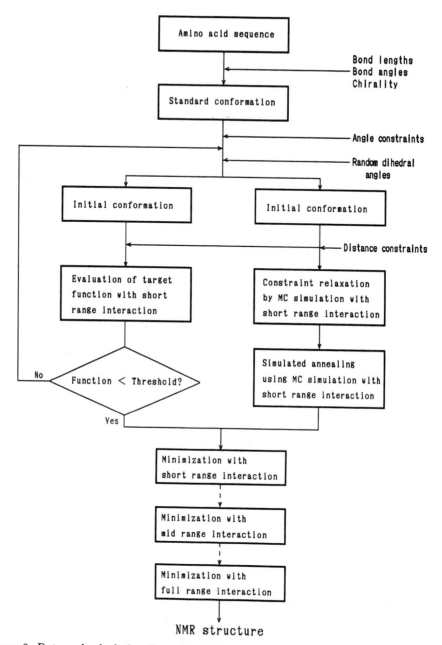

Figure 2. Data and calculation flow with DADAS. Either the selection of initial structures satisfying the short range distance constraints (left side) or the Monte Carlo simulated annealing (right side) is followed by minimization of the variable target function to obtain NMR structures.

4 Å are also identified. The following five distance sets are then prepared. A distance set I consists of the upper and lower bounds assigned to exact distances in the reference structure. This is a set of ideal distance constraints except for the ambiguities introduced by the use of pseudoatoms. A distance set II consists of only upper bounds assigned to exact distances. It should be noticed here that the possible distance between a pair of atoms in 1-4 positions is fairly limited. If the simulated distance of these pairs is close to the upper limit, one should use it as a lower bound rather than an upper bound. Accordingly, in a distance set III some of the constraints in the set II are replaced to lower bounds when they pertain to a 1-4 interaction pair and their restrained distance is close to its maximum value. A distance set IV has relaxed distances of the set I whose upper bound values are 125% of the exact distances of set I and lower ones are 80%. The total number of atom pairs with identified distance constraints in the sets I to IV is 1060. A distance set V has the actual constraints observed by NMR[15] which consist of 228 distance constraints and 43 dihedral angle constraints.

All calculations were done on a Stardent TITAN 1500 with 4 CPUs or a TITAN 3000 with 2 CPUs.

RESULTS AND DISCUSSION

Convergence to the Exact Structure for Precise Distance Information

The complexity of behaviors of computational methods used to determine protein conformations from NMR data does not warrant us to interpret calculated conformations naively. Most of the existing methods produce a number of conformations, each of which more-or-less satisfies input distance constraints. We have to examine behavior of the methods critically from points of view such as: (a) Does the mean of a number of calculated conformations corresponds to the mean of actually fluctuating conformations? (b) Does the extent of conformational diversity among calculated conformations correspond to the extent of real fluctuations? As a starting point of the critical examination we think that it should be elucidated whether a distance geometry algorithm is clean, i.e., whether it can reconstruct the exact structure from precise distances. The metric matrix algorithm is clean in this sense, because the exact structure can be reconstructed analytically if exact distances are given for all pairs[17].

We calculated thirty structures by the SA+VMIN2 procedure shown in Table 2 for each of the distance constraint sets I, II, III and IV. Calculations were started from 30 randomly generated structures. Fig. 3 shows the quality of calculated 120 structures in terms of two criteria: the root mean square

Table 2. An example of the command procedure in DADAS90 (SA+VMIN2 procedure)

title BPTI NMR experimental distance set; Wagner et al. (1987)								
rmol	bpti.mol							
rrdt	bpti.rdt							
rrag	bpti.rag							
dfiang	RANDOM							
doconf	30							
stiang								
aang	bpnmini.ang							
setp	temp 1000							
mc	2	-1	1.	0.	5.	0.	10000	-0.0
aang	bpnmmc1.ang							
mc	2	1	1.	1.	5.	0.	2000	-0.002
aang	bpnmmc2.ang							
min2	0	0	40.0	4.0	200.0	0.0	100	-4.
min2	1	1	40.0	4.0	200.0	0.0	100	-4.
min2	2	2	40.0	4.0	200.0	0.0	100	-4.
min2	3	3	40.0	4.0	200.0	0.0	100	-4.
min2	4	4	40.0	4.0	200.0	0.0	100	-4.
min2	5	5	40.0	4.0	200.0	0.0	100	-4.
min2	6	6	40.0	4.0	200.0	0.0	100	-4.
min2	7	7	40.0	4.0	200.0	0.0	100	-4.
min2	8	8	40.0	4.0	200.0	0.0	100	-4.
min2	10	10	40.0	4.0	200.0	0.0	100	-4.
min2	13	13	40.0	4.0	200.0	0.0	100	-4.
min2	22	22	40.0	4.0	200.0	0.0	100	-4.
min2	26	26	40.0	4.0	200.0	0.0	100	-4.
min2	29	29	40.0	4.0	200.0	0.0	100	-4.
min2	34	34	40.0	4.0	200.0	0.0	100	-4.
min2	36	36	40.0	4.0	200.0	0.0	100	-4.
min2	40	40	40.0	4.0	200.0	0.0	100	-4.
min2	48	48	40.0	4.0	200.0	0.0	100	-4.
min2	67	67	40.0	4.0	200.0	0.0	100	-4.
min2	72	72	40.0	4.0	200.0	0.0	100	-4.
min2	90	90	40.0	8.0	200.0	0.0	100	-4.
min2	107	107	40.0	10.0	200.0	0.0	100	-4.
min2	155	155	40.0	20.0	200.0	0.0	100	-4.
min2	173	173	40.0	40.0	40.0	0.0	200	-12.
aang	bpnmr.ang							
enddo								
exit								

Table 3. The degree of convergence of structures obtained with different minimization procedures

procedure name	mean RMSD[a]	no. of structures	no. of minimization	maximum iteration number in each minimization
VMIN2	3.10	30	24	100 (200)[b]
VMIN1	2.95	30	24	500 (2000)[b]
MC+VMIN2	2.50	30	(MC+)24	100 (200)[b]
MC+VMIN1	2.59	30	(MC+)24	500 (2000)[b]
SA+VMIN2	2.43	30	(SA+)24	100 (200)[b]
DISMAN[c]	2.92	10[d]	24	500 (5000)[b]

[a]The degree of convergence is represented by the mean of pairwise RMSD between 5 structures with the least target function value. [b]The number in the parenthesis is the maximum iteration number in the final step of the minimization. [c]This results was obtained by Wagner et al. (1987). [d]These structures were selected by the sampling of local ϕ and ψ angles so that distance constraints of sequential NOE, $d_{\alpha N}$ and d_{NN}, may be satisfied.

distance (RMSD) of main chain atoms from the reference structure and the sum of violations for distance constraints used. From now we call this graph a violation-RMSD plot. In the distance constraint sets I, II and III the VMIN2 minimization procedure was terminated in about half of the cases by locating a point with an essentially vanishing gradient and non-negative hessian. In another half of the cases, minimization was carried out up to the pre-specified number of 200 steps and then was terminated. In these cases further MIN2 minimization of 200 steps was carried out but points in Fig. 3 moved only by insignificant amounts. Therefore, the procedure in Table 2 is efficient and practically sufficient.

The cleanness of the algorithm should be examined at first for the set I, the set with the most precise distance constraints among the sets studied. In this case 16 out of 30 calculated structures converged within 0.05 Å RMSD from the reference structure. For practical purposes, this is a very good convergence. However, in terms of residual violations for distance constraints the algorithm is not completely clean, i.e., it has finite sums of residual violations less than 10 Å. The situations are similar in the cases of both distance constraint set II and III. In the case of set III, 17 out of 30 calculated structures converged within 0.13 Å RMSD with the sums of residual violations less than 7 Å. The eight structures with the least violations for the set III are shown in Fig. 4A in their best fit superposition to the reference structure. In this figure it is observed that the main structural differences are in side chains. The main chain structures are very similar,

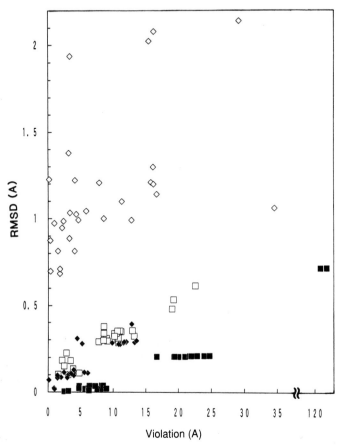

Figure 3. The violation-RMSD plot of structures obtained by the SA+VMIN2 procedure of DADAS90 with several kinds of distance constraint sets I (■), II (□), III (◆), and IV (◇). RMSD for each structures is calculated from the X-ray reference structure used to generate the distance sets I to IV. Detailed description of distance constraints in the text.

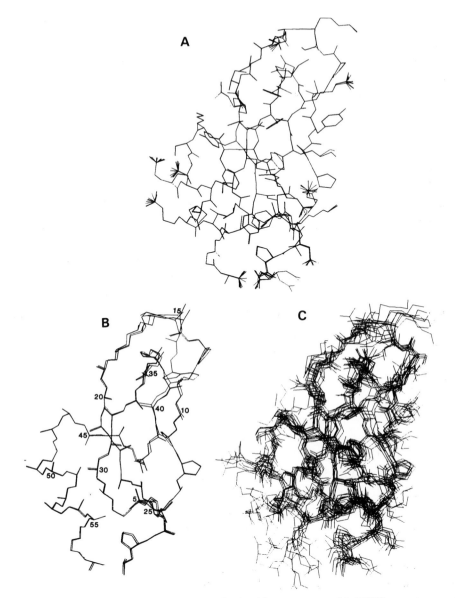

Figure 4. (A) Best fit superposition of the backbone atoms of 8 BPTI structures with the least violation among 30 structures obtained by DADAS90 using the distance set III, the exact upper distance constraints derived from the X-ray reference structure. (B) The superposition of 3 structures with the least violation, 3 structures with 0.2 Å RMSD and 2 structures with 0.75 Å RMSD among 30 structures obtained using the distance set I, the exact upper and lower constraints, as shown by (■) in Fig. 3. (C) The superposition of 8 structures with the least violation from 30 structures obtained using the distance set IV, the distance constraints relaxed at the 25% accuracy of the exact distances.

reflected to the small value of the main chain RMSD. This situation is even more clearly so in the 16 structures mentioned above in the case of set I.

The reconstruction of the essentially exact main chain structure with varied conformations of some side chains can be understood as a result of the use of pseudoatoms in the related side chains. When distance constraints are given for pseudoatoms, more than one side chain conformation are often consistent with a set of intra-residue distance constraints, for example, as in lysine with four methylene groups. Then, after the SA procedure with only the short-range information in the target function, the side chains may assume one of these conformations more-or-less arbitrarily. The strategy of the use of variable target function in the DADAS minimization algorithm fails to rectify these wrongly chosen side chain conformations as can be seen in Fig. 4A. Therefore the basic strategy to avoid this difficulty of defining side chain conformation should be to carry out NMR measurements in such a way that one can assign side chain atoms stereo-specifically and also one can obtain coupling constants so that information about side chain dihedral angles can be obtained. If the present calculation were done for a truly ideal distance set of all methylene protons, sufficient to fix side chain conformations to exact ones from intra-residue distance information, the truly exact conformations including side chains should have been obtained by the calculation using the SA+VMIN2 procedure.

In Fig. 3 the calculated structures for distance constraint sets I, II and III are seen to be clustered. For set I, 12 structures are seen to be clustered and displaced from the reference structure by about 0.21 Å and have sums of residual violations in the range of 16-25 Å. The mutual main chain RMSD among these 12 structures is less than 0.05 Å. This means that these 12 structures are indeed clustered in a small conformational subspace. Differences among them are again observed mainly in some side chains. A similar situation is observed also for the two structures with RMSD being about 0.75 Å and sum of violations being about 120 Å. Differences of the main chain structures of the three clusters is shown in Fig. 4B. As is clearly seen, the difference is localized. A similar localized main chain difference in structures calculated from distance information was already observed previously[3].

When the distance information is very precise as in sets I, II and III, calculated structures are clustered as just mentioned. The best cluster corresponds to the precise main chain fold. Clusters corresponding to locally misfolded main chain structures have distinctly larger values of the sum of residual violations of the distance constraints. Therefore, by judging from the value of the sum of violations, structures with misfolded main chain can be discarded.

A conclusion from the above is that the algorithm of DADAS90 with the procedure of SA+VMIN2 is clean, i.e., is able to reconstruct the exact structure from precise distance information, if precise distance information corresponding to stereospecific assignment is used and structures with residual violations larger than a certain value are discarded.

When the precision of the distance information is reduced as in set IV, the clustering nature of calculated structures becomes less conspicuous. It should be noted that, in sets III and IV, structures with virtually no violations are obtained. These are perfectly clean structures. Their RMSD values indicate that the extent of conformational diversity corresponding to the conformation subspace in which the target function vanishes exactly is about 0.05 Å and 1.0 Å in terms of RMSD values for sets III and IV, respectively. The eight structures with the lowest violations are shown in Fig. 4C. A detailed analysis of these structures indicates that the conformational deviations from the reference structure, from which the simulated data are constructed, is not biased as was shown previously for the structures determined by the older version of DADAS[15].

Comparison of Various Minimization Procedures

Five types of minimization procedures were employed and examined for distance constraint set III, for which many structures with low violations were obtained in Fig. 3. Each type of minimization procedure was applied for 30 randomly generated initial structures. The five types of procedures are; (1) VMIN2, minimization of the variable target function by MIN2, (2) VMIN1, the same as VMIN2 but by MIN1, (3) MC+VMIN2, the Monte Carlo simulation with constant temperature followed by VMIN2, (4) MC+VMIN1, the same as MC+VMIN2 but followed by VMIN1, and (5) SA+VMIN2, the simulated annealing followed by VMIN2 as detailed in Table 2. Fig. 5 shows the violation-RMSD plot of the calculated 150 structures with the same criteria as in Fig. 3.

VMIN1 and MC+VMIN1 procedures were terminated at a prespecified number of 2000 MIN1 minimization iterations. As in Fig. 3 about half of the minimizations by MIN2 in VMIN2, MC+VMIN2 and SA+VMIN2 procedures were terminated by locating a point with a vanishing gradient and non-negative hessian. Average CPU times used in each of the five procedures is roughly equal. Because terminated points by MIN1 minimization are not minimum points of the objective function, further minimizations of either 4000 steps of MIN1 or 200 steps of MIN2 were carried out by starting from structures obtained by MC+VMIN1 procedures shown in Fig. 5. CPU time for 4000 steps of MIN1 minimization is approximately equal to the time for 200 steps of MIN2 minimization in the current calculation. Results of these further minimizations are shown in Fig. 6.

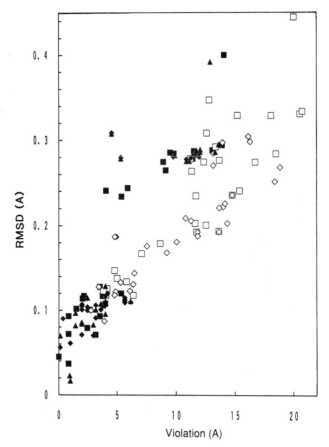

Figure 5. The violation-RMSD plot of structures obtained by DADAS90 with the distance set III using various kinds of procedures VMIN2 (■), VMIN1 (□), MC+VMIN2 (◆), MC+VMIN1 (◇), and SA+VMIN2 (▲). These symbols are explained in the text.

In Fig. 5 filled symbols are structures minimized by MIN2 and open ones by MIN1. It is clear that minimization by MIN2 is more efficient at obtaining better structures. This is again also clear in Fig. 6. In the practical point of view, we conclude that it is difficult to find real minimum points only by MIN1 because all further MIN1 minimizations were also terminated at the prespecified number of 4000 steps.

Structures Obtained for Experimental NMR Constraints

The same five types of minimization procedures as in Fig. 5 were carried out for the constraint set V consisting of NMR experimental information[15]. Each type of minimization procedure was applied again for 30 randomly generated initial structures. Fig. 7 shows the quality of calculated 150 structures. There are some differences in Fig. 7 from Figs. 3 and 5. The main chain

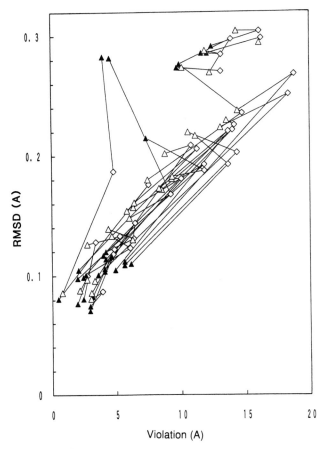

Figure 6. The violation-RMSD plot of structures further minimized by 200 steps of MIN2 (▲), or by 4000 steps of MIN1 (△) from the structures obtained by the MC+VMIN1 procedure in Fig. 4 (◇).

RMSD was measured from a new reference structure, NMR reference structure, which was obtained by MC+VMIN2 procedure starting from the X-ray reference structure with the constraint set V. The abscissa represents the violation of constraints by the obtained structure. Since the constraint set V contains both distance and angle constraints, the violation of constraints can be expressed by a single number only as a weighted sum of distance and angle violations. As such a quantity the value of the target function itself is employed. The value of the target function of the NMR reference structure is 7.4, which is the smallest among those of other structures calculated from the randomly generated structures.

The quality of the obtained structures is comparable to that in the previous report[15], in which the older program, DISMAN, was used. The results are summarized in Table 3. In the present calculation 30 structures were calculated for each type of procedure, while in the previous report 10 struc-

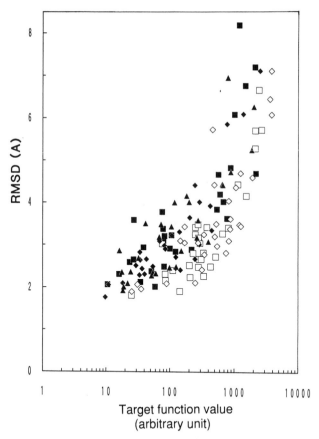

Figure 7. The quality of structures obtained by DADAS90 with the distance constraint set V obtained from NMR, using the procedures VMIN2 (■), VMIN1 (□), MC+VMIN2 (◆), MC+VMIN1 (◇), and SA+VMIN2 (▲).

tures were obtained. However initial structures of these 10 structures were generated from ϕ and ψ angles chosen randomly from a range consistent with the short range distance constraints, $d_{\alpha N}$ and d_{NN}, inferred from sequential NOE. The mean value of pairwise RMSD between the 5 best structures in terms of the residual value of the target function in the present calculation is roughly equal or slightly less than that in the previous report. The mean RMSD value calculated for the 5 best structures among the first 10 structures (instead of the 30 structures) obtained by MC+VMIN2 procedure is 2.90 (instead of 2.50 in Table 3). This value is very close to that the reported previously, i.e., 2.92. This fact indicates that the sampling of initial structures by Monte Carlo simulation with the short range interaction is quantitatively equivalent to the selection of initial structures consistent with the short range information.

Practical comparison among the results obtained by several minimization procedures is also made. Fig. 7 again claims that in general MIN2 gives better structures than MIN1 as observed in Fig. 5. All MIN1 minimizations in the final step of the VMIN1 procedure were here also terminated at the end of pre-specified number of 2000 steps, while about 1/3 of the MIN2 minimizations were converged within 200 steps in the final step of VMIN2. For the distance set V the SA+VMIN2 procedure seems to give better results than VMIN2 or MC+VMIN2 procedures. However these three procedures provided almost the same quality in obtained structures as given by the distance set III shown in Fig. 5. SA or MC become more effective at convergence of structures when distance constraints are more relaxed.

Behavior of the new version of DADAS should further be tested for constraints sets of different quality. Experience of such tests should be accumulated as different procedures of minimization fitted best for constraint sets of different quality. We are now making efforts along this direction.

REFERENCES

1. W. Braun, C. Bösch, L. R. Brown, N. Gō, and K. Wüthrich, *Biochim. Biophys. Acta* **667**, 377-396 (1981).
2. G. M. Crippen, *J. Comp. Chem.* **10**, 896-902 (1989).
3. W. Braun, and N. Gō, *J. Mol. Biol.* **186**, 611-626 (1985).
4. T. Noguti, and N. Gō, *J. Phys. Soc. Japan* **52**, 3685-3690 (1983).
5. H. Abe, W. Braun, T. Noguti, and N. Gō, *Comp. Chem.* **8**, 239-247 (1984).
6. R. Kaptein, E. R. P. Zuiderweg, R. M. Scheek, R. Boelens, and W. F. van Gunsteren *J. Mol. Biol.* **182**, 179-182 (1985).
7. A. T. Brünger, G. M. Clore, A. M. Gronenborn, and M. Karplus, *Proc. Natl. Acad. Sci. USA* **83**, 3801-3805 (1986).
8. M. Nilges, G. M. Clore, and A. M. Gronenborn, *FEBS Lett.* **229**, 317-324 (1988).
9. T. F. Havel and K. Wüthrich, *Bull. Math. Biol.* **46**, 673-698 (1984).
10. M. F. Summers, T. L. South, B. Kim, and D. R. Hare, *Biochemistry* **29**, 329-340 (1990).
11. T. Noguti, and N. Gō, *Biopolymers* **24**, 527-546 (1985).
12. F. A. Momany, R. F. McGuire, A. W. Burgess, and H. A. Scheraga, *J. Phys. Chem.* **79**, 2361-2381 (1975).
13. G. Némethy, M. S. Pottle, and H. A. Scheraga, *J. Phys. Chem.* **87**, 1883-1887 (1983).
14. H. Wako, and N. Gō, *J. Comp. Chem.* **8**, 625-635 (1987).
15. G. Wagner, W. Braun, T. F. Havel, T. Schaumann, N. Gō, and K. Wüthrich, *J. Mol. Biol.* **196**, 611-639 (1987).
16. K. Wüthrich, M. Billeter, and W. Braun, *J. Mol. Biol.* **169**, 949-961 (1983).
17. G. M. Crippen and T. F. Havel, *Acta Cryst.* **A34**, 282-284 (1978).

AN AMATEUR LOOKS AT ERROR ANALYSIS IN THE DETERMINATION OF PROTEIN STRUCTURE BY NMR

Jeffrey C. Hoch

Rowland Institute for Science
100 Cambridge Parkway
Cambridge, MA 02142

ABSTRACT

Error analysis constitutes an important part of experimental physical chemistry, yet many of the concepts taught at the undergraduate level have yet to be systematically applied to the determination of protein structure by NMR. At first glance, the application of such simple concepts to the complex problem of structure determination might appear quixotic. However, with straightforward extensions, these concepts can yield valuable insights into the precision of NMR based structures. Some of these concepts are explored, using illustrative examples from the data of Driscoll et al. (*Biochemistry* **20**, 2188 (1989)) for the 43 residue protein BDS-I.

ELEMENTARY ERROR ANALYSIS

Precision vs. Accuracy

As Professor Jardetzky is fond of reminding us, the precision of an observation is a measure of how reproducible it is when the measurement is repeated. Accuracy is a measure of how closely a result reflects the truth. Precision reflects random errors, while accuracy reflects both random and systematic errors. The magnitude of random errors can be assessed by making repeated observations. Systematic errors are more difficult to assess, unless the true result is somehow known.

Measures of Precision

The distribution of a set of independent observations o_i of some parameter x,

$$o_i = x + \epsilon_i \tag{1}$$

Computational Aspects of the Study of Biological Macromolecules by Nuclear Magnetic Resonance Spectroscopy, Edited by J.C. Hoch *et al.*, Plenum Press, New York, 1991

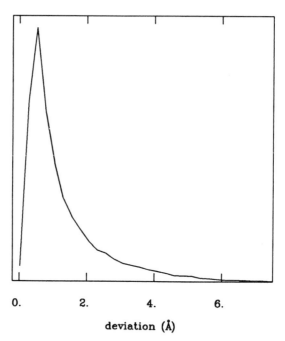

0.   2.   4.   6.

deviation (Å)

Figure 1. Histogram of the atomic deviations from the mean for the 42-member ensemble of structures for BDS-I by Driscoll et al..

where the ϵ_i are the random errors, provides an indication of the precision of the measurement. Common ways of characterizing the distribution of ϵ_i are the range, the average deviation, and the variance. The variance $S^2 = \frac{1}{N-1} \sum (o_i - \bar{o})^2$ has the advantage of being insensitive to sample size; the average deviation is sensitive to sample size. When the ϵ_i are normally distributed, $\bar{\epsilon} = 0$ and the variance alone is sufficient to completely describe the distribution. When the distribution of the observations o_i about the mean $\bar{o}$ is not normal, additional parameters may be necessary to completely characterize the distribution, such as third or higher moments about the mean.

A normal distribution of errors is often assumed, a priori. A distribution that is not normal, for example a highly skewed distribution, may be indicative of systematic rather than random error. In any case, it is always a good idea to examine the distribution of residuals $o_i - \bar{o}$.

Fig. 1 shows the distribution of atomic deviations from the mean structures for an ensemble of 42 structures obtained by Driscoll et al. for the protein BDS-I using a hybrid distance geometry-dynamical simulated annealing (DG-DSA) protocol.[1] The shape of this distribution is clearly not Gaussian, and is due in part to the least squares criterion used to (rigidly) rotate two structures into registration before computing the rms deviation.

Propagation of Error

When the desired result R of an experiment is derived from the observations o_i,

$$R = f(o_i) \tag{2}$$

the error in the result δR due to infinitesimal errors in the observations δo_i is

$$\delta R = \sum_{i=1}^{N} \frac{\partial f}{\partial o_i} \delta o_i \tag{3}$$

which is simply a first order Taylor expansion. For finite errors Δo_i, small enough that the partial derivatives do not vary significantly over the range, the finite error ΔR is

$$\Delta R = \sum_{i=1}^{N} \frac{\partial f}{\partial o_i} \Delta o_i. \tag{4}$$

The average of the squared error (the variance) is

$$S^2(R) = \sum_{i=1}^{N} \left(\frac{\partial f}{\partial o_i}\right)^2 S^2(o_i), \tag{5}$$

assuming that the errors Δo_i are independent, so that cross terms of the type $\sum \Delta o_j \Delta o_k$ vanish for $j \neq k$. A particularly readable discussion of error propagation can be found in ref. 2. To be able to conduct error analysis by propagation of error, we need to know not only about the errors in the observations, but we also need to know about the function f.

SALIENT FEATURES OF NMR STRUCTURE DETERMINATION

The structure $\vec{r}$ of a protein determined from NMR data is a derived quantity that depends not only on the experimental NMR observations, but also on other prior knowledge about the structure, such as the sequence and the nature of chemically plausible structures. To call the structure a *function* (in the mathematical sense) of the observations is a bit of an oversimplification: there may be a function f, which, applied to the observations, yields the structure, but we don't know what it is or if it exists. Instead, we try to solve the inverse problem by generating trial solutions and testing to see if they are consistent with the observations.

There are a number of different methods for solving this inverse problem, including distance geometry[3] (DG), restrained molecular dynamics[4] (RMD), heuristic methods,[5] and hybrid methods. While the underlying principles vary considerably, there are some similarities between the methods. Most methods can be described as consisting of two phases: *discovery*, in which approximate structures satisfying the criteria of plausibility and consistency with the experimental data are determined, and *refinement*, in which relatively small perturbations are introduced to improve the plausibility and consistency.

A constrained optimization paradigm serves to illustrate the distinction between discovery and refinement. Suppose we have some measure $E(\vec{r})$ that gives the plausability of a structure $\vec{r}$, and a measure $C(\vec{r})$ of how well a structure agrees with the experimental observations. The smaller the values of E and C, the more plausible the structure and the better the agreement with experiment, respectively. The constrained optimization seeks the most plausible structure that is consistent with the data,

$$\min\ E(\vec{r})\ s.t.\ C(\vec{r}) \le C_o \tag{6}$$

where C_o represents the limit corresponding to experimental uncertainty. This can be converted to an unconstrained optimization by introducing a Lagrange multiplier λ to create the objective function

$$O(\vec{r}) = E(\vec{r}) + \lambda C(\vec{r}) \tag{7}$$

where the value of λ is chosen so that $C(\vec{r}) \le C_o$. The constrained optimization problem is thus equivalent to finding the minimum of the function $O(\vec{r})$. Unfortunately, the surface defined by $O(\vec{r})$ is quite complicated, with many local minima. While descent methods (e.g., steepest descent, conjugate gradients) will converge to a local minimum, there is no formal test for deciding whether a local minimum is the global minimum. The only way of knowing for sure is to search over all the degrees of freedom, which is computationally impractical. Hence the importance of the discovery phase: to locate new "basins of attraction." The refinement phase then locates the local minimum of the basin (Fig. 2). A good discovery algorithm is one that is biased toward the best basins of attraction.

The local minima of $O(\vec{r})$ satisfy the condition

$$\frac{\partial O(\vec{r})}{\partial r_i} = 0 \text{ for all } i. \tag{8}$$

Two distinct types of solutions exist. There are those for which

$$\frac{\partial E}{\partial r} = -\lambda \frac{\partial C}{\partial r} \ne 0 \tag{9}$$

and those for which

$$\frac{\partial E}{\partial r} = \frac{\partial C}{\partial r} = 0 \quad . \tag{10}$$

256

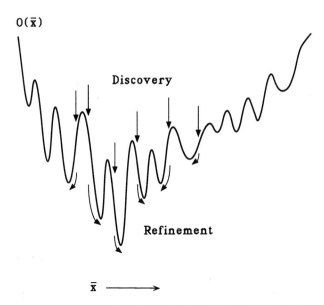

O(x̄)

Discovery

Refinement

x̄ ⟶

Figure 2. Searching the objective function $O(\bar{x}) = E(\bar{x}) + \lambda C(\bar{x})$. The discovery phase locates basins of attraction, and the refinement phase locates the (local) minimum.

I will refer to these as non-trivial and trivial solutions, respectively. Trivial solutions occur whenever there is a local minimum of E that coincides with a local minimum of C. Non-trivial solutions occur when the gradients of E and C point in opposite directions, and the ratio of their magnitudes

$$\frac{\bar{\nabla} E}{\bar{\nabla} C} = \lambda \quad . \tag{11}$$

The most widely used measure $E(\vec{r})$ of the chemical plausibility of a structure is the potential energy, based on an empirical force field such as CHARMM[6], AMBER[7], GROMOS[8], or ECEPP[9]. The measure of consistency with the experimental data, the "restraint" in methods such as restrained molecular dynamics, is usually based not on the directly observed NOE intensities, but on derived quantities, the approximate distance between proton pairs. More recently, analytic expressions for the gradient of the (multispin) nuclear Overhauser effect with respect to nuclear coordinates have allowed refinement using the NOE intensities themselves as restraints.[10]

The mathematical form of the restraint term can affect the type of solutions found for the constrained optimization problem. One widely used form utilizes upper and lower distance bounds, with the constraint statistic (the pseudo potential energy) zero when the restrained distance falls between these bounds and increasing quadratically outside the bounds (Fig. 2). The derivative of the restraint energy also vanishes when the restrained distance falls between the upper and lower bounds. Consequently any maximum of $E(\vec{r})$ that falls within the hypervolume defined by

$$d^i_{lower} \leq d^i \leq d^i_{upper} \tag{12}$$

for all restrained distances d^i is a trivial solution of the constrained problem.

When E is the potential energy, the existence of trivial solutions to the constrained optimization means that there is a local minimum of the potential energy surface that also satisfies the experimental constraints. It is by no means obvious, however, that such solutions should exist. Another way of describing a trivial solution is that it represents a single chemically plausible configuration of the protein atoms that is consistent with the experimental data. Yet the experimental observations represent averages over an ensemble of molecules, and all of the configurations of the atoms that are thermally accessible (on the time scale of the experiment). The average structure, and structures that give rise to the average NMR parameters, do not necessarily correspond to chemically plausible structures. Nor does the most probable structure (corresponding to the global minimum of the potential energy surface) necessarily give rise to average values of the NMR parameters. Concrete examples of the differences between NMR parameters of average structures and most probable structures have been given elsewhere.[11] For the purpose of the present discussion, we simply note that the existence of a trivial solution to the constrained optimization problem, using experimental restraints based on lower and upper distance bounds, implies that the *range* of distances is large enough to account for the influence of dynamical averaging in addition to other sources of experimental uncertainty.

Since the feasible hypervolume (Eq. 12) increases as the differences between the upper and lower bounds increase, and since

$$\frac{\partial C}{\partial \vec{r}} = 0 \tag{13}$$

within these boundaries, the likelihood that a trivial solution can be found increases as the range between the upper and lower bounds is increased. The likelihood of obtaining trivial solutions when other forms for the restraint potential are used may be lower than when using a flat bottom potential. For example, Prestegard et al.[12] have employed a restraint potential of the form

$$E_{NOE} = k \left[\left(\frac{1}{r_{ij}^3} - \frac{1}{r_{o_{ij}}^3} \right)^2 - \frac{1}{r_{o_{ij}}^6} \right] \tag{14}$$

(Fig. 3b) which was chosen to mimic more closely the actual uncertainty in NOE-derived distance restraints. The likelihood of obtaining trivial solutions for this restraint potential is lower than for a square bottom potential, since

$$\frac{\partial C}{\partial \vec{r}} = 0 \qquad (15)$$

only when

$$d^i = d^i_{exp} \text{ for all } i \qquad . \qquad (16)$$

Regardless of whether or not trivial solutions can be found, it is important that the potential energy of the constrained structure be low enough for it to be considered chemically plausible.

AN EXAMPLE: ESTIMATES OF THE PRECISION OF THE NMR DERIVED SOLUTION STRUCTURE OF BDS-I

A widely-used measure of the precision of NMR-based protein structures is the rms deviation between a family of structures and the mean structure. For the family of 42 structures obtained for BDS-I by Driscoll et al. using a hybrid distance geometry-dynamical simulated annealing protocol,[1] the RMSD's are 0.6 Å for the backbone atoms and 0.9 Å for all atoms.

What do these numbers mean? The average RMSD is not an infallible measure of the width of the distribution of the deviations, for a number of reasons, one of which is apparent from the shape of the distribution in Fig. 1. Only 52% of the deviations are below 0.9 Å. The limit below which 95% of the values fall is 3.8 Å. The shape of this distribution is due in part to the least squares criterion used to (rigidly) rotate two structures into registration before computing the rms deviation. Havel has recently discussed some further aspects of the RMSD and its limitations as an indicator of precision.[13]

The goal of error analysis, ultimately, is to determine the range of chemically plausible structures that are consistent with the experimental data. Protocols which conclude by minimizing an objective function $O = E + \lambda C$, including the protocol used by Driscoll et al., result in an ensemble of structures that are local minima of O. This will represent the full range of structures only if the local minima span the entire hypervolume of chemically plausible and experimentally consistent structures. The influence of thermally accessible conformations that are not local minima of O, and the possibility that the local minima are not distributed out to the edges of the hypervolume (the surfaces corresponding the chemical plausibility, $E \leq E_o$, and experimental consistency, $C \leq C_o$), even if they could all be sampled, means that the distribution of structures probably overestimates the true precision of the structure.

How then should we assess precision? The point is not that there may be other simple measures of precision that are better than the RMSD, but that unless the distribution of deviation is particularly well-behaved (for example having zero mean and symmetry about zero), then no *single* number will accurately describe the distribution. We must avoid the temptation to treat the RMSD as if it were a standard deviation for a normal (Gaussian) distribution. Instead of relying on a single number to characterize the precision, we should look at the actual distribution of structures. More importantly, we should seek methods that will faithfully propagate the error: the numbers used to characterize a distribution of structures matters little if the means used to generate that distribution is systematically biased.

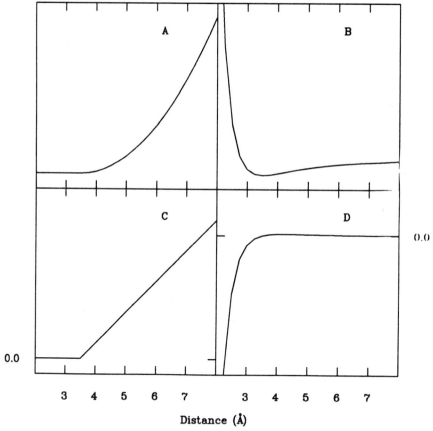

Figure 3. Distance restraint potentials and their derivatives.
A) $E_{NOE} = K_{NOE}(r - r_{upper})^2$, $r_{upper} = 3.5$ Å.
B) E_{NOE} of Eq. 9, $r_o = 3.5$ Å.
C) $\frac{dE}{dr}$ corresponding to A).
D) $\frac{dE}{dr}$ corresponding to B). Note that $\frac{d\epsilon}{dr} = 0$ for $r_{lower} \leq r \leq r_{upper}$ in C) and only at $r = 3.5$ Å in D.

A narrowly distributed family of structures that are chemically feasible and consistent with the experimental data does not necessarily reflect high precision. A narrow distribution may also result if the algorithm used to generate structures is significantly biased. It may, for example, explore only one basin of attraction. Until it becomes feasible to systematically explore all of conformation space, we are faced with the reality that our confidence in the *uniqueness* of any structure depends on the number of sampled conformations and the belief that our sampling is not biased. Sampling bias in distance geometry-based methods has been explored recently,[13,14] using a number of diagnostics for sampling. Among these are the end-to-end distance of the polypeptide chain and the distribution of ϕ and ψ angles. Another diagnostic is the range of distances between atoms for which there are experimental distance restraints. The upper and lower restraint distances usually reflect the experimental uncertainty in the restrained distance: any value between the upper and lower is deemed consistent with the experimental observations. For any family of feasible and consistent structures, it is unlikely that the distribution of restrained distances will span the entire range between the upper and lower limits, for at least two reasons. One is that the repulsive interactions between atoms will preclude some distance values, and another is that correlations between restrained distances due to the covalent structure of the molecule will render some distance values for one atom pair incommensurate with some distance values for another atom pair.

In general, for two populations of structures that are consistent with the experimental constraints, the one with the larger range(s) is the one with the least sampling bias. Furthermore, the ultimate goal of error analysis is to provide a lower bound to the precision, or an upper bound to the uncertainty of the structure. Fig. 4 illustrates, via histograms, the range Δr_i for the family of 42 structures obtained by Driscoll et al.[1] Panels B and C depict histograms of Δr_i obtained from restrained molecular dynamics trajectories at 300° K and 600° K, respectively, using the Dreiding I potential energy function implemented in NMRgraf[15] together with distance restraints identical to those described by Driscoll et al. The 1-4 van der Waals interactions were empirically scaled to give total van der Waals energies comparable to those obtained using the CHARMM force field. Each trajectory was computed, in vacuo, for 50 psec: 25 for equilibration and 25 for analysis, starting from the average of the 42 structures obtained by Driscoll et al. Molecular dynamics provides a means of sampling individual basins of attraction, and basins that are separated by barriers low enough to be surmounted by thermal energy, but is not terribly efficient at sampling regions that are only (thermally) accessible via highly correlated fluctuations. Examples in-

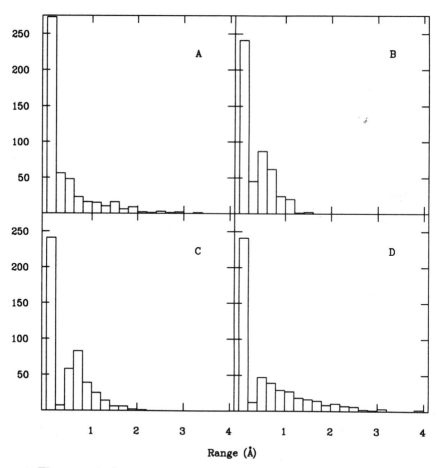

Figure 4. The range Δr between the shortest and longest distance values for which there are experimental restraints (BDS-I). A) The 42 hybrid DG-DSA ensemble obtained by Driscoll et al. B) restrained MD, 300° K, C) restrained MD, 600° K D) restrained MD, 300° K, no non-bonded terms.

clude ring flips (tyrosine and phenylalanine) and "crankshaft motions" of side chains. The barriers to these types of motions arise from the impenetrability of matter and the compact nature of folded proteins: condensed matter physicists use the word "frustration". This "frustration" can be relieved somewhat by turning off the non-bonded terms in the force field. In practice, it is also necessary to "turn off" the electrostatic and hydrogen bond terms (if any) in the force field so that the protein will not collapse on itself. The trajectories that result are similar in spirit to "distance bounds driven dynamics", described by Scheek et al.[4] A histogram of the ranges Δr_i for such a trajectory are shown in Fig. 4D. While many of the structures in such a trajectory will not be chemically plausible (due to atomic

262

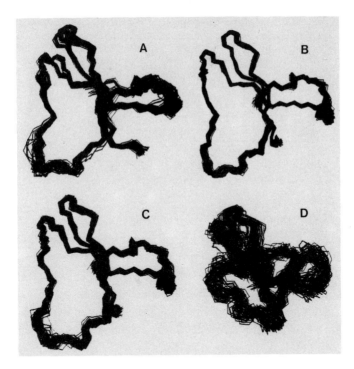

Figure 5. Superimposed backbones of the ensembles of BDS-I structures described in Fig. 4.

overlap), the ensemble of structures can more confidently be considered as an estimate of the upper bound on the uncertainty of the structure.

Comparison of panels A through D of Fig. 4 at first glance suggests that the sampling achieved by the hybrid procedure of Driscoll et al. is more extensive than that of simple restrained molecular dynamics at 300° K or 600° K, and is comparable to that of restrained molecular dynamics without non-bonded interactions at a "temperature" of 300° K. However, it turns out that for a majority of the restrained distances, the range Δr_i for the hybrid DG-DSA ensemble is less than the range for all of the dynamics trajectories: 55%, 60%, and 82% for the 300°, 600°, and 300°-non "non-bond" trajectories, respectively. A somewhat different picture emerges when comparing the RMSD values for each ensemble or the average distance variation (for restrained distances); the values for the "non-nonbonded" ensemble are significantly larger than the hybrid DG-DSA ensemble. The superimposed backbone traces of the four ensembles illustrate the statistics in Table 1 somewhat more dramatically (Fig. 5).

Further insight into the sampling properties of the ensembles comes from examining individual distances. Figs. 6-8 give some illustrative examples.

Table 1. Statistical comparison of four BDS-I ensembles.

	rms deviation		rms restraint	distance
	backbond	overall	violation	variation
DG-DSA	0.67	1.13	0.09	0.20
RMD, 300°	0.35	0.44	0.05	0.14
RMD, 600°	0.49	0.62	0.06	0.19
RMD, 300°, no-NB	1.07	1.33	0.03	0.27

Fig. 6 mirrors the general trend observed for Δr_i. There are, however, numerous exceptions to this trend. Fig. 7 illustrates a case where the Δr is lower for the hybrid DG-DSA ensemble than that for the restrained MD ensemble, and much lower than that for the restrained MD trajectory without non-bonded interactions. Fig. 8 illustrates a peculiar bimodal distribution for the hybrid ensemble; one might conclude that repulsive interactions or incommensurate restraints preclude the region near 3 Å. Yet the restrained MD trajectory starting from the average of the hybrid structures samples this region extensively. Distributions such as those in Figs. 7 and 8 suggest the possibility of algorithmic bias, when compared with the other distributions. Analysis of the distribution of Δr conducted as an integral part of the structure determination protocol, can help to ensure that the observed distributions accurately reflect the precision of the structure.

Energetics

Failure to identify "trivial" solutions to the constrained optimization problem, that is, where the gradients of the empirical potential energy and the experimental restraint both vanish, may simply be due to failure of the algorithm to locate such solutions, or it may be because trivial solutions don't exist. Trivial solutions might not exist for a number of reasons including

i) there are incorrect assignments or erroneous distance restraints;

ii) there is motional averaging or structural heterogeneity in solution, so that the assumption that a single chemically plausible structure can account for the experimental observations is invalid; and

iii) the empirical potential energy function is in error.

Identification of a trivial solution does not, however, guarantee that a chemically plausible structure has been found. The total potential energy must also be considered. Unfortunately, the energies obtained using different

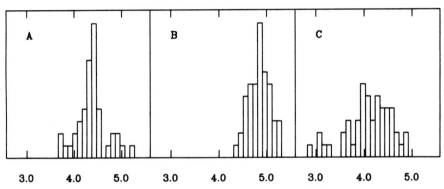

Figure 6. Histograms of the distance between the ϵ protons of Tyr 41 (averaged over H^ϵ, and H^{ϵ_2}) and H^N of Gly 13.
A) Hybrid DG-DSA.
B) restrained MD, 600° K.
C) restrained MD, 300° K, no non-bonded terms.

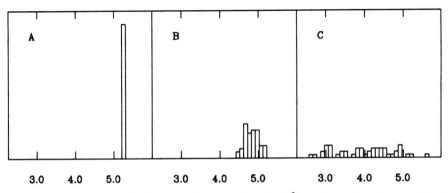

Figure 7. Histograms of the distance between Cys 40 H^{β_1} and the β protons of Cys 22 (averaged over H^{β_1} and H^{β_2}). The histograms correspond to the same ensembles as in Fig. 6.

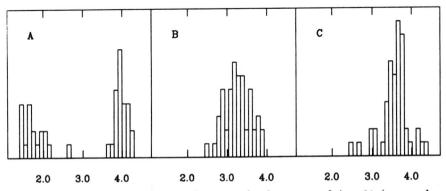

Figure 8. Histogram of the distance between the δ protons of Asn 31 (averaged over H^{δ_1} and H^{δ_2}) and Asn 31 H^N.

force fields vary sufficiently that it is difficult to say precisely what constitutes a chemically plausible potential energy. Experience with the particular force field (applied to native protein structures) must be the guide.

CONCLUDING REMARKS

Classical error propagation can be applied to the determination of biomolecular structure by NMR. The multiple-minimum problem means that analytic expressions for error propagation will only represent the error associated with the local minimum of a basin of attraction. The total error will include contributions from other basins that are thermally accessible. The practice of generating families of structures — either by DG or hybrid DG-DSA protocols — provides one means of determining the error due to different basins, although when the final step consists of minimization, the distribution of the structures should not be viewed as the full range of structures that are consistent with the experimental data. Constant temperature restrained MD and restrained MD without non-bonded terms provide additional means of propagating error. These protocols differ in their sampling properties on the distance constrained potential energy surface. Short of performing a complete conformational search, comparison of the distributions obtained by more than one sampling protocol can provide a more reliable estimate of the uncertainty in a structure.

A reliable method for numerically propagating error is just one of the requirements for realistic error analysis. It is also important to have well-defined criteria for the chemical plausibility and for consistency of candidate structures with experimental results. Progress toward an "NMR R-factor"[15,16] addresses the need for a measure of the consistency. Unfortunately, the prospect for a universal criterion for chemical plausibility is hampered by the variety of force fields used by different laboratories.

Finally, reliable estimates of the experimental uncertainty and of the derived distances are required (if direct refinement against the NOE intensities is not used). Multispin treatments of NOE's such as MARDIGRAS[16] and IRMA[17] have recently improved the accuracy with which distances can be derived from NOE's, and consequently improved the reliability of error propagation from experimental uncertainty to derived distances. This capability can and should be used to supplant the practice of *ad hoc* overestimation of the uncertainties in distances, if realistic error analysis is to be achieved. While certain aspects of experimental uncertainty have been examined,[18,19] a classical analysis of the statistical uncertainty in NOE measurements through repeated experiments has, apparently, not yet been performed.

ACKNOWLEDGEMENTS

Marius Clore and Angela Gronenborn graciously provided the ensemble of structures they obtained for BDS-I. I am grateful to Lesley Pew, MaryAnn Nilson, and Jay Scarpetti for assistance in preparing the manuscript. I am indebted to Barry Olafson, Alan Stern, and Abraham Szöke for useful discussions, Molecular Simulations Inc., for a software grant of Biograph, and the Rowland Institute for Science for unwavering support.

REFERENCES

1. P. Driscoll, A. M. Gronenborn, L. Beress, and G. M. Clore, *Biochemistry* **28**, 2188 (1989).
2. D. P. Shoemaker, C. W. Garland, and J. I. Steinfeld, "Experiments in Physical Chemistry," McGraw-Hill, New York, 1974.
3. I. D. Kuntz, J. F. Thomason, and C. M. Oshiro, *Methods Enzym.* **177**, 159 (1989).
4. R. M. Scheek, W. F. van Gunsteren, and R. Kaptein, *Methods Enzym.* **177**, 204 (1989).
5. R. B. Altman and O. Jardetzky, *Methods Enzym.* **177**, 218 (1989).
6. B. R. Brooks, R. E. Bruccoleri, B. D. Olafson, D. J. States, S. Swaminathan, and M. Karplus, *J. Comp. Chem.* **4**, 187 (983).
7. S. J. Weiner, P. A. Kollman, D. T. Nguyen, and D. A. Case, *J. Comp.* **1**, 230 (1986).
8. GROMOS (by W. F. Van Gunsteren and H. J. C. Berendsen) is available from BIOMOS b.v., Nijenborgh 16, 9747 AG Gronengen, The Netherlands.
9. M. J. Sippl, G. Némethy, and H. A. Scheraga, *J. Phys. Chem.* **88**, 6231 (1984).
10. P. Yip and D. A. Case, *J. Magn. Reson.* **83**, 643 (1989).
11. J. C. Hoch, C. M. Dobson, and M. Karplus, *Biochemistry* **21**, 1118 (1982).
12. T. A. Holak, J. H. Prestegard, and J. D. Forman, *Biochemistry* **26**, 4652 (1987).
13. T. F. Havel, *Biopolymers* **29**, 1565 (1990).
14. W. J. Metzler, D. R. Hare, and A. Pardi, *Biochemistry* **28**, 7045 (1989).
15. Biograf Users Manual, Biodesign, Inc., Pasadena, CA (19??).
16. C. Gonzalez, J. A. C. Rullmann, M. J. J. Bonvin, R. Boelens, and R. Kaptein, *J. Magn. Reson.* **91**, 659 (1991).
17. P. D. Thomas, V. J. Basus, and T. L. James, *Proc. Natl. Acad. Sci. USA* **88**, 1237 (1991).
18. R. Boelens, T. M. G. Konig, and R. Kaptein, *J. Mol. Struct.* **173**, 299 (1988).
19. T. A. Holak, J. N. Scarsdale, and J. H. Prestegard, *J. Magn. Reson.* **74**, 546 (1987).

STRUCTURAL INTERPRETATION OF NMR DATA IN THE PRESENCE OF MOTION

J. H. Prestegard and Yangmee Kim

Department of Chemistry
Yale University
New Haven, CT, 06511

ABSTRACT

Proteins exhibit a variety of internal motions that are, in many cases, necessary for function. Yet, most approaches used to convert NMR data into a three dimensional structure assume a rigid model in translating cross-relaxation data into proton-proton distance constraints. For some types of motion this can lead to substantial errors in the final structure determined. Means of recognizing the presence of significant internal motions and means of minimizing the impact on the structures determined are discussed. Data on a small protein important in fatty acid biosynthesis, acyl carrier protein from spinach, are used to illustrate the procedures.

INTRODUCTION

Most approaches to the determination of the structure of proteins from NMR data rely on the assumption that 2D-NOE cross-peak intensities can be interpreted on the basis of a structural model which represents a protein as a rigid sphere tumbling in a viscous medium[1]. For cases in which data are acquired with a very short cross-relaxation time, this leads to an ability to convert cross-peak intensities to distance estimates using simple proportionality factors. When a more quantitative evaluation is desired these factors are taken to be the inverse sixth root of intensity ratios between cross peaks arising from a proton pair at an unknown distance and a cross peak arising from a proton pair at a known distance. The above assumption also leads to an ability to define, in a straightforward way, elements used in a complete relaxation matrix treatment of NMR data[2]. Such analyses allow one to use longer mixing time data which is usually of a considerably higher signal to noise ratio. Yet, intuition would suggest that the assumption of a rigid model is incorrect, and that one should have to consider the consequences

Computational Aspects of the Study of Biological Macromolecules by Nuclear Magnetic Resonance Spectroscopy, Edited by J.C. Hoch *et al.,* Plenum Press, New York, 1991

of violation of this assumption in selection of structure determination protocols or methods for evaluation of the structures produced. Here, we hope to demonstrate the violation of the rigid molecule assumption in one class of small protein, acyl carrier proteins (ACPs), and suggest ways in which the undesirable effects of internal protein motions on structure determination might be minimized.

Internal motions in proteins can occur with a variety of time scales and with a variety of different amplitudes. Surprisingly, the more rapid and generally widespread oscillatory motions of short segments within proteins cause relatively small departures from the NOEs that would be predicted on the basis of a rigid model. This can be understood on the basis of a formalism introduced by Lipari and Szabo[3]. This formalism uses two correlation times, one characterizing internal motion,τ_i, and one characterizing overall motion, τ_o, in addition to a generalized order parameter, S, to describe relaxation effects for a dipolar coupled pair of protons. The cross-relaxation rate, σ, important for the establishment of NOE cross peaks between a pair of coupled spins in a large molecule can be taken to be proportional to the zero quantum spectral density, $J(0)$. $J(0)$ is in turn given by the following expression:

$$J(0) = 2/5 \left(S^2 \tau_o + (1 - S^2)\tau_i \right). \tag{1}$$

In the limit where internal motions are very fast compared to overall tumbling ($\tau_i < 10^{-9}s$) and are of moderate amplitude ($S^2 > 0$), expression (1) reduces to an expression showing a simple proportionality to the overall correlation time with the scaling constant being the square of the generalized order parameter, S^2. Thus, all the effects of internal motion are contained in the order parameter.

In practical terms, S^2 can be evaluated as an average over the interproton distance, r, and second order spherical harmonics, $C_{2m}(\theta, \phi)$. The average is taken for a time period long compared to τ_i but short compared to τ_o.

$$S_2 = \sum_{m=-2}^{2} \left| \langle C_{2m}(\theta, \phi)/r^3 \rangle \right|^2 . \tag{2}$$

From Eq. 2 it is clear that both distance variations and angular variations will contribute to the size of the scaling factor, S^2. What is less clear is that these two types of variation compensate one another. Random variations in the internuclear distance, r, will lead to an increase in S^2 over what S^2 would have been, had r been fixed at the mean internuclear distance. Random variations in θ and ϕ will lead to a decrease in S^2 over what S^2 would have been, had no internal motions been present. As shown by LeMaster et al.[4], if the motions of the two protons are uncorrelated and sample a centrosymmetric distribution about their mean position, the compensation is exact. Rapid random internal motions, therefore, result in very small departures of NOEs from those expected based on a pair of protons rigidly fixed at their mean positions, and, as in a rigid model, cross-relaxation rates can be taken to be simply proportional to $1/r^6$.

Unfortunately, this fortuitous situation no longer exists when internal motions become slow compared to the overall tumbling time for a protein ($\tau_i > 10^{-8}$ s). As long as the timescale of motion is short compared to the mixing time used for the NOE experiment ($\tau_i < 10^{-1}$ s), the effective relaxation rate for a pair of interacting spins will appear to be a simple average of the relaxation rates for a series of models representing points sampled along the trajectory of internal motions.

$$\sigma = \sum_i p_i \sigma_i. \tag{3}$$

Here, p_i is the population of a discreet model, and the individual σ_i are proportional to $1/r^6$.

$1/r^6$ is a very steep function of distance and can easily lead to the overall cross-relaxation rate being dominated by distances of closest approach. For example, a simple model in which two equally populated states exist, one with a proton pair separation of 3 Å, and one with a proton pair separation of 6 Å, leads to an NOE only a few percent greater than half the intensity expected for a model with full occupation of the 3 Å site. If the NOE were mistakenly interpreted on the basis of a single rigid model, the distance would be scaled by $(1/2)^{-1/6}$, and an interproton distance only 10% greater than 3 Å would be predicted. For a single pair of protons this is not a problem; one of the two states is just better modeled than the other. But, close approaches of one proton to a series of protons constrained to be distant from one another can easily give rise to a set of geometrically inconsistent constraints. For example, the first proton cannot be simultaneously be 3 Å from two other protons constrained to be 9 Å apart.

EVIDENCE FOR MOTION IN PROTEINS

Do motions on time scales which can cause this problem exist in real protein systems? And, can we easily recognize cases where they do exist? We will examine these questions using data on acyl-carrier-protein-I from spinach in its reduced form (ACP-I). This is a protein of modest size (82 amino acids) responsible for carrying a growing fatty acid chain during its synthesis by the fatty acid synthetase system[5]. We have previously determined structures for the homologous protein from *Escherichia coli* but have been somewhat dissatisfied with the quality of fit of a rigid model to the available NOE data[6]. The spinach protein appears to be a better test case in that direct evidence for the existence of slowly interconverting conformational states is more readily obtained[7,8].

One source of evidence can be NOE spectra. While normally cross peaks in these spectra indicate cross relaxation between proton pairs in a single structure, they can also indicate magnetization transfer between conformational states having discrete resonances for a given proton in the two states. In cases where such transfer can be demonstrated the intensity of these cross peaks can be used to evaluate an exchange time.

Below we show a segment of a one dimensional proton spectrum and the corresponding two dimensional NOE spectrum showing connectivities among aromatic protons in spinach ACP-I. This protein has just two aromatic residues, Phe 31 and Phe 52. The upfield resonances in the one dimensional spectrum, C1, C2, and D, are easily assigned to Phe 31, and the downfield resonances, A and B2 are easily assigned to Phe 52 using conventional 2D sequential assignment strategies (Kim et al., unpublished). B1 has a shape and intensity that would be consistent with assignment to the para proton of Phe 52. However, RELAY spectra do not show a connectivity to the ortho protons, peak A. Moreover, if B1 were the para proton, peak B2 would have too much intensity for the remaining meta protons.

On examining Fig. 1, several expected NOE cross peaks are seen. A strong connection of peak D to peak C2 is, for example expected because of the spatial proximity of the para and meta protons on the aromatic ring of Phe 31. The strong connection of peak A and B2 is also expected based on the spatial proximity of ortho and meta protons in Phe 52. However, if B1 were the para proton of Phe 52, the strong connection to D could only be explained if Phe 31 and Phe 52 were spatially proximate. Given the other anomalies associated with assigning B1 to the para proton of Phe 52, a far more likely explanation is that this peak arises from magnetization transfer by exchange between conformational states. In this case the peak would be a resonance from the meta protons of Phe 31 in a minor conformational form of spinach ACP-I. This explanation has, in fact, been confirmed using rotating frame NOE experiments which can distinguish cross-relaxation peaks and exchange peaks based on their sign[8]. A similar explanation applies to cross peaks between C1, C2, and B1, and peaks under or near B2.

The intensities of the exchange derived cross peaks and their corresponding diagonal peaks as a function of mixing time, τ, can be used to estimate exchange rates in the system. The magnetization associated with the diagonal peaks, M_d, and the cross peaks, M_c, in a system where transfer by both exchange and cross relaxation occur, can be approximated as follows:

$$M_d = M_o \exp\left[-(\rho + k)\tau\right] \tag{4}$$

$$M_c = M_o \exp(-\rho\tau)\left[1 - \exp(-k\tau)\right]. \tag{5}$$

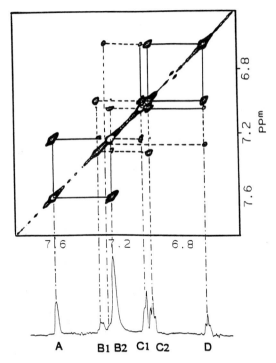

Figure 1. 500 MHz 1H 2D NOESY spectrum and 1D spectrum of the aromatic region of a 6 mM spinach ACP sample at pH 6.0, 25° C, in 50 mM D_2O phosphate buffer. The mixing time was 180 ms and acquisition required 43 hours.

In the above equations it has been assumed that the back exchange rate constant is small compared to the forward exchange rate constant, k, and it has been assumed that the total cross-relaxation rate, ρ, is the same for both states. The symbol M_o represents the total magnetization associated with the originating site.

Fig. 2 shows an idealized plot of intensity versus mixing time for a pair of diagonal and cross peaks under exchange conditions. Clearly, measurement of intensities of diagonal peaks and exchange derived cross peaks for resonances such as that labeled D in Fig. 1 can provide the data needed to evaluate the rate constants in Eqs. (4) and (5). Analysis of experimental points for a sample in the presence of 12 mM Mg^{2+} at 25° C and pH 6.0 was carried out using data at mixing times of 80 and 180 ms. The Mg^{2+} reduces the population of the minor component to approximately 4% and insures

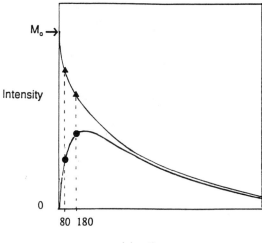

Figure 2. Idealized intensities of diagonal and exchange peaks versus mixing time. The triangles and squares represent experimental points for the diagonal and exchange peaks respectively at 80 and 180 ms mixing times.

a small back exchange rate constant. Under these conditions the forward exchange rate was found to be 5 s^{-1} and the backward rate was found to be 0.2 s^{-1}.

The above rates clearly are approaching a range where they could be effective in averaging cross-relaxation effects seen in the two conformers. Moreover, it is easy to imagine similar conversions occurring at much more rapid rates. Relatively small changes in molecular energies (1.5 kcal) can change rate constants by more than an order of magnitude if they serve to reduce an activation energy barrier.

These more rapid rates can compound the problem with NOE interpretation because they are frequently rapid enough to produce a single resonance at the average of the chemical shifts for a given proton in the two states. Under these conditions there will be no extra resonances or extra cross peaks in NOESY data to warn of potential motional effects. Thus, one should always be sensitive to the possibility of motions on time scales which can significantly contribute to the observed NOE pattern.

Recognition of the kinds of incompatibilities in NOE distance constraints described above can be non-trivial. NOE distance constraints are almost always too few in number to independently determine the spatial coordinates of all atoms in a protein. The deficiency in number of constraints is usually overcome by incorporating chemical information about bonding and non-bonding interactions among atoms. Structures determined by typical NMR structure determination protocols are, therefore, a combined result of our knowledge of preferred amino acid geometries and our knowledge of experimental distance constraints. If experimental constraints are incorporated as heavily weighted harmonic or square-well potentials, and only violation of these constraints are used to assess the quality of a structure, one can easily suppress inconsistencies in the data. A set of solutions converging to a well defined single structure consistent with NMR constraints can be found. Only if one examines the molecular energy associated with the structures does one find that satisfaction of distance constraints has been achieved at the expense of distorted bonding geometry.

An alternate approach that can highlight possible inconsistencies in structural data without resort to quantitative evaluation of molecular energies is the use of highly asymmetric pseudopotentials to represent distance constraints[6]. This is particularly effective if the potential is designed to impart a small penalty for violation of distance constraints when distances are too long. Since motional processes on an intermediate time scale tend to produce extra NOEs with apparent distances that are too short for a given conformational state, an ability to violate these constraints by pulling a proton pair apart is precisely the process which allows accurate modeling of one or another of the equilibrating structures. The family of structures produced by a search constrained with such asymmetric potential functions will span all conformational sets, and in each set constraints more appropriate to another set will be violated. This pattern of systematic violation is easily recognized. We have applied a procedure incorporating an asymmetric potential to 2D-NOE data on acyl carrier protein from spinach (Kim et al, unpublished). Data were collected on a 6 mM sample at 25° C, pH 6.0, containing two molar equivalents of Mg^{2+}. These conditions produce spectra that appear to give a single set of resonances in the aromatic region of the spectrum. However, as detailed above, we know the potential for some averaging of NOE effects still exists. The total number of assignable NOE constraints (about 200) falls short of the number normally desired for a complete structure determination on a molecule of this size. However, the observed constraints are concentrated in the last two thirds of the protein with the first third appearing to be largely unstructured. Also, there are regions in the last two thirds that appear particularly well constrained.

The procedure normally employs the first two stages of a distance geometry program to produce crude structures that are useful starting points for refinement under the influence of the potentials described above[6]. The structures are refined with constraint potentials added to the normal molecular force field of the model building program AMBER[9]. Results showing distance constraint violations involving Ile 67 are presented in Table 1 for the 9 structures which showed an overall rms distance violation of less than 1 Å. The nomenclature uses one letter codes for the amino acid type, followed by a residue number, followed by a code for the constrained atom. It

Table 1. NOE Constraint Violations in Structural Models for Spinach ACP[a]

MODEL	ATOM 1	ATOM 2	DISTANCE
MMD S1	T66HN	I67CG2	2.8
MMD S2	T66HN	I67CG2	2.1
MMD S3	T66HN	I67CG2	2.3
	F52CG	I67CG2	6.9
MMD S4	I67HN	I64CB	3.0
	F52CG	I67CG2	6.6
MMD S5	F52CG	I67CG2	8.4
MMD S10	I67HN	I64CB	2.4
	F52CG	I67CG2	3.9
MMD S11	I67HN	I64CB	3.4
	F52CG	I67CG2	7.4
MMD S12	I67HN	I64CB	2.7
	F52CG	I67CG2	3.3
MMD S13	T66HN	I67CG2	3.7
	F52CG	I67CG2	3.4

[a]Only constraint violatons involving I67 and on those over 2 Å are listed. Constraint violations are given in Å.

is clear that Ile 67 is constrained by contacts with Phe 52, Ile 64, and Thr 66, among other residues. It, however, appears that at most two of these contacts can be satisfied within 2 Å for any one of the structures. This is precisely the pattern which we might expect for a protein showing the effects of motional averaging.

METHODS FOR DEALING WITH MOTION

Given that there appear to be cases of motional averaging among discrete conformational states in proteins, it is useful to formulate a structure determination approach which can explicitly recognize these processes. One such approach begins by representing constraints in the refinement procedure as classical error functions. For this purpose we can define a pseudoenergy term, E_{NOE}, as a weighted square deviation between measured values of cross-relaxation rates, $\sigma - abo$, and values calculated on the basis of distances in an appropriate structural model, σ_{ab}.

$$E_{NOE} = W \left(\sqrt{\sigma_{ab}} - \sqrt{\sigma_{abo}} \right)^2 . \qquad (6)$$

The use of a square root of the observable in this error function is not strictly justified, but it makes the potential a little more gradual and may improve convergence properties. W is a weighting function chosen to make violation of a 3 Å distance constraint approximately equal to 30 kcal. When

276

we are dealing with a rigid structure model, σ_{ab} is simply proportional to $1/r^6$, and the function has the useful symmetry properties discussed above. When we are dealing with multiple conformations, σ_{ab} can be expressed as an average of cross-relaxation rates in discrete states as follows:

$$\sigma_{ab} = \sum_i p_i \sigma_{abi}. \tag{7}$$

Constraint functions of this form have been used in a two state model for the structure of acyl carrier protein from *Escherichia coli*[10]. While computationally demanding because of the necessity of modeling coordinates for two complete sets of protein atoms, the approach does provide one possibility for dealing simultaneously with the structural and motional properties of molecules which show evidence of motional averaging.

ACKNOWLEDGEMENT

We are grateful to J. B. Ohlrogge for the preparation of the spinach ACP, and to the NIH for support through grant GM32243

REFERENCES

1. K. Wüthrich, *Meth. Enzymol.* **177**, 125-131 (1989).
2. G. A. Borgias and T. L. James, *J. Magn. Reson.* **79**, 493-512 (1988).
3. G. Lipari and A. Szabo, *J. Am. Chem. Soc.* **104**, 4546-4559 (1982).
4. D. M. LeMaster, L. E. Kay, A. T. Brünger, and J. H. Prestegard, *FEBS Lett.* **236**, 71-76 (1988).
5. J. B. Ohlrogge, "The Biochemistry of Plants 9", P. K. Strumpf, ed., pp. 137-157 (1987).
6. T. A. Holak, S. K. Kearsley, Y. Kim, and J. H. Prestegard, *Biochemistry* **27**, 6135-6142 (1988).
7. Y. Kim, J. B. Ohlrogge, and J. H. Prestegard, *Biochem. Pharm.* **40**, 7-13 (1990).
8. Y. Kim and J. H. Prestegard, *J. Am. Chem. Soc.* **112**, 3707- 3709 (1990).
9. U. C. Singh, P. K. Weiner, D. A. Case, J. Caldwell, and P. A. Kollman, AMBER 3.0 (A program obtained through a licensing agreement with the Regents of the University of California at San Francisco) (1986).
10. Y. Kim and J. H. Prestegard, *Biochemistry* **28**, 8792-8797 (1989).

NEW INTERACTIVE AND AUTOMATIC ALGORITHMS FOR THE ASSIGNMENT OF NMR SPECTRA

Martin Billeter

Institut für Molekularbiologie und Biophysik
ETH Hönggerberg
8093 Zürich, Switzerland

ABSTRACT

The assignment of the resonances in ^{1}H-NMR spectra to individual protons in small proteins relies on three types of magnetization transfers: along three chemical bonds, i.e., between protons attached to neighboring heavy atoms (e.g., COSY), within a spin system (e.g., TOCSY), and through space covering short distances (e.g., NOESY). Several aspects of the assignment problem favor the use of computers: Due to the large number of cross peaks and of initially possible assignments, bookkeeping is best done by computer programs. Furthermore, increased reliability of the results is obtained by systematically checking all assignments for consistency with respect to all other data. Finally, the need of good interactive graphics software is obvious since the basic data is usually provided in a graphical form, i.e., as multi-dimensional spectra.

INTRODUCTION

In our laboratory, several program packages are currently in use or under development. The program EASY (Eccles et al., to be published) allows the display of one or several spectra and their manipulation with a wealth of commands (e.g., real-time zoom and change of plot levels, creation and updating of peak lists, proton lists and spin system descriptions, joint integration of overlapping cross peaks). To these a number of automated routines are added that perform peak picking of anti-phase and in-phase cross peaks, and delineate possible spin systems.

Recent efforts aim at the assignment of a significant percentage of the resonances in a completely automatic way. "Expected" cross peaks (determined from the protein sequence) are related to the complete set of observed cross peaks from COSY, TOCSY and NOESY in a single data structure. This allows at all times the combined use of all three types of spectra, and

Computational Aspects of the Study of Biological Macromolecules by Nuclear Magnetic Resonance Spectroscopy, Edited by J.C. Hoch *et al.,* Plenum Press, New York, 1991

thus the simultaneous delineation of spin systems and their sequential connections. The advantage over the splitting up of the assignment process into several independent routines (e.g., for spin system delineation by COSY, modification of these by TOCSY, and sequential connections by NOESY) is that in the latter procedure it is difficult to correct in subsequent steps the uncertainties, errors and missing data emerging from an earlier step. The algorithm can readily be modified to the use of other types of spectra (e.g., heteronuclear spectra or 3-D spectra), often by simple modification of a program library defining the expected peaks.

The determination of the three-dimensional structure of a protein in solution by ^{1}H-NMR may be considered as the succession of the following steps: First, the protein has to be isolated, purified, and dissolved in solution in a sufficiently high concentration. The next step consists of the actual NMR experiments, i.e., of recording various spectra that show magnetization transfer within individual spin systems (e.g., COSY, TOCSY), or through space (e.g., NOESY). This is then followed by the assignment of resonances to individual protons, and finally by the collection of structural data and the structure calculation. With increasing size of the proteins, the calculational part becomes more CPU intensive. But more important in terms of human effort is the increase in time spent for the sequence-specific assignments of the spectra, because these still involve to a significant amount human interactions.

Efforts to speed up this part of the structure determination generally aim at using computers for bookkeeping and consistency checks of the large lists of cross peaks and of possible assignments, and of introducing computer graphics to the work on multidimensional spectra. Two computer programs are described in the following. The first, short description is about a fully interactive graphics package called EASY (ETH Automated Spectroscopy). The second program, named CASANS (Complete Automatic Sequence-specific Assignments of NMR Spectra), aims at obtaining significant parts of the sequence-specific assignments in a completely automatic way.

THE PROGRAM EASY

The EASY program (Eccles et al., to be published) is implemented on SUN workstations. It offers well over a hundred commands to display NMR spectra, to assign resonances and to extract structure-relevant information. General purpose tools include the display of a spectrum, or part of it, as an intensity plot or using contour lines, zoom and shift commands for the selection of displayed regions, viewing of cross sections and others. The display of intensity plots is particularly useful because it allows real-time adjustments

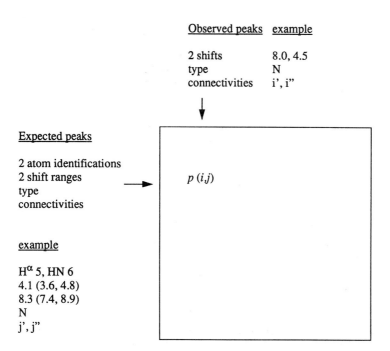

Observed peaks example

2 shifts   8.0, 4.5
type   N
connectivities   i', i"

Expected peaks

2 atom identifications
2 shift ranges   p (i,j)
type
connectivities

example

H$^\alpha$ 5, HN 6
4.1 (3.6, 4.8)
8.3 (7.4, 8.9)
N
j', j"

Figure 1. Schematic display of the peak table used in the program CASANS. The text under the headings "Observed peaks" and "Expected peaks" describes the elements i and j, respectively, of the two peak lists (see text) for which the table entry $p(i, j)$ is the probability of correspondence. Under the headings "example", possible entries for these peaks are given. Here, "N" stands for NOESY, i.e., the peak is observed in a NOESY spectrum.

of the noise level, and thus helps to identify weak peaks. The assignment strategy proposed to the user follows the classical strategy; thus, special menus with commands for the following topics are implemented: spectrum handling (e.g., calibration), peak analysis (e.g., peak picking), spin system identification, sequential assignments, and NOESY peak integration. A few commands require on the order of one minute of CPU time (e.g., antiphase peak picking on a large spectral region), but in general all commands are designed to be used in a fully interactive way, i.e., without delay times for the user. Figs. 4-6 give some display examples of situations occurring when working with EASY.

THE PROGRAM CASANS

The goal of the program CASANS is to obtain, starting from a set of NMR spectra (e.g., COSY, TOCSY and NOESY), a significant amount of sequence-specific resonance assignments in a purely automatic way. The present version does not work on the spectra directly, but rather reads lists of cross peaks from all spectra. The underlying ideas of CASANS are (1)

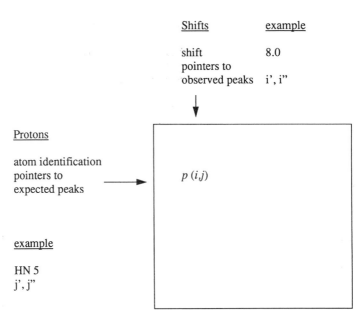

Figure 2. Schematic display of the proton table used in the program CASANS. The text under the headings "Shifts" and "Protons" describes the elements i and j, respectively, of the proton lists for which the table entry $p(i,j)$ is the probability of correspondence. Under the headings "example", possible entries for these list elements are given.

that it works with the basic spectral information, i.e., with cross peaks, (2) that it uses the data from the different spectra simultaneously, and (3) that it creates a list of expected peaks based on the sequence of the protein which is then compared and matched to the list of observed peaks.

The simultaneous use of all spectral information replaces the stepwise procedure in which cross peaks are first collected to delineate spin system and the latter are then connected in their sequential order. It thus avoids that errors of the earlier step are carried over to the latter one during which they can hardly be corrected. The list of observed peaks consists of the two chemical shifts given for every peak plus a code indicating in which types of spectra this peak is observed. In addition, pointers are given to other observed peaks with one chemical shift in common. The list of expected peaks is obtained using the protein sequence and a library indicating for each residue type what kind of COSY, TOCSY and NOESY (only sequential and intraresidue) peaks are expected. It contains for every expected peak the identification of the two atoms that are related by the peak, for each of these atoms a range in which its chemical shift is expected, the type of peak (e.g., COSY, TOCSY, NOESY) and pointers to other expected peaks with one atom in common.

Two basic data structures are used within CASANS. A schematic description of the first one, the peak table, is given in Fig. 1. Each entry

Table 1. CASANS example run

Input:	
Sequence:	RPDFCLE (res. 1-7 of BPTI)
Shift accuracy:	0.01 ppm
Number of expected peaks:	97
Number of protons:	43
Number of observed peaks*:	69
Number of shifts:	39
Result:	
Number of unique shift assignments:	28
Number of shifts with 2 assignments:	5
Number of shifts with > 2 assignments:	6 (max. 4 assignments)
Unassigned proton:	1 (degenerate)
Time (Sun 4/390):	76 s

*COSY, TOCSY and NOESY peaks connecting the same two shifts are counted as one peak.

$p(i,j)$ in this table gives the probability that the observed peak i corresponds to the expected peak j. Initially, all these entries are set to the same value. Then a number of rules are applied to the table in order to identify peak correspondences (i,j) with higher probability $p(i,j)$. The following are examples of these rules: Chemical shifts depend on the type of the proton, e.g., shifts of amide protons are larger than 6 ppm. Observed COSY (TOCSY) peaks can only correspond to expected COSY (TOCSY) peaks. For NOEs this statement has to be expressed in a weaker form since long range NOEs appear only in the list of observed peaks. The pattern of connectivities of an observed peak, i.e., the number and types of peaks having one shift in common with the observed peak under consideration, should match the connectivity pattern of the corresponding expected peak. The observation of two COSY peaks from an α proton to shifts smaller than 6 ppm exclude the residue types Ala, Ile, Thr and Val (here, α protons are identified by having a COSY peak to an amide proton). As a final example consider the case where two spin systems are connected by two NOEs. This indicates with a high probability that the two spin systems are neighbors in the protein sequence.

The second important data structure is a proton table, shown schematically in Fig. 2. Here, the protons determined by the protein sequence and the observed shifts are listed along the two axes, and the entry $p(i,j)$ gives the probability that the shift i corresponds to the proton j. This table is obtained from the above described peak table by using all high probabilities of correspondences between observed and expected peaks to determine pos-

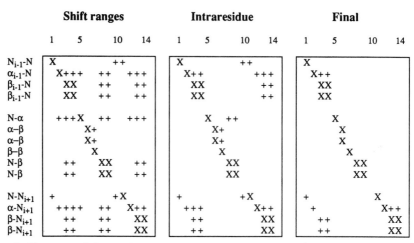

	Shift ranges				Intraresidue				Final			
	1	5	10	14	1	5	10	14	1	5	10	14
N_{i-1}-N	X		++		X		++		X			
α_{i-1}-N	X+++	++	+++		X++		+++		X++			
β_{i-1}-N	XX	++	++		XX		++		XX			
β_{i-1}-N	XX	++	++		XX		++		XX			
N-α	+++X	++	+++		X	++			X			
α–β		X+				X+				X		
α–β		X+				X+				X		
β–β		X				X				X		
N-β	++	XX	++			XX				XX		
N-β	++	XX	++			XX				XX		
N-N_{i+1}	+		+X		+		+X		+		X	
α-N_{i+1}	++++	++	X++		+++		X++		+		X++	
β-N_{i+1}	++	++	XX		++		XX		++		XX	
β-N_{i+1}	++	++	XX		++		XX		++		XX	

Figure 3. Extracts of the peak table at different stages of an example run. The extracts are selected to show on the vertical axis all expected peaks involving protons of Cys 5, and on the horizontal axis all observed peaks for this residue. The numbering of the latter (from 1 to 14) was arranged for this table after the run was finished. Probabilities exceeding a certain threshold are shown as "X" in the case of a correct assignment, and as "+" for additional assignments that the program still considers at the various stages.

sible shifts for individual protons of the protein. Again, a number of rules can be formulated for this proton table, some of which are translations of similar rules for the peak table: Chemical shifts depend on the type of the protons, (e.g., amide protons). Protons have only one shift. This rule is very trivial but nonetheless powerful; it can be extended to pairs or triples of protons (e.g., a pair of protons can have only two shifts).

AN EXAMPLE RUN OF CASANS

The first application presented here is an example run using the N-terminal heptapeptide of BPTI together with the observed peaks for this polypeptide. The first part of Table 1 summarizes the input. Fig. 3 shows extracts of the peak table at various stages of the run. The left panel describes the situation after comparing the observed shifts versus the expected shift ranges. The middle panel reports the situation after checking correspondences of all COSY and TOCSY peaks, and the right panel is the result of the application of all peak table rules to this example. Table 2 shows an extract of the proton table with the protons of Phe 4, Cys 5 and Leu 6. The correct assignment (column 2) was always found. Additional assignment possibilities given by the program are listed in column 3. Some have an obvious explanation, e.g., the shift of the γ proton of Leu 6 is degenerate with the shift of one of the β protons; this results then in ambiguous assignments of the δ methyl groups. The protons in column 4 are assignments that were

Table 2. Proton table extract for Phe 4 — Leu 6 before and after optimization

Shift	Assigned Proton		Before Optimization
	Correct	In Addition	
7.83:	HN4	Qδ4, Qε4, Hζ4	HN5
4.59:	Hα4	Hα3	
3.35:	Hβ4		Hα4, Hβ5
2.94:	Hβ4		Hβ5, Hα4, Hβ5
7.00:	Qδ4	Qε4, Hζ4, HN4	
7.37:	Qε4	Qδ4, Hζ4	
7.31:	Hζ4	Qδ4, Qε4	
7.46:	HN5		Hε1, HN4, Qδ4, Qε4, Hζ4
4.35:	Hα5		
2.74:	Hβ5		Hδ1, Hβ3, Hβ4
2.86:	Hβ5		Hδ1, Hβ3, Hβ4
7.57:	HN6		
4.49:	Hα6		
1.85:	Hβ6		Hβ1, Hγ1, Hβ2, Hγ2, Hγ6
1.70:	Hγ6, Hβ6		Hβ1, Hγ1, Hβ2, Hγ2
0.94:	Hδ6	Hβ6, Hγ6	Hβ1, Hγ1, Hβ2, Hγ2
0.85:	Hδ6	Hβ6, Hγ6	Hβ1, Hγ1, Hβ2, Hγ2

still present after "projecting" the peak table into the proton table but were eliminated by the application of the proton table rules. A summary of the result for this heptapeptide is given in the second part of Table 1. 28 out of 39 shifts could be assigned uniquely; one degenerate proton was not assigned to any shift.

APPLICATION OF CASANS TO BPTI

The second application discussed here is the use of CASANS to assign cross peaks in a complete protein, namely in BPTI. The input is summarized in the first part of Table 3, and Figs. 4-6 show extracts of the spectra used for this run of CASANS. The 826 expected peaks include all COSY and TOCSY peaks that could possibly be observed in the real spectra, and thus many of those do not show up in the observed peak list. On the other hand, the 573 observed cross peaks include also long range NOEs that are not considered in the list of expected peaks. The major obstacles that prevent a completely automatic assignment of parts of the protein are the relatively large shift differences observed for identical protons in different spectra. In particular, the shifts in the TOCSY spectrum show significant differences, most likely due to heating effects of the strong irradiation during the pulse sequence. For the present run of CASANS this problem was avoided by manually

Table 3. Application of CASANS to BPTI

Input:	
Sequence:	BPTI (58 residues)
Shift accuracy:	0.01 ppm
Number of expected peaks:	826
Number of protons:	338
Number of observed peaks:	573
Number of shifts:	221
Result:	
Residues with assigned backbone:	41 (70%)
Residues with β-H assignments:	26 (50%)
Assigned backbone shifts:	66
Assigned β shifts:	36
Assigned side chain shifts:	13
Shifts with multiple assignments:	75
Shifts with no assignments:	16
Wrong assignments:	15
Time (Sun 4/390):	1 h 20 min

resetting the peak positions so that differences of proton shifts between different spectra do not exceed 0.01 ppm (the same value as assumed for the accuracy of shift measurements within each spectrum). The adjusted peak positions are marked for the peaks shown in Figs. 4-6. The second part of Table 3 and Fig. 7 summarize the result: 70% of the residues have at least one backbone proton assigned uniquely; the corresponding number for β positions is 50%. Additional unique assignments are found in several side chains. Only 15 shifts do not have the correct proton in the list of possible assignments; these affect mostly side chain protons.

DISCUSSION

The program CASANS yields a substantial number of resonance assignments prior to any manual or interactive work, i.e., the information obtained is free (provided that the CPU costs for one or few hours on a workstation are not counted). Because the resulting assignments are spread over the entire protein (see Fig. 7), the manual task of filling the assignment gaps should be much less complex than the problems encountered before any assignments are made. For several of the shifts that CASANS could not assign uniquely to a proton, the program offers a strongly reduced list of possibilities, which should further help in completing the assignment on a manual basis. The major unsolved problem concerns the differences in

286

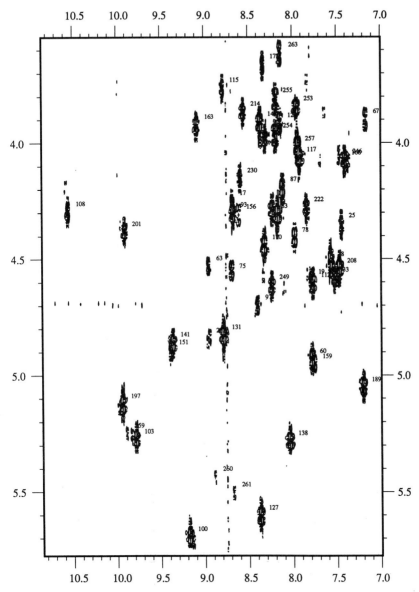

Figure 4. Region of a COSY spectrum of BPTI. Adjusted peak positions entered in the peak list (see text) and used for the application of CASANS to BPTI are marked with a cross and labeled.

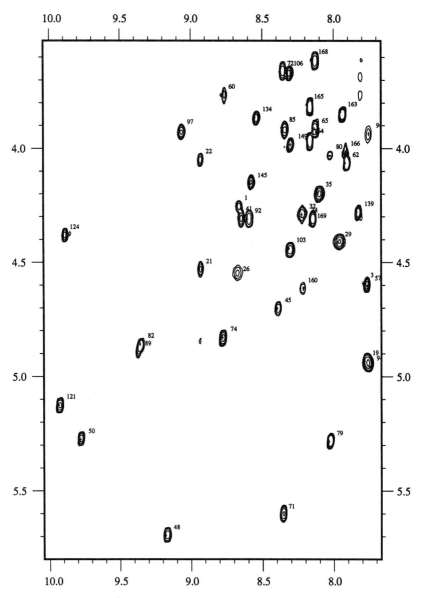

Figure 5. Region of a TOCSY spectrum of BPTI. Adjusted peak positions entered in the peak list (see text) and used for the application of CASANS to BPTI are marked with a cross and labeled.

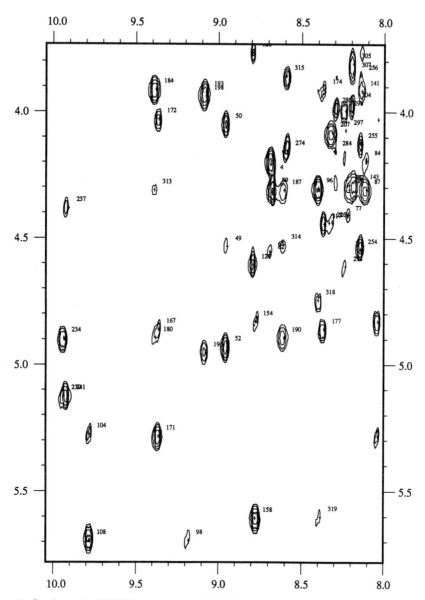

Figure 6. Region of a NOESY spectrum of BPTI. Adjusted peak positions entered in the peak list (see text) and used for the application of CASANS to BPTI are marked with a cross and labeled.

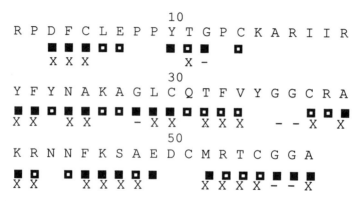

Figure 7. Result of the application of CASANS to BPTI. Below the sequence of BPTI, residues with unique assignments for both backbone protons (filled squares) or for one backbone proton only (empty squares) are identified. The sign "X" further marks those residues for which at least one β proton was assigned.

chemical shifts for identical protons in different spectra. Improvements will certainly include better consideration of this problem during the recording of the spectra. Automatic pattern matching of peak sets that should show up in an identical way in various spectra, e.g., COSY cross peaks are expected in the NOESY, might also reduce the number of necessary manual adjustments of peak positions.

The applications presented here are both based on the use of a COSY, a TOCSY and a NOESY spectrum. The use of other types of spectra, e.g., of heteronuclear spectra, requires in most cases no change of the program code but only the modification of a library that describes the expected peaks for each residue type. The extension of the code to 3D spectroscopy is straightforward, but one might run into time and computer memory problems if the number of peaks increases too much, e.g., in homonuclear 3D NMR.

ACKNOWLEDGMENTS

This project was supported by the Schweizerischer Nationalfunds (project 31.25174.88). Thanks go to C. Eccles, who wrote most of the program EASY, and to K. Wüthrich for support and encouragement.

OUTLINE OF A COMPUTER PROGRAM FOR THE ANALYSIS OF PROTEIN NMR SPECTRA

Mogens Kjær, Kim Vilbour Andersen, Svend Ludvigsen,
Hengyi Shen, Dan Windekilde,[†] Bo Sørensen,[†] and
Flemming M. Poulsen*

Carlsberg Laboratorium
Kemisk Afdeling
Gamle Carlsberg Vej 10
DK-2500 Valby
Copenhagen, Denmark
and
[†]Axion A/S
Bregnerødvej 133
DK-3460 Birkerød, Denmark

INTRODUCTION

It was realized early on in the work process preceeding the determination of three-dimensional structures of proteins in solution by ^{1}H NMR spectroscopy that the very large amount of data in the analysis required the use of computers, primarily as computational devices for 'number-crunching'; however, the potential of the computer as a bookkeeper and an automation device was also anticipated. The result has been the development of a large number of computer programs, that can assist the analysis of protein NMR data at all levels, from Fourier transformation, automated cross peak identification, amino acid spin system recognition, and sequential and stereospecific assignments.[1-10] The development in this field has been promising. In particular, recent analyses of three- and four-dimensional spectra have shown that the larger dispersion of these favors automation.[10] However, the full automation of the analysis of protein two-dimensional NMR data involving complete assignment of both NOESY, TOCSY and COSY spectra, full integration of peak volumes, and accurate determination of

Computational Aspects of the Study of Biological Macromolecules by Nuclear Magnetic Resonance Spectroscopy, Edited by J.C. Hoch *et al.,* Plenum Press, New York, 1991

291

coupling constants has not yet been achieved. The major reason for this is partly the complexity of the problem, partly the coincidental overlap of resonances, and partly the line broadening of nuclei in certain regions of proteins.

From our experience with the analysis in four completed protein NMR studies it has become clear that full automation of the analysis of two-dimensional protein ^{1}H NMR spectra has very limited possibilities for success. Still, after the application of automation procedures a number of steps in the process of analysis has to be carried out by a human expert. Therefore, in our work we have been concerned with the analysis of protein two-dimensional ^{1}H NMR spectra using a combination of automated procedures with a set of tools for interactive analysis and subsequent human decision making at each level of the spectrum analysis. The intention for designing this computer program has been to create an environment in which all results at all levels of analysis can be stored and retrieved, respectively, in and from one master database, and at the same time be linked to all the experimental spectra used for the analysis. Here we describe the lay-out for the design of such a system and a number of software automation tools which are required for a protein NMR analysis computer program. The production of a computer program based on this design is presently in progress, and several of the modules have already been tested; however, the purpose of this paper is merely to present the idea of a data base operated protein NMR assignment tool.

THE DATABASE

The database consists of a set of catalogs. Each catalog is designed to contain a set of information. The catalogs can be opened for entering of information and for retrieving of the information present. Entries in one catalog can be linked to entries in the preceding and the subsequent catalogs in the hierarchy, except catalogs at the bottom and top which have only one-way links. The bottom catalog is the spectrum catalog. The spectrum catalog contains a list of the spectra in the analysis, and for each spectrum the catalog can contain information such as type of spectrum, recording dates, sample conditions, contouring parameters, etc. The next catalog is the cross peak catalog, where the entry is the individual cross peak, and this is linked to the spectrum of origin in the spectrum catalog and contains information about the chemical shifts, the intensity, and the active and passive coupling constants. The next step up is the coupling catalog. The entry here is the coupling between two specific atoms in a spin system as encountered in the different types of spectra in the analysis. When assigned the appropriate atom names can be given to the coupling. The coupling catalog is linked to the cross peak catalog, where the individual cross peaks

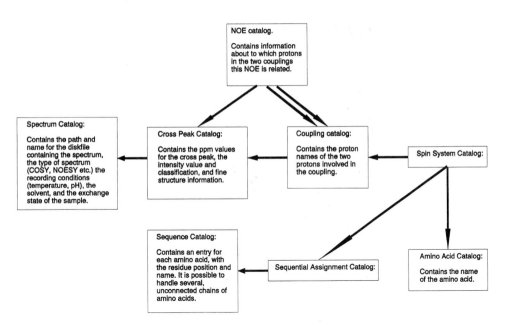

Figure 1. Schematic presentation of the database design.

are marked as members of the coupling. Upwards, the coupling catalog is linked to the spin system catalog. The spin system catalog entry is a spin system, and the information in this catalog is about the entries in the coupling catalog, the couplings, that the spin system is made from. Furthermore the spin system catalog can contain information about the amino acid type or category. The sequential assignment catalog contains strings of spin systems which are potential candidates for being sequentially connected. This is a transit catalog where strings of spin systems are entered for examination by the sequential assignment maker (see later) prior to the final sequential assignment. Each entry in the system can be linked to the sequence catalog. The sequence catalog is the top catalog, the entries here are the individual amino acid residues in the sequence. Enquiries at this entry level can retrieve the relevant entries in the underlaying catalogs via the links in the database. However, in the process of assignment and after completion the reverse analysis is also possible, going from a cross peak in the cross peak catalog to the amino acid residue in the sequence catalog. The database has two auxiliary catalogs. One is the amino acid catalog where each of the common amino acids is an entry. The entry contains the full name and the one-letter and three-letter names and the atom names of the amino acid; other less common amino acids, prosthetic groups, substrates and ligands can be entered into this catalog. The other catalog contains all the atom names encountered for the common amino acids in proteins.

It is important to have the ability to print the content of the database and summaries of the results of the analysis made by the system. The system

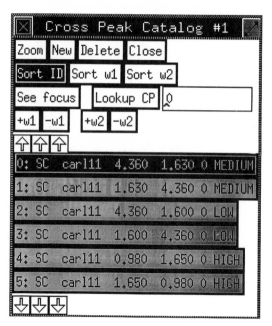

Figure 2. The cross peak catalog window; the operational buttons are the 'zoom', 'new', 'close', 'sort ID', 'sort w1', 'sort w2', 'see focus', 'look up CP' and the arrows up and down buttons; the functional buttons are the entries each carrying an ID the 'used in' information, where S indicates used in a spin system and C used in a coupling, the spectrum information, the chemical shift information, the coupling information, and the intensity information; the current peak is high-lighted.

therefore has a reporting facility for the printing of the contents of the spin system catalog, for plotting the sequential NOE diagram, for plotting contoured spectra with assignment labels and crosses, and a reporting facility listing the NOE assignment in various formats, in particular formats suited for input to the programs for structural computations. Another set of facilities which has proven to be necessary and useful is the checking routines. These routines are applied to the system after a working session, where they make a consistency examination of the contents of the major database catalogs, and report the found inconsistencies, such as wrong atoms in a given amino acid type, more than one chemical shift for the same resonance, etc.

THE USER INTERFACE

The user interface is designed to assist the protein NMR analysis in two ways, the forward way, which is the bottom-to-top analysis, starting with the recorded NMR spectrum and assisting the user in most aspects encountered in the process from cross peak identification to the sequential assignment and the NOE assignments. The reverse analysis which is the top-to-bottom analysis, where the results of an analysis can be checked starting for instance

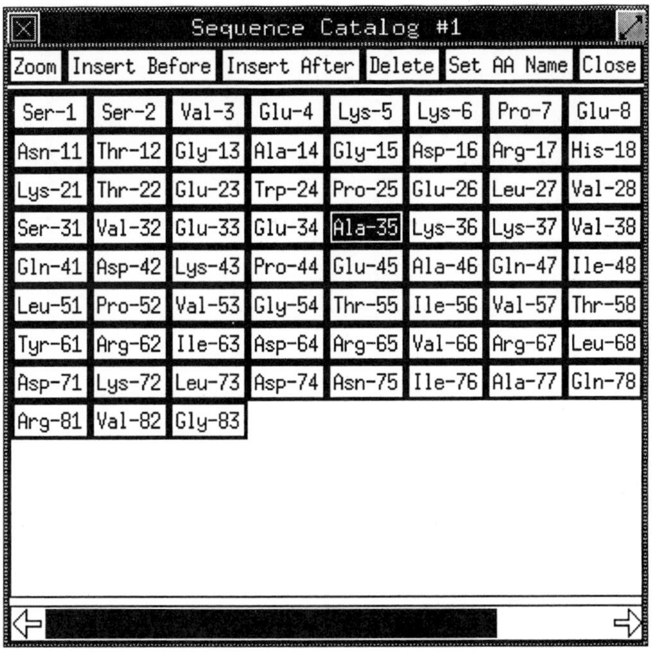

Figure 3. The amino acid sequence catalog window showing the full sequence; in this window the individual residues are shown with the three-letter code; each residue is an entry and a functional button; the high-lighted field is the current sequence position which activates the corresponding entry in the spin system catalog.

with one amino acid in the sequence, and subsequently in a set of steps through the data base catalogs the information that led to the assignment of the NMR resonances to the atoms of the selected amino acid residue can be retrieved from the database.

The user interface is designed using X-Windows. The windows contain a set of operational buttons and a set of function buttons. The operational buttons activates common operations such as opening new windows, scrolling within the present window, addition or deletion of information. Furthermore, some windows have specific operational buttons for operations that are required for the window function, for instance sorting, look up, and merging of information. The functional buttons, mainly used in the database catalogs, are connected with the individual entries and from these database functions can be activated. Many of the operational buttons are used in the bottom-to-top analysis, in particular for entering the contents of the data base catalogs. When the content has been entered a functional entry button is created in the database catalog window. In these the functional buttons can activate a command to the database manager. The functional buttons are the keys to the navigation in the system between the experimental data to the results achieved and stored in the catalogs analysis both in the top-

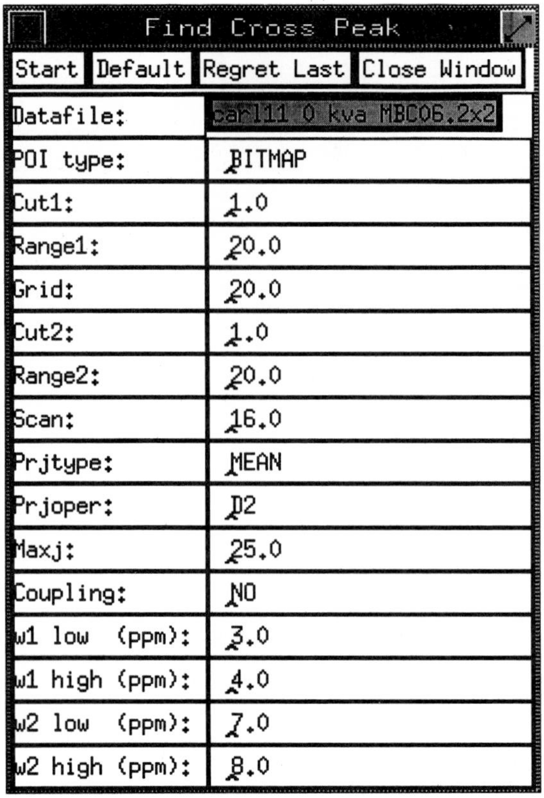

Figure 4. The set-up window for the cross peak finder.

to-bottom analysis and the reverse. Two examples of catalog windows are shown in Figs. 2 and 3.

The system has a master menu window, which is the entrance to the system. From this there is a database entrance, an entrance to each of the associated tools, and an entrance to the error window. From the master menu new windows can be opened, for instance from the database button a new database menu window can be opened. The database menu has a set of operational buttons that can open each of the database windows separately and one that opens all in one step. Each catalog in the database has a catalog window. For display of the NMR data there is a contouring window, there is a window for the various reporting facilities, a window for the internal consistency checking facilities, each of the analytical tools have a set-up window, and finally there is an error window. The tool windows are set-up windows which have entries for the various parameters required for the individual tools, Figs. 4, 5, and 7.

THE ANALYTICAL TOOLS

The system is designed with the flexibility to suit the individual user's preferred working approach to the assignment process. This means that the user can do the analysis in a very similar fashion to the one used for the analysis of spectra plotted on paper with the exception though that all spectra can be displayed on the computer display with only a few commands, and results are entered directly to the database on the computer with a minimum amount of typing. This in itself should be an advantage over the so-called manual methods previously used. In addition however, the protein NMR analysis system has been equipped with four tools that should make it possible to save even more time in the analysis. These are the cross peak finder, the spin system maker, the sequential assignment maker, and the NOE assignment program, four computer programs that can perform their specific tasks in an automated fashion. However, the ambiguities, encountered in two-dimensional spectra of proteins prevents these tools from being fully automated, and each of them is therefore equipped with an inspection tool that allows the user to examine all the results produced by the programs. All the results of these analytical tools are entered into the respective catalogs of the database after the user's inspection and/or acceptance.

The Cross Peak Finder

This is the first tool to be used in the analysis. It is a computer program that finds the position of cross peaks in two-dimensional NMR spectra and enter these as pairs of chemical shifts defining the cross peak position to the cross peak catalog of the database. The program can analyze COSY, TOCSY, NOESY, and 2QS types of spectra. For each type of spectra a particular algorithm is used for the cross peak finding, however common to them is the application of symmetry recognition[11]. The cross peak finder can be applied to the entire spectrum, to regions of the spectrum, or to just one chemical shift value in one of the two chemical shift axis. The result of the analysis can be checked in several ways; one is to display a region of the spectrum in the contouring window where all the found peaks will be marked. In some instances cross peaks will for one reason or another not be found by the program. Such peaks can only be identified by a review of the spectrum. When these peaks are identified, they can be entered to the cross peak catalog using a pointer and a 'new' button in the cross peak catalog. In regions of the spectrum where artifacts occur the program may not be able to distinguish real and false peaks; however, such peaks are readily removed from the database using a delete button in the cross peak catalog window. The cross peak finder program has a number of facilities that permit the

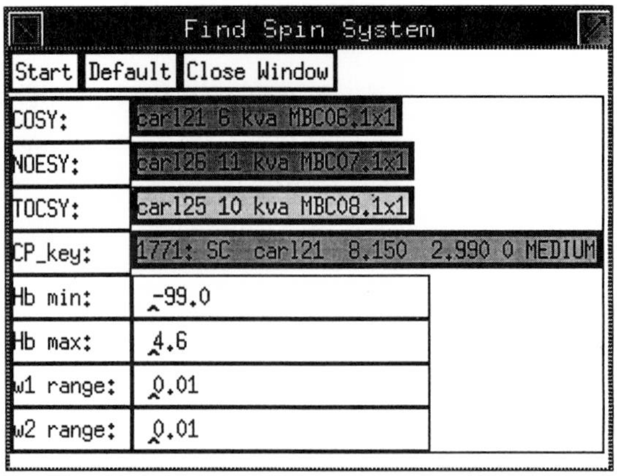

Figure 5. The set-up window for the spin system finder.

setting of filters that will allow the user to optimize the cross peak finding before the visual inspection is necessary.

Additional tools are available for further analysis of the identified cross peaks. For instance integration of the cross peaks of NOESY spectra can be performed[12], and the intensity of the integrated peak can be entered in the cross peak catalog. The passive coupling constants of E. COSY peaks can be determined using the symmetry recognition approach[13]. Also when the sequential assignment has been achieved the method for determining $^3J_{HNH\alpha}$ that uses both COSY and NOESY spectra can be applied[14]. Also the coupling constant information can be stored in the cross peak catalog.

The Spin System Finder

The spin system finder is a search program that builds spin system networks on the basis of the chemical shift information in the cross peak catalog. The program requires primarily data from NOESY, COSY, and TOCSY spectra. However, other spectra can be included. The spin system network is built from one starting point, typically an $H^N - H^\alpha$ cross peak, but any cross peak can be chosen. The procedure is based on the observation that an amino acid spin system in a protein NMR spectrum is often fully represented in a NOESY spectrum. Therefore the first step involves a network building procedure that selects the NOESY cross peaks related to the two chemical shifts of the starting cross peak. Subsequently the nodes of this network are checked for coinciding correlation cross peaks of COSY, TOCSY and other correlation spectra. On the basis of this information a spin system is generated.

Associated with the spin system finder is an inspection tool that allows immediate inspection of the result of the analysis in the contouring window showing in extracts of the spectra the regions around the nodes of the network proposed by the analysis. An example of this is shown in Fig. 6. The contoured cross peaks can be displayed showing the same region of several spectra superimposed. This inspection tool offers an efficient method for eliminating nodes in the spin system network, which are somehow inconsistent. When the nodes of the spin system have all been confirmed, the entire spin system can be entered to the spin system catalog and the assignment to the amino acid type or category can be written to the catalog including the chemical shifts of the atom resonances.

The spin system finder is particularly useful for the analysis of spin systems where partial overlap with parts of others hampers the analysis. The combination of an automated tool that propose solutions to the analysis and the inspection tool has been proven to be efficient and a time saver. For spin systems where no overlap problems exist the spin system finder provides the solution automatically.

The Sequential Assignment Program

The sequential assignment program identifies spin systems of neighboring residues in the amino acid sequence. The prerequisites of the analysis are: a set of amino acid spin systems stored in the spin system catalog of which a substantial fraction must be assigned to an amino acid type or category; the amino acid sequence in the corresponding database catalog; and a complete list of identified NOE cross peaks in the database cross peak catalog. The program firstly identifies in the cross peak catalog all the NOEs that could possibly be of the sequential types $H^{\alpha}(i)$ to $H^{N}(i+1)$, $H^{N}(i)$ to $H^{N}(i+1)$ or $(i-1)$, and $H^{\beta}(i)$ to $H^{N}(i+1)$, where i is the sequence position. Subsequently it uses this information to examine pair wise all the combinations of spin systems for being sequentially connected by either one or several of the possible sequential NOEs. The strings of sequentially connected spin systems are then matched with the amino acid sequence and all the matching strings are written to the sequential assignment catalog in the database as possible solutions to the analysis. The inspection tool permits examination in the spectra of the validity of the proposed solutions of sequential strings of spin systems. This tool is a set up procedure for the contouring window that allows contouring of the individual sequential NOEs that were used to make the sequential connectivity of the two spin systems.

The NOE Assignment Program

The NOE assignment program is a set of interactive procedures which assists the user in the assignment of the NOE cross peaks in the spectrum. The prerequisite for the assignment is the sequential assignment and the

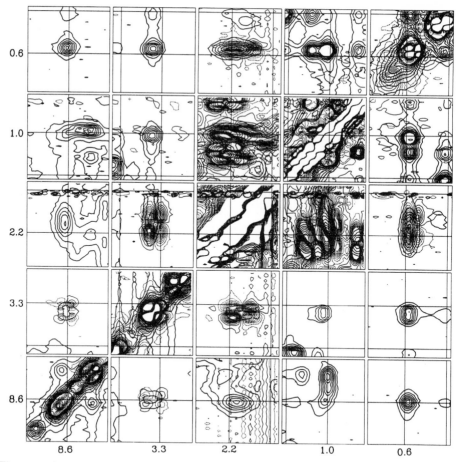

Figure 6. An amino acid spin system from the acyl Coenzyme A binding protein, Val-77. Each square is an extract of the spectrum centered around the position of a spin system coupling. The positive and negative contours of the DQF-COSY spectrum are shown with thin lines, the contours of the NOESY cross peaks are shown with a thicker line.

information accumulated in the database in the process that led to this. To build the NOE catalog the database management has to operate with pointers in the following catalogs: the sequence catalog where it can point to a current atom; the spin system catalog where the pointers can be to all couplings involving the current atom; the coupling catalog where the pointers involve the NOESY couplings of the current atom; the cross peak catalog where there are pointers to the NOESY cross peaks in the spin system couplings; and to the spectrum catalog where there are pointers to the current spectra. In other words the database manager can from the catalogs create an entry which combines the current atom, the current chemical shift, the current NOESY spectrum, the current intraresidue NOE coupling, and a current NOESY cross peak that can serve as a reference for an alignment in

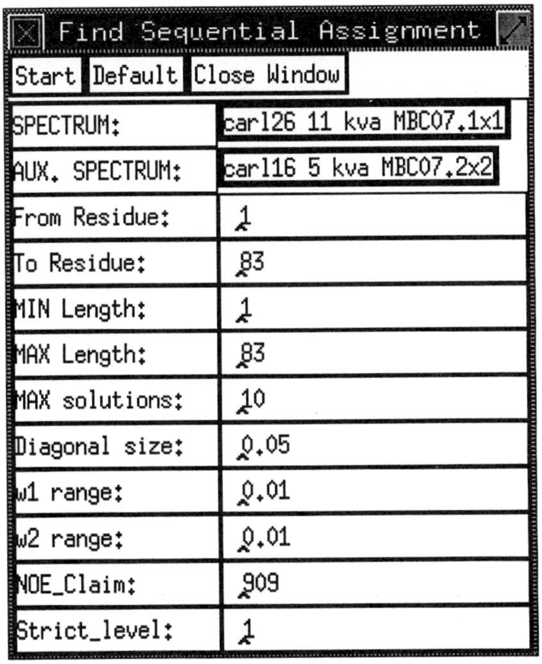

Figure 7. The set-up window for the sequential assignment finder.

the contour diagram. The database can create this information for each of the atoms in the protein. However, for the NOE assignment the database can also point from a current chemical shift to all those atoms which have a resonance at this value, and subsequently the pointers can activate the catalog entries relevant for these individual atoms. With these relatively simple database management operations the program permits the assignment procedure to be initiated either by stepping through each individual NOESY cross peak in the spectrum or by a systematic assignment stepping from atom to atom and making the assignment of the effects aligned with the peak. On the basis of the atom assignment list the system propose the possible assignment for a given NOE and the contouring window receives a parameter set-up for contouring the reference cross peaks of the atoms involved.

RESULTS AND CONCLUSIONS

The major purpose of this paper has been to describe the design of a computer based protein NMR analysis tool, as however, modules of this program have already been produced according to the described design plan, some experience with the performance already exists. The first testing of these modules has been applied to the two-dimensional NMR spectra of the barley serine proteinase inhibitor CI-2 for which almost complete resonance assignment exists[15] and the structure has been determined both in crystal[16] and solution[17]. The modules which have been tested so far are the database, the cross peak finder, the spin system finder, and the sequential assignment finder. These tests have proven that the analysis of the 65 residue protein can be performed much more efficiently and faster than by the traditional methods.

ACKNOWLEDGEMENTS

The present work is a result of an Eureka collaboration. Both parties acknowledge the research and development subsidies provided by the Danish Eureka authorities.

REFERENCES

1. K. P. Neidig, H. Bodenmüller, and H. R. Kalbitzer, *Biochem. Biophys. Res. Commun.* **125**, 1143 (1984).
2. M. Billeter, V. J. Basus, and I. D. Kunz, *J. Magn. Reson.* **76**, 400 (1988).
3. C. Cieslar, G. M. Clore, and A. M. Gronenborn, *J. Magn. Reson.* **80**, 119 (1988).
4. P. L. Weber, J. A. Malikayil, and L. Mueller, *J. Magn. Reson.* **82**, 419 (1989).
5. C. D. Eads and I. D. Kuntz, *J. Magn. Reson.* **82**, 467 (1989).
6. P. J. Kraulis, *J. Magn. Reson.* **84**, 627 (1989).
7. G. J. Kleywegt, R. M. J. N. Lamerichs, R. Boelens, and R. Kaptein, *J. Magn. Reson.* **85**, 286 (1989).
8. F. J. M. van de Ven, *J. Magn. Reson.* **86**, 633 (1990).
9. P. Catasi, E. Carrara, and C. Nicolini, *J. Compt. Chem.* **11**, 805 (1990).
10. C. Cieslar, T. A. Holak, and H. Oschkinat, *J. Magn. Reson.* **87**, 400 (1990).
11. J. C. Hoch, H. Shen, M. Kjær, S. Ludvigsen, and F. M. Poulsen, *Carlsberg Res. Commun.* **52**, 111 (1987)
12. H. Shen and F. M. Poulsen, *J. Magn. Reson.* **89**, 585 (1990).
13. H. Shen, S. Ludvigsen, and F. M. Poulsen, *J. Magn. Reson.* **90**, 346 (1990).
14. S. Ludvigsen, K. V. Andersen, and F. M. Poulsen, *J. Mol. Biol.*, (in press).
15. M. Kjær, S. Ludvigsen, O. W. Sørensen, L. A. Denys, J. Kindtler, and F. M. Poulsen, *Carlsberg Res. Commun.* **52**, 355 (1987).
16. C. A. McPhalen and M. N. G. James, *Biochemistry* **26**, 261 (1987).
17. C. M. Clore, A. M. Gronenborn, M. Kjær, and F. M. Poulsen, *Protein Eng.* **1**, 305 (1987).

ASSIGNMENT OF THE NMR SPECTRA OF HOMOLOGOUS PROTEINS

Christina Redfield and James P. Robertson

Oxford Centre for Molecular Sciences
University of Oxford
South Parks Road
Oxford OX1 3QR England

The assignment of resonances in the complex spectrum of a protein is still one of the bottlenecks in the application of NMR spectroscopy to the study of the structure and dynamics of protein molecules in solution. During the last few years there has been great interest in the automation of the sequential assignment procedure. It is clear from several of the papers contained in this volume that progress is being made in the automation of many aspects of the spectral-assignment process. In this paper a specific aspect of automated assignment, dealing with the assignment of the spectra of homologous proteins, will be discussed.

Once the spectrum of a protein has been assigned by sequential methods is it possible to use the information from this spectrum to assign the spectra of mutant and other homologous proteins in a rapid and reliable manner? The interest in this problem derives from our extensive work on the NMR spectrum of hen lysozyme and its homologous proteins[1,2]. An automated assignment procedure for homologous proteins would be useful for several reasons. It would provide a fast, easy and reliable procedure for assigning the NMR spectra of mutant proteins and naturally occuring homologous proteins that could be used by researchers who are not experts in spectral assignment. Such a procedure could also provide an aid in the assignment of the spectra of homologous proteins which prove difficult by standard sequential assignment methods because of poor quality spectra due to, for example, a limited amount of sample. Full sequential assignments are available for three homologous lysozymes. The sequences of hen and bobwhite quail lysozymes differ at only four amino acid positions, whereas those of hen and human lysozymes differ at 51 positions including an amino acid insertion at position 48 in human lysozyme. The availability of assignments

Computational Aspects of the Study of Biological Macromolecules by Nuclear Magnetic Resonance Spectroscopy, Edited by J.C. Hoch *et al.,* Plenum Press, New York, 1991

303

Table 1. Comparison of NMR Parameters for Hen, Bobwhite Quail and Human Lysozymes [a]

PARAMETER	HEN — BWQ	HEN — HUMAN
CHEMICAL SHIFTS		
NH RMSD	0.084 ppm	0.35 ppm
αCH RMSD	0.051 ppm	0.27 ppm
NH EXCHANGE		
% with same classification	94%	88%
SEQUENTIAL NOE's		
% of residues with same NOE		
NH-NH	99%	84%
αCH -NH	97%	82%
NH-αCH COUPLING CONSTANTS		
RMSD	0.71 Hz	0.90 Hz

[a]The NMR parameters for hen and bobwhite quail lysozyme are compared in the second column, the parameters for hen and human lysozyme are compared in the third column. For the chemical shift and coupling constant data the root mean square differences (RMSD) are given. The exchange rates of the amides of all three lysozymes have been qualitatively divided into slow and fast categories; the percentage of residues with the same hydrogen exchange category in the two proteins is shown. The percentage of residues for which the same type of sequential NOE effect is observed in two proteins is given.

for these three homologous proteins allows an assessment to be made of the reliability of the various NMR parameters in an automated procedure. The NMR parameters which will be compared for the three proteins are the NH and αCH chemical shifts, hydrogen exchange rates, sequential NOE information and NH-αCH coupling constants.

The chemical shift, hydrogen exchange, NOE and coupling constant data for hen, bobwhite quail, and human lysozymes are compared and summarized in Table 1. The root mean square difference (RMSD) between the chemical shifts of hen and bobwhite quail lysozymes is 0.08 ppm for the NH resonances and 0.05 ppm for the αCH resonances. The largest shifts observed between these two proteins are less than 0.4 ppm and more than 90% of the shifts are less than 0.1 ppm. The strong correlation of chemical shifts between these proteins can be seen from a comparison of the two fingerprint regions shown in Figs. 1 and 2. Much larger RMSD values are obtained from the comparison of the chemical shifts of hen and human lysozymes; an RMSD of 0.35 ppm is observed for the NH resonances and a value of 0.27 ppm for the αCH resonances. Shifts as large as 1.0 ppm are observed for a couple of NH resonances and these can be correlated with changes in the hydrogen bonding pattern observed in the two X-ray structures[2]. However, 85% of residues have shifts of less than 0.4 ppm and

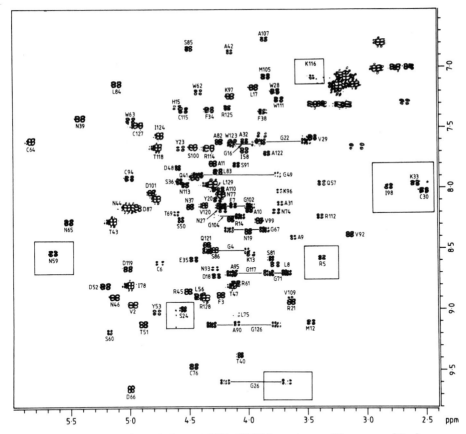

Figure 1. Fingerprint region of the 500 MHz COSY spectrum of hen egg-white lysozyme.

60% have shifts of less than 0.2 ppm. The pattern of hydrogen exchange observed for the three lysozymes is very similar. In all three proteins about 90% of the residues have the same qualitative hydrogen exchange behavior. This is a result of the very similar pattern of hydrogen bonds found in the protein structures[2]. A very similar pattern of sequential NOE effects is observed for hen, bobwhite quail and human lysozymes; for about 80% of residues the same sequential NOE (NH-NH or αCH-NH) is observed in hen and human lysozymes whereas for almost 100% of residues the same NOE is observed in hen and bobwhite quail. The close similarity of NOE patterns is a consequence of the close structural homology of these three proteins. NH-αCH coupling constants have been measured for the majority of residues in the three homologous proteins[2,3]. Comparison of the hen and bobwhite quail coupling constants gives an RMSD of 0.71 Hz, whereas comparison of the hen and human values gives an RMSD of 0.90 Hz. For the majority of residues the difference in coupling constant is less than 1.0 Hz.

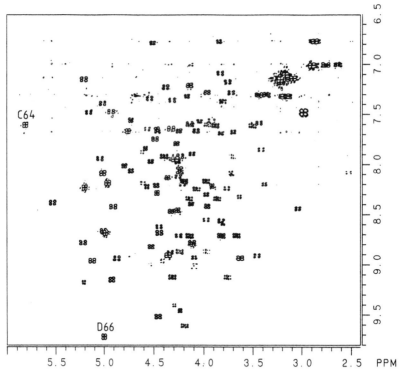

Figure 2. Fingerprint region of the 500 MHz COSY spectrum of bobwhite quail lysozyme. The cross peaks of C64 and D66 are labeled.

There is a strong correlation between the chemical shifts, hydrogen exchange rates, NOE effects and coupling constants for the three lysozyme proteins studied. It is, therefore, likely that all of these parameters will be useful in an automated assignment procedure which is based on sequence homology. The correlation of NMR parameters is better for hen and bobwhite quail lysozyme than it is for hen and human lysozyme indicating that the success of a homologous assignment procedure is likely to decrease as the sequence homology decreases.

It is possible to envisage several different ways of utilizing the NMR parameters discussed above in a homologous assignment procedure. The strategy that has been adopted so far relies on the chemical shifts and NOE effects as the primary parameters and uses information about hydrogen exchange and coupling constants to confirm or discard potential assignments. There are two major comparisons between the homologous proteins on which proposed assignments are based. The first uses the NH and αCH chemical shift information and spin system classification information as the basis for assignment. The second uses the chemical shifts, spin system classifications, and sequential NOE information as the basis. These two methods are described below. Examples of their application to the assignment of

306

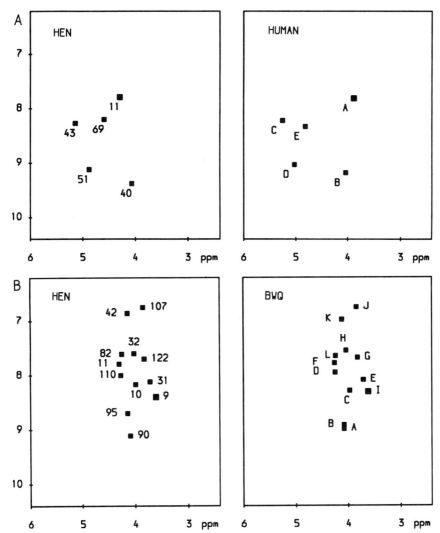

Figure 3. a) Schematic representation of the fingerprint region cross peaks of the threonine residues found in human lysozyme and the corresponding residues in hen lysozyme. The hen lysozyme peaks are labeled with their residue assignments, the human lysozyme peaks are given arbitrary labels. b) Schematic representation of the fingerprint region cross peaks of the alanine residues found in hen and bobwhite quail lysozymes. The hen lysozyme peaks are labeled with their residue assignments, the bobwhite quail lysozyme peaks are given arbitrary labels.

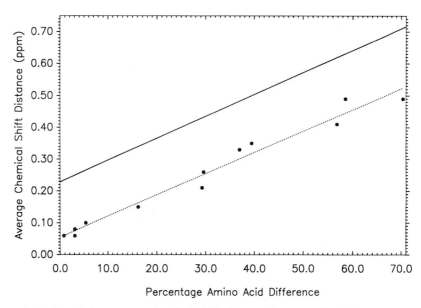

Figure 4. A plot of the average chemical shift distance between the NH-αCH cross peaks of a pair of homologous proteins versus the percentage of amino acid differences for that pair of homologous proteins. The circles represent the data points for proteins including lysozymes, plastocyanins, cytochromes c and growth factors. The broken line is a least squares fit to these data points. The solid line is used to determine the cutoff distance appropriate for a particular pair of homologous proteins. For example, a cutoff distance of 0.5 ppm is used for hen and human lysozymes which differ in 40% of their amino acid sequence. If less conservative cutoff distances are desired then the solid line should be redrawn with a smaller vertical offset from the broken line.

the bobwhite quail and human lysozyme spectra on the basis for the hen lysozyme spectrum are given.

The first method used in the homologous assignment approach involves the comparison of the chemical shifts of the fingerprint region cross peaks in the spectra of the two homologous proteins. The underlying assumption in this approach is that the NH and αCH chemical shifts of the two proteins will be very similar. For individual cross peaks which are very well separated from all other peaks in the fingerprint region, that is those with unique NH and αCH chemical shifts, assignments can often be made by a simple visual comparison of the spectra. For example, the cross peaks of D66 and C64 can be assigned in the spectrum of bobwhite quail lysozyme, shown in Fig. 2, by comparison with the spectrum of hen lysozyme, shown in Fig. 1. However, for the majority of cross peaks, assignment by this simple comparison is not possible because of the close proximity of many cross peaks in the spectrum. The problem of crowding of cross peaks can be overcome, to some extent, if spin system information is used. NH-αCH cross peaks arising from Gly, Ala, Thr, Val and Ile residues can usually be identified on the basis of cross peak fine structure or long-range connectivities found in RELAY, DOUBLE-RELAY, and HOHAHA spectra. For example, the NH and αCH chemical shifts of the threonine residues found in hen and human lysozymes are compared in Fig. 3a. Peaks such as residue 11, which are located in crowded regions of the full spectrum, become isolated when the fingerprint region is simplified on the basis of spin system information. The assignment of the threonine spin systems in human lysozyme can now be obtained by a simple comparison with the hen lysozyme spectrum. The alanine NH and αCH chemical shifts of hen and bobwhite quail lysozyme are compared in Fig. 3b. In this case simplification of the spectrum does not lead to unambiguous assignment of all alanines. The cross peaks of A11, A32, A82, A110 and A122 are still crowded together in the spectrum. Because of the small shifts that occur as a result of amino acid substitutions it would be possible to make an incorrect assignment of these alanine residues of bobwhite quail lysozyme on the basis of a simple comparison with the hen spectrum. Thus, in some cases spin system information can lead to a dramatic simplification of the assignment problem whereas in other cases it does not. It is clear that some criterion needs to be introduced in order to assess when assignments can be made reliably and when assignments based on a chemical shift comparison might be incorrect. This criterion is based on a comparison of the relative values of the expected shifts in the spectrum due to amino acid substitutions and the proximity of cross peaks of a particular amino acid type to each other in the spectrum. Before the comparison of the chemical shifts in the spectra of the two different proteins is made the chemical shifts of resonances of a particular amino acid type in the spectrum of the assigned protein (in this case hen lysozyme) must be analyzed. For each peak of a particular amino acid type the chemical shift distance to the closest peak of that type is calculated. The chemical shift distance between peaks i and j in the spectrum of the assigned protein is defined as

$$\text{distance} = \left[(\delta_{NHi} - \delta_{NHj})^2 + (\delta_{\alpha CHi} - \delta_{\alpha CHj})^2 \right]^{1/2}.$$

This distance is then compared with a cutoff distance for the unassigned protein. If the chemical shift distance to the closest peak for a particular residue is greater than the cutoff distance then this residue can be used for assignment. If the distance is less than the cutoff then it is possible that the changes in chemical shifts due to the amino acid substitutions will lead to an incorrect assignment and this residue is, therefore, not used in the assignment process. The cutoff distance is determined by the degree of sequence homology between the two proteins and increases as the homology decreases. Cutoff distances can be obtained from a plot of the average chemical shift distance between two proteins versus the percentage of amino acid differences, such as that shown in Fig. 4.

Table 2. Summary of Nearest Neighbor Distances for the Threonine Residues of Hen Lysozyme [a]

RESIDUE NUMBER	NEAREST NEIGHBOR RESIDUE NUMBER	CHEMICAL SHIFT DISTANCE (PPM)
11	69	0.52
40	51	0.85
43	69	0.54
51	40	0.85
69	11	0.52

[a]The chemical shift distance between a threonine peak i in the hen spectrum and its nearest neighbor threonine peak j in the hen spectrum is defined as distance $= \left[(\delta_{NHi} - \delta_{NHj})^2 + (\delta_{\alpha CHi} - \delta_{\alpha CHj})^2\right]^{1/2}$.

The procedure can be illustrated for the threonine residues of hen and human lysozymes. The distances between nearest neighbor threonine pairs in hen lysozyme are summarized in Table 2. The threonine peaks are well separated in the spectrum of hen lysozyme and the smallest nearest neighbor distance is 0.52 ppm. Hen and human lysozymes differ in 40% of their sequences; a cutoff distance of 0.5 ppm is obtained from Fig. 4. All the nearest neighbor distances are greater than this cutoff and as a result it should be possible to assign all the threonine cross peaks in human lysozyme unambiguously. The chemical shift distance between each threonine peak in hen lysozyme and the two nearest peaks in the spectrum of human lysozyme are summarized in Table 3. In all cases the distance to the closest peak is less than the 0.5 ppm cutoff, and the distance to the next closest peak is greater than the 0.5 ppm cutoff, in all cases but one. On the basis of chemical shift similarity all the threonine peaks in the human lysozyme spectrum can be assigned to the residues which correspond to the closest peak in the hen spectrum; peaks A, B, C, D and E are assigned to T11, T40, T43, T52 and T70, respectively. In order to confirm these assignments information about hydrogen exchange and coupling constants can be used as shown in Table 3. All the threonine residues have the same hydrogen exchange classification for the proposed set of assignments and the coupling constants for all these residues agree to within 0.6 Hz.

This procedure can also be applied to the alanine residues of hen and bobwhite quail lysozymes. The distances between nearest neighbor Ala pairs in hen lysozyme are summarized in Table 4. The distances obtained for the alanines are all smaller than those obtained above for the threonines and this reflects the greater crowding of the alanine cross peaks in the spectrum. Hen and bobwhite quail lysozymes differ at only 4 amino acid positions and, therefore, a much smaller cutoff can be used for hen and bobwhite quail than

Table 3. Assignment of the threonine residues of human lysozyme on the basis of a comparison with the hen lysozyme spectrum[a]

Hen Peak	Closest Peak				Next Closest Peak			
	Human Peak	Dist ppm	NH Exch	Coupling Constant	Human Peak	Dist ppm	NH Exch	Coupling Constant
11	A	0.41	Y	0.6	E	0.76	N	4.5
40	B	0.20	Y	0.5	D	1.02	Y	4.4
43	C	0.13	Y	0.1	E	0.33	Y	0.0
51	D	0.17	Y	0.0	E	0.79	Y	0.5
69	E	0.26	Y	0.0	C	0.66	Y	0.1

[a]The threonine peak in the human lysozyme spectrum with the smallest distance to each of the hen peaks is listed in column 2; the peak with the second smallest distance is listed in column 6. The hydrogen exchange categories for the hen and human peaks are compared in columns 4 and 8; a Y indicates that the two peaks are assigned to the same exchange category whereas an N indicates different exchange categories. The difference between the hen and human NH-αCH coupling constants measured for each pair of peaks is listed in columns 5 and 10. On the basis of the data shown here cross peaks A, B, C, D, and E in human lysozyme are assigned to T11, T40, T43, T52, and T70, respectively.

was used for hen and human lysozymes. A cutoff value of 0.25 ppm is obtained from Fig. 4. The nearest neighbors of A11, A32, A82, A110, and A122 are all closer than the 0.25 ppm cutoff value and, therefore, these residues can not be used, with confidence, as a basis for assignment using the chemical shift comparison. Assignments for A9, A10, A31, A42, A90, and A95 can be attempted on the basis of the chemical shift comparison and confirmed or rejected using the hydrogen exchange and coupling constant data.

Using the procedure outlined above assignments can be made for the Gly, Ala, Thr, and Val residues of bobwhite quail and human lysozymes by comparison with the known assignments of hen lysozyme. In the case of bobwhite quail lysozyme 23 of a possible 38 assignments can be made whereas for human lysozyme only 9 of a possible 39 assignments can be made. This result is in accord with the expectation that the success of a homologous assignment procedure will decrease as sequence homology decreases.

The second method used in the homologous assignment procedure uses chemical shift, spin system, and NOE information. The underlying assumption in this approach is that the chemical shifts of the two proteins will be very similar and that the pattern of sequential NH-NH and αCH-NH NOE effects will also be the same. Each NH-NH or αCH-NH NOE effect can be considered as connecting two peaks, i and j, in the fingerprint region. An NOE effect observed in the spectrum of hen lysozyme is very similar to an effect observed in the spectrum of a homologous protein if it

Table 4. Summary of Nearest Neighbor Distances for the Alanine Residues of Hen Lysozyme

RESIDUE NUMBER	NEAREST NEIGHBOR RESIDUE NUMBER	CHEMICAL SHIFT DISTANCE (PPM)
9	31	0.30
10	31	0.27
11	82	0.19
31	10	0.27
32	122	0.21
42	107	0.31
82	11	0.19
90	95	0.41
95	90	0.41
107	42	0.31
110	11	·0.21
122	32	0.21

involves the same type of backbone protons (NH-NH or αCH-NH), and if the sum of the chemical shift distances between the pairs of i peaks and j peaks is small. The sum of the chemical shift distances between peaks i and j in one spectrum and peaks i' and j' in the other spectrum is defined as

$$\text{distance} = \left[(\delta_{NHi} - \delta_{NHi'})^2 + (\delta_{\alpha CHi} - \delta_{\alpha CHi'})^2 \right]^{1/2}$$
$$+ \left[(\delta_{NHj} - \delta_{NHj'})^2 + (\delta_{\alpha CHj} - \delta_{\alpha CHj'})^2 \right]^{1/2}.$$

In this method each NH-NH or αCH-NH NOE effect observed in the spectrum of the assigned protein, hen lysozyme, is compared with each NH-NH or αCH-NH NOE effect observed in the spectrum of the homologous protein. If the spin system classifications of the peaks i and j in the spectrum of the homologous protein are consistent with those of hen lysozyme (taking sequence differences into account) then the sum of the chemical shift distances of the pairs of peaks i and j are calculated. The NOE effect in the homologous protein which has the smallest distance to the hen NOE effect can then be tentatively assigned to the pair of residues which give rise to the NOE effect in the hen spectrum. Some NOE effects in the hen lysozyme spectrum connect cross peaks with unique chemical shifts. However, other NOEs connect peaks in crowded regions of the spectrum and, therefore, errors in the assignment of NOE peaks could be made. As in the comparison of chemical shifts described above, a criterion for assignment must be applied in the analysis of NOE effects. Again the criterion is based on how unique a particular NOE effect is in the spectrum of the

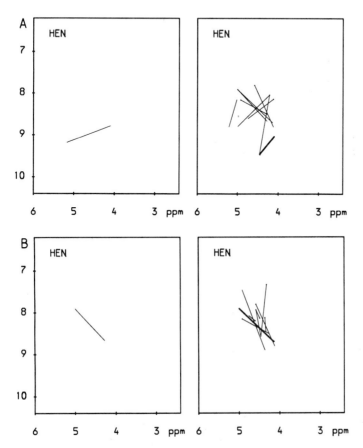

Figure 5. a) Comparison of the hen NH-NH NOE between S60 and R61 (left) and the nine
most similar hen NH-NH NOEs (right). NOE effects are represented schematically by a
line connecting the fingerprint region cross peak positions of the two residues involved.
The NOE effect which gives the smallest sum of chemical shift distances is shown in bold
on the right; this NOE effect involves L75 and C76 and has a distance of 1.01 ppm. b)
Comparison of the hen NH-NH NOE between N93 and C94 (left) and the nine most similar
hen NH-NH NOEs (right). The NOE effect which gives the smallest sum of chemical shift
distances is shown in bold on the right; this NOE effect involves C94 and A95 and has a
distance of only 0.15 ppm. Eight of the nine NOE effect shown have distances of less than
1 ppm.

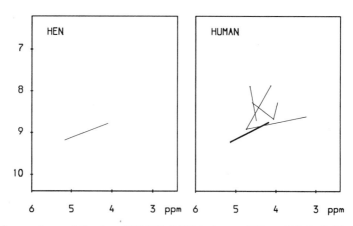

Figure 6. Comparison of the hen NH-NH NOE between S60 and R61 (left) and the six most similar human NH-NH NOEs (right). The NOE effect which gives the smallest sum of chemical shift distances is shown in bold on the right; this human lysozyme NOE effect can be assigned to residues S61 and R62 of human lysozyme. The second closest NOE effect has a distance of 1.38 ppm.

assigned protein, hen lysozyme. Each NOE effect in the spectrum of hen lysozyme is compared with all the other NOE effects of the same kind in hen lysozyme and the NOE effect with the smallest sum of the chemical shift distances is found. The NOE effects which have a large distance to the nearest neighbor NOE effect can be used as the basis for assignments whereas those with a small distance are not used for assignment. The number of NOE effects on which assignments can be based will depend on the cutoff distance chosen and, therefore, on the degree of sequence homology. For example, the NH-NH NOE effect between S60 and R61 in hen lysozyme is compared with the nine most similar NH-NH NOEs in hen in Fig. 5A. The next closest NOE effect involves L75 and C76 and is at a distance of 1.01 ppm. Thus, this NH-NH NOE effect in hen involves unique chemical shifts and will be a good basis for assignment even in proteins, such as human lysozyme, with many amino acid differences. By contrast the NH-NH NOE effect between N93 and C94 does not involve unique chemical shifts as illustrated in Fig. 5B; the closest NOE effect is at a distance of only 0.15 ppm and there are 9 NOE effects at a distance of less than 1.0 ppm. Therefore, this NOE would not be a good basis for assignment even in a protein such as bobwhite quail lysozyme which has a high degree of sequence homology.

The procedure can be illustrated for the comparison of hen and human lysozymes; these two proteins differ in 40% of their sequences. A cutoff value of 1.0 ppm is used. This cutoff is twice the value used above in the first method because a pair of peaks is being considered. There are 6 NH-NH NOE effects in hen lysozyme which have a nearest neighbor distance of greater than 1.0 ppm; one of these is the NOE between S60 and R61 described above. A comparison of this hen NOE with all of the NH-NH NOEs in human lysozyme yields 36 human NOEs which have the correct spin system classification. The 6 NOE effects in human lysozyme which give the smallest distance to the hen peaks are illustrated in Fig. 6. The closest NOE effect in human lysozyme differs from the hen NOE by a distance of only 0.14 ppm, the next closest NOE is at a distance of 1.38 ppm. Thus, the NOE effect with the smallest distance in human lysozyme can be assigned to the residues which correspond to S60 and R61 in the sequence of human lysozyme (S61 and R62). The assignments can be confirmed by comparing the hydrogen exchange classifications and coupling constants for these residues. This procedure is repeated for the other 5 NH-NH NOE effects which are above the 1.0 ppm cutoff and for the αCH-NH NOE effects above the cutoff. This analysis yields assignments for 14 residues. A similar analysis carried out for bobwhite quail lysozyme using a smaller cutoff value of 0.50 ppm yields a total of 42 assignments. Again the number of assignments made decreases as the sequence homology decreases.

Using the methods described above a number of assignments can be made for a homologous protein such as bobwhite quail or human lysozyme. These assignments will have a high degree of confidence because they have been based on very conservative cutoff values and have been confirmed using hydrogen exchange and coupling constant data. It is important that the assignments made at this stage are correct because they will serve as the basis for additional assignments. Therefore, any assignment for which the hydrogen exchange or coupling constant data does not fit should be discarded at this stage.

The third stage of assignment is based on analysis of the NOE data using the assignments made using the two methods described above. NOE effects in the spectrum of the assigned protein which involve one of the residues assigned earlier for the homologous protein are now considered even if they fall below the cutoff distance. Instead of considering all the NOE effects observed for the homologous protein, however, only those which involve the assigned cross peak need to be considered. For example, the human lysozyme cross peaks of S61 and R62 were assigned on the basis of the S60-R61 NOE effect in hen. The NH-NH NOE involving R61 and W62 in hen lysozyme was not considered above because the nearest neighbor pair distance of 0.59 ppm was well below the cutoff value of 1.0 ppm used for human lysozyme. This hen lysozyme NOE effect can now be compared with the human NOE effects; there are 5 human NOEs that are within 1.0 ppm. However, only one of these NOEs involves the peak assigned above to R62. Therefore, the cross peak corresponding to Y63 in human lysozyme can be assigned on the basis of the earlier assignment of R62. In this way a number of additional assignments can be made in the spectrum of the homologous protein on the basis of the high confidence assignments already obtained.

Assignments for a total of 51 residues human lysozyme are obtained using the homologous assignment procedures detailed above. For bobwhite quail lysozyme 96 residue assignments are obtained. These assignments are entirely consistent with the sequential assignments made previously[2]. There are several reasons why a complete set of assignments is not obtained for either homologous protein. In order to avoid confusion between αCH and βCH resonances only those αCH-NH NOE effects involving αCH resonances downfield of the water peak were used in the analysis. Some of the NH-NH or αCH-NH NOE effects observed in hen lysozyme are not observed in bobwhite quail or human lysozymes because of chemical shift overlap or a change in backbone conformation. Some of the peaks in the spectra of the homologous proteins shift relative to those in hen as a result of amino acid substitutions and, therefore, the chemical shift distances obtained for these peaks from the two methods of comparison will be too great to allow an unambiguous assignment. However, the large number of assignments obtained using the homologous assignment method can serve as a starting point for further assignments based on the sequential approach and will dramatically simplify and speed up the remaining assignment problem. The methods described above are just a couple of the ways in which information from the spectrum of an assigned protein can be used to suggest assignments in the spectrum of an unassigned homologous protein. We are continuing to develop new tools to aid in the homologous assignment procedure.

REFERENCES

1. C. Redfield and C. M. Dobson, *Biochemistry* **27**, 122-136 (1988).
2. C. Redfield and C. M. Dobson, *Biochemistry* **29**, 7201-7214 (1990).
3. L. J. Smith, M. J. Sutcliffe, C. Redfield, and C. M. Dobson, *Biochemistry* **30**, 986-996 (1990).

INCORPORATION OF INTERNAL MOTION IN NMR REFINEMENTS BASED ON NOESY DATA

Ping F. Yip and David A. Case

Department of Molecular Biology
Research Institute of Scripps Clinic
La Jolla, California 92037

ABSTRACT

One promising way of carrying out the final refinements for solution structures of proteins and nucleic acids involves an automated fitting of the observed NOESY peak intensities to those calculated from a multi-spin relaxation model. We have previously discussed how simulated annealing procedures could be used to carry out such calculations (Yip and Case, *J. Magn. Reson.* **83**, 643-648 (1989)). We now extend this initial model to allow discrete jump models of internal motions (assumed to be uncoupled from overall rotational diffusion), and compare these results to those from static models or effective order parameters. Preliminary results for plastocyanin suggest that this level of theory will be sufficiently accurate in many cases to be used as a basis for structural refinement.

INTRODUCTION

The past decade has seen remarkable progress in the ability to determine approximate solution structures of proteins and nucleic acids based upon NMR data. The general procedure is by now fairly well established[1-4]: distance constraints based on nuclear Overhauser effect (NOE) intensities, and angular constraints based on coupling constants, are used to first deduce and then refine a three-dimensional structure consistent with both the general stereochemical constraints derived from the covalent structure and with the particular experimental data for the molecule in question. Although there is little question that the overall three-dimensional structures of biomolecules determined in this way are correct (at least in the absence of large-amplitude conformational disorder,) the extent to which one can rely upon the details of such structures is less clear. One attractive starting point for making such judgments would involve measures of the extent to which simulated and observed spectra agree with each other; such comparisons form the basis for

Computational Aspects of the Study of Biological Macromolecules by Nuclear Magnetic Resonance Spectroscopy, Edited by J.C. Hoch *et al.,* Plenum Press, New York, 1991

refinement methods in x-ray crystallography, for example[5]. We recently described a way in which such a scheme might work for nmr spectroscopy[6], and other refinement techniques that involve comparisons between calculated and observed NOESY intensities have also been described[7-11].

It is clear that such refinement procedures can be no more accurate than the models that are used to compute spectral intensities from model structures; their accuracy may also be limited by noise and systematic errors in experimental intensity estimations. Progress is being made in both areas, although many problems remain. In this paper we focus on the computational side of the problem, i.e., on the development of models for estimation of NOESY intensities from assumed structures and motional models.

The simplest models for magnetization transfer in a NOESY experiment posit a rigid set of spins undergoing isotropic rotational diffusion. In the past few years a number of groups have developed methods for going beyond the two-spin approximation, allowing relaxation among the entire pool of coupled spins. A logical next step is to incorporate what we know about anisotropic diffusion and internal motions into the computational models, and to test the resulting theory against experimental data. Discrete jump models offer an attractive means for considering effects of internal motion since they involve relatively few additional parameters (in some cases none,) and can cover a wide range of physical situations, from fast rotation of methyl groups to problems of chemical exchange[12-15]. Here we outline the formulation of general jump models for internal motion, and the computation of the derivatives of the resulting intensities with respect to atomic positions. Our preliminary applications to plastocyanin (reported elsewhere) suggest that these models are accurate enough to serve as useful target functions for nmr refinements. In the final section we compare these ideas to other approaches that can be used to model internal motions in macromolecules.

DISCRETE JUMP MODELS

Our refinement procedure is an elaboration of one we proposed earlier[6], in which penalty functions based on NOE intensities are added to a molecular mechanics potential function, and the resulting combined function is optimized by a simulated annealing procedure. A key ingredient for carrying out this procedure involves the calculation of the gradient of the NOE intensity with respect to the structural parameters (in our case the Cartesian coordinates) of the molecule. In addition to allowing for (isotropic) rotational diffusion, our current model adds internal motions for methyl groups (where the three equivalent proton sites are assumed to interchange via a jump model with a rate constant fast compared to the overall molecular rotational diffusion time) and for chemically-exchanging δ and ϵ protons on phenylalanine and tyrosine (which also exchange through a jump model, but now on a time scale slow compared to the rotational diffusion time.) In these limits, the effect of the internal motion is independent of the actual values for the residence times in each state, so that no additional parameters are required to model these motions. The basic relaxation formulas for these models for internal motion are given by Tropp[13].

Consider the longitudinal relaxation of protons coupled by dipolar inter-actions. Suppose the protons can be divided into N magnetically distinct species with population $n_i, i = 1, \ldots N$. The longitudinal magnetization m_i above thermal equilibrium for each species then follows the Bloch equations.

$$\frac{dm_i}{dt} = -\sum_{j=1}^{N} R_{ij} m_j. \tag{1}$$

The relaxation rate matrix $\mathbf{R}$ is determined by the spectral densities,

$$R_{ii} = k(n_i - 1)\left(J_{ii}^1(\omega) + 4J_{ii}^2(2\omega)\right) + k \sum_{j \neq i} n_j \left(\frac{1}{3}J_{ij}^0(0) + J_{ij}^1(\omega) + 2J_{ij}^2(2\omega)\right). \tag{2}$$

And for $i \neq j$,

$$R_{ij} = kn_i(2J_{ij}^2(2\omega) - \frac{1}{3}J_{ij}^0(0)). \tag{3}$$

Here $k = (6\pi/5)\gamma^4\hbar^2$, and the spectral densities J^0, J^1 and J^2 depend on the geometry and the motion of the protons. For distinct species i and j,

$$J_{ij}^n(\omega) = \int_0^\infty \langle \frac{Y_{2n}(\phi_{ij}^{lab}(0))Y_{2n}^*(\phi_{ij}^{lab}(t))}{r_{ij}^3(0)r_{ij}^3(t)} \rangle \cos(\omega t)dt. \tag{4}$$

$\phi_{ij}^{lab}(t)$ denotes the set of polar angles of the vector from species i and j at time t in the laboratory frame. $r_{ij}(t)$ is the distance between them. The angular bracket denotes ensemble average. The spectral densities J_{ii}^n occur only in the diagonal elements of the rate matrix. They differ from J_{ij}^n only in that the distance and the angles are those of the vector from one proton to another within the same group.

In the case when neither species i nor j has internal rotation, the distance r_{ij} is a constant and thus can be taken out of the average and the integral to give the usual r_{ij}^{-6} factor and a Lorentzian dependence on ω:

$$J_{ij}^n(\omega) = \frac{1}{4\pi} \cdot \frac{1}{r^6} \cdot \left(\frac{\tau_c}{1 + \omega^2\tau_c^2}\right). \tag{5}$$

As with other models below, in the case of isotropic tumbling the right-hand-side of Eq. (5) is independent of n. This is the simplest motional model for relaxation, in which all of the spins are rigidly attached to the rotating macromolecule. It is probably approximately correct for protein backbones, but is clearly inadequate for methyl groups (where internal rotation is almost always fast compared to overall tumbling for macromolecules) and for many other side-chain spins; the extent to which a single correlation time can account for motions in oligonucleotides is still unclear.

When either i or j suffers internal rotation, however, r_{ij} can change and the ensemble averaging in Eq. (4) should include averaging over internal motion as well. In our approach, these are modelled as instantaneous jumps between discrete conformations. We now show how to calculate Eq. (4) in the case of jump models. First express the spherical harmonics in the molecular frame:

$$Y_{2n}(\phi_{ij}^{lab}) = \sum_{m=-2}^{m=2} D_{nm}^2(\Omega) Y_{2m}(\phi_{ij}^{mol}). \tag{6}$$

The D_{nm}^2 is the Wigner rotation matrix, Ω denotes the set of Euler angles between the laboratory frame and the molecular frame, and ϕ_{ij}^{mol} is the set of polar angles of the vector between i and j in the molecular frame. Substituting Eq. (6) into Eq. (4) we have,

$$J_{ij}^n(\omega) = \sum_{m,m'} \int_0^\infty \Big\langle D_{nm}^2(\Omega(0)) D_{nm'}^{2*}(\Omega(t))$$

$$\frac{Y_{2m}\left(\phi_{ij}^{mol}(0)\right) Y_{2m'}^*\left(\phi_{ij}^{mol}(t)\right)}{r_{ij}^3(0) r_{ij}^3(t)} \Big\rangle \cos(\omega t) dt. \tag{7}$$

To make Eq. (7) tractable, we assume that the overall molecular tumbling and the internal rotation are uncorrelated. Thus one can evaluate the average of the product in Eq. (4) as the product of the averages. We further assume that the overall molecular tumbling is isotropic with a correlation time τ_c. We then have[12,16]:

$$\Big\langle D_{nm}^2(\Omega(0)) D_{nm'}^2,(\Omega(t)) \Big\rangle = \frac{1}{5} \delta_{mm'}, \exp(-t/\tau_c). \tag{8}$$

Eq. (7) then becomes,

$$J_{ij}^n(\omega) = \frac{1}{5} \sum_m \int_0^\infty \langle \frac{Y_{2m}\left(\phi_{ij}^{mol}(0)\right) Y_{2m}^*\left(\phi_{ij}^{mol}(t)\right)}{r_{ij}^3(0) r_{ij}^3(t)} \rangle \exp(-t/\tau_c) \cos(\omega t) dt. \tag{9}$$

Since the right-hand-side is independent of n, we can henceforth remove this index from J. We now make use of the jump model to calculate the average in Eq. (9). Suppose the vector from i to j suffers jumps among N states.

The probability P_μ that the system is in the μ-th conformer satisfies the differential equation,

$$\frac{dP_\mu}{dt} = \sum_{\nu=1}^{N} A_{\mu\nu} P_\nu \tag{10}$$

where $A_{\mu\nu}$ characterizes the rate of jumps among sites. The solution to Eq. (10) can be written down immediately.

$$P_\mu(t) = \sum_{\nu=1}^{N} \exp(\mathbf{A}t)_{\mu\nu} P_\nu(0), \tag{11}$$

where $\exp(\mathbf{A}t)$ is the matrix exponential operator of the matrix $(\mathbf{A}t)$. At large t, the probability P_μ tends to the equilibrium $\langle P_\mu \rangle$. The average in Eq. (9) becomes

$$\left\langle \frac{Y_{2m}\left(\phi_{ij}^{mol}(0)\right) Y_{2m}^*\left(\phi_{ij}^{mol}(t)\right)}{r_{ij}^3(0)r_{ij}^3(t)} \right\rangle =$$

$$\sum_{\mu,\nu=1}^{N} \langle P_\nu \rangle \exp(\mathbf{A}t)_{\mu\nu} \frac{Y_{2m}\left(\phi_{ij,\nu}^{mol}\right) Y_{2m}^*\left(\phi_{ij,\mu}^{mol}\right)}{r_{ij,\nu}^3 r_{ij,\mu}^3}. \tag{12}$$

Explicit indices are inserted to denote site dependence, i.e., $r_{ij,\nu}$ is the ij distance in conformer ν. Finally substituting Eq. (12) into Eq. (9), we obtain the spectral densities desired.

$$J_{ij}(\omega) = \frac{1}{5} \sum_m \sum_{\mu,\nu=1}^{N} \frac{Y_{2m}\left(\phi_{ij,\nu}^{mol}\right) Y_{2m}^*\left(\phi_{ij,\mu}^{mol}\right)}{r_{ij,\nu}^3 r_{ij,\mu}^3}$$

$$\int_0^\infty \langle P_\nu \rangle \exp(\mathbf{A}t)_{\mu\nu} \exp(-t/\tau_c) \cos(\omega t) dt. \tag{13}$$

Although the general solution of Eq. (13) requires the knowledge of $\mathbf{A}$, and is rather involved[13], there are important cases where it becomes quite simple.

Case 1 — Slow Jumps Among Equally Probable Sites.

In this case, all $\langle P_\mu \rangle = 1/N$. The jump rate $\mathbf{A}$ is slow compared to $1/\tau_c$. We can then approximate $\exp(\mathbf{A}t)$ by the identity matrix in Eq. (12) to obtain

$$J_{ij}(\omega) = \frac{1}{5N} \sum_m \sum_{\mu=1}^{N} \frac{Y_{2m}(\phi_{ij,\mu}^{mol})Y_{2m}^*(\phi_{ij,\mu}^{mol})}{r_{ij,\mu}^6} \int_0^\infty \exp(-t/\tau_c) \cos(\omega t) dt \tag{14}$$

$$= \frac{1}{5N} \left(\frac{\tau_c}{1+\omega^2\tau_c^2}\right) \sum_m \sum_{\mu=1}^{N} \frac{Y_{2m}\left(\phi_{ij,\mu}^{mol}\right) Y_{2m}^*\left(\phi_{ij,\mu}^{mol}\right)}{r_{ij,\mu}^6}. \tag{15}$$

321

Making use of the addition theorem for spherical harmonics, we obtain

$$J_{ij}(\omega) = \frac{1}{4\pi N} \left(\frac{\tau_c}{1 + \omega^2 \tau_c^2} \right) \sum_{\mu=1}^{N} \frac{1}{r_{ij,\mu}^6}. \tag{16}$$

Thus in this case, the spectral densities are simple averages of the spectral densities of different sites. This is expected to be the case, for example, for the δ and ϵ protons of phenylalanine and tyrosine rings, which are commonly in fast exchange on the nmr time scale, but flip more slowly than the 10^{-9} sec tumbling times of small proteins. Since the "jump" involves a 180°rotation about χ_2 to an equivalent position, the net spectral density is just the average of values calculated for the two sides of the ring.

Case 2 — Fast Jumps Among Equally Probable Sites.

Again, we have $\langle P_\mu \rangle = 1/N$. The assumption of fast jumps implies that $\langle P_\nu \rangle \exp(\mathbf{A}t)_{\mu\nu} = \langle P_\nu \rangle \langle P_\mu \rangle$. Therefore Eq. (13) becomes in this case,

$$J_{ij}(\omega) = \frac{1}{5N^2} \left(\frac{\tau_c}{1 + \omega^2 \tau_c^2} \right) \sum_m \sum_{\mu,\nu=1}^{N} \frac{Y_{2m}\left(\phi_{ij,\mu}^{mol}\right) Y_{2m}^{*}\left(\phi_{ij,\nu}^{mol}\right)}{r_{ij,\mu}^3 r_{ij,\nu}^3}$$

$$= \frac{1}{5} \left(\frac{\tau_c}{1 + \omega^2 \tau_c^2} \right) \sum_m \left| \frac{1}{N} \sum_{\mu}^{N} \frac{Y_{2m}\left(\phi_{ij,\mu}^{mol}\right)}{r_{ij,\mu}^3} \right|^2. \tag{17}$$

Again, one can simplify Eq. (17) by using the addition theorem.

$$J_{ij}(\omega) = \frac{1}{4\pi N^2} \left(\frac{\tau_c}{1 + \omega^2 \tau_c^2} \right) \sum_{\mu,\nu=1}^{N} \frac{P_2 \left(\cos(\theta_{ij,\mu\nu}) \right)}{r_{ij,\mu}^3 r_{ij,\nu}^3}, \tag{18}$$

where $\theta_{ij,\mu\nu}$ is the angle between the vectors from i to j in site μ and site ν, and P_2 is a Legendre polynomial. For taking derivatives, it helps to further rewrite this as:

$$J_{ij}(\omega) = \frac{1}{8\pi N^2} \left(\frac{\tau_c}{1 + \omega^2 \tau_c^2} \right) \sum_{\mu,\nu=1}^{N} \frac{1}{r_{ij,\mu}^5 r_{ij,\nu}^5} \left(3(\mathbf{r}_{ij,\mu} \cdot \mathbf{r}_{ij,\nu})^2 - r_{ij,\mu}^2 r_{ij,\nu}^2 \right), \tag{19}$$

where as before, $\mathbf{r}_{ij,\mu}$ stands for the vector from i to j for site μ.

One application of Eq. (19) would be to methyl groups, which may be modelled as jumping among three equally probable sites with a time constant that is short compared to molecular tumbling[17,18]. This should be an acceptable approximation for most macromolecules, overall rotational correlation times are greater than 2 nsec. Methyl correlation times are expected to be in the range of 0.02 to 0.2 nsec[19,20]. As with aromatic ring rotations, the sites are equally probable, but now the angular factors do not disappear. Although formally the distance dependence that is averaged in Eq. (17) is r^{-3} (compared to r^{-6} for the slow jump model) we show below that it is generally *not* a good approximation to use an r^{-3} average of the various conformers while ignoring the angle changes, since angle and distance variations often work in opposite directions, partially cancelling each other[21,22].

For cross relaxation between two methyl groups, this model can be expanded to a 9-state model[17]; there are then 81 terms in the summation in Eq. (19), but this poses little computational burden since the time-consuming steps of our refinement are located elsewhere, as we illustrate below.

Case 3 — Combination of Slow and Fast Jumps.

Finally, we have the case where the vector r_{ij} exhibits slow jumps among M groups of sites, where within each group the jumps are fast. In this case, the spectral densities are given by a simple average over the M groups of the spectral densities computed using Eq. (18) or Eq. (19) for each individual fast jumping group. More discussion of this general case is given elsewhere[23].

Derivatives of the Rate Matrix Elements

In order to carry out automated refinements, we also need the derivatives of the spectral densities with respect to the cartesian coordinates of the protons. Since the slow-jump case is straightforward, we present here just the derivatives of Eq. (19). We start with the derivative with respect to the coordinates of the interproton vector in conformational state σ. (In the following the subscripts ij will be suppressed for clarity, it being understood that both the spectral density and its derivative correspond to the same ij.) First, simplifying Eq. (19), we get

$$J(\omega) = \frac{1}{8\pi N^2} \left(\frac{\tau_c}{1 + \omega^2 \tau_c^2} \right) \sum_{\mu,\nu=1}^{N} \left(\frac{1}{r_\mu^5 r_\nu^5} 3 \left(\sum_{\alpha=1}^{3} r_m u^\alpha r_\nu^\alpha \right)^2 - \frac{1}{r_\mu^3 r_\nu^3} \right). \quad (20)$$

The coordinates of $\mathbf{r}$ are explicitly shown above. To take the derivative with respect to r_σ^α, we note that

$$\frac{\partial r_\mu}{\partial r_\sigma^\alpha} = \frac{r_\sigma^\alpha}{r_\sigma} \delta_{\sigma\mu}. \quad (21)$$

We then have,

$$\frac{\partial J(\omega)}{\partial r_\sigma^\alpha} = \frac{3}{8\pi N^2} \left(\frac{\tau_c}{1 + \omega^2 \tau_c^2} \right)$$

$$\times \sum_{\mu,\nu=1}^{N} \left[\frac{-5 r_\sigma^\alpha \left(\sum_{\beta=1}^{3} r_{\mu\beta} r_\nu^\beta \right)^2}{r_\sigma{}^7} \left(\frac{\delta_{\mu\sigma}}{r_\nu^5} + \frac{\delta_{\nu\sigma}}{r_\mu^5} \right) \right.$$

$$\left. + \frac{2 \left(\sum_{\beta=1}^{3} r_\mu^\beta r_\nu^\beta \right) \left(r_\nu^\alpha \delta_{\sigma\mu} + r_\mu^\alpha \delta_{\sigma\nu} \right)}{r_\mu^5 r_{\nu_5}} + \frac{r_\sigma^\alpha}{r_\sigma^5} \left(\frac{\delta_{\mu\sigma}}{r_\nu^3} + \frac{\delta_{\nu\sigma}}{r_\mu^3} \right) \right]. \quad (22)$$

To obtain the derivative with respect to the atomic coordinates, we have

$$r_{ij,\sigma}^\alpha = x_{j,\sigma}^\alpha - x_{i,\sigma}^\alpha. \quad (23)$$

Therefore,

$$\frac{\partial J_{ij}(\omega)}{\partial x_{j,\sigma}^\alpha} = -\partial J_{ij}(\omega)\partial x_{i,\sigma}^\alpha = \frac{\partial J_{ij}(\omega)}{\partial r_{ij,\sigma}^\alpha}. \quad (24)$$

Finally,

$$\frac{\partial J_{ij}(\omega)}{\partial x_j^\alpha} = \sum_\sigma \frac{\partial J_{ij}(\omega)}{\partial x_{j,\sigma}^\alpha}. \quad (25)$$

Eqs. (22) through (25) give the desired expression for the derivative of the spectral density with respect to the Cartesian coordinates of the protons. Once these are available, the derivatives of the NOESY intensities can be determined by the methods we outlined earlier[6]. The integrated peak intensities in an idealized NOESY spectrum are given by the exponential of $\mathbf{R}$ times τ_m, the mixing time. This is most conveniently treated by using the eigenvectors $\mathbf{L}$ and eigenvalues[24] $\mathbf{\Lambda}$ of $\mathbf{R}$:

$$\exp\left(-\mathbf{R}\tau_m\right) = \mathbf{L}^T \exp(-\mathbf{\Lambda}\tau_m)\mathbf{L}. \quad (26)$$

This gives results equivalent to numerical integration[10,25] of Eq. (1), but is generally more efficient if many peaks are to be computed, and further is amenable to analytical differentiation[6]:

$$\nabla_\mu \left[\exp(-\mathbf{R}\tau)\right]_{ij} = \sum_{rstu} L_{ir}^T L_{rs}(\nabla_\mu \mathbf{R})_{st} L_{tu}^T L_{uj} \left[\frac{\exp(-\lambda_r \tau_m) - \exp(-\lambda_u \tau_m)}{(\lambda_r - \lambda_u)} \right].$$

$$(27)$$

Here μ represents any parameter that affects the elements of the matrix $\mathbf{R}$; generally, these will be the coordinates of the protons, but could also be scaling factors or correlation times as well. Once the intensity derivatives are available, refinement can take place via simulated annealing techniques like those in common use in crystallography[5].

Comparison to Other Motional Models

One powerful way of representing the effects of internal motion on NMR relaxation is the "model-free" approach developed by Lipari and Szabo[23]. Here the potentially complex spectral density is replaced by a two-term expression in which the Lorentzian term representing overall motion (see Eq. 5) is reduced by a generalized order parameter S^2, and a second Lorentzian term (weighted by $1 - S^2$) represents the averaged effects of internal motion. When both angular and distance fluctuations are present, the revised spectral density becomes:

$$J_{ij}(\omega) \frac{1}{4\pi} \langle r_{ij}^{-6} \rangle \left(\frac{S^2 \tau_c}{1 + \omega^2 \tau_c^2} + \frac{(1 - S^2)\tau}{1 + \omega^2 \tau^2} \right) \tag{28}$$

where

$$\tau^{-1} \equiv \tau_c^{-1} + \tau_e^{-1} \tag{29}$$

and τ_e is an effective internal correlation time. It is possible to relate τ_e to the individual rate constants in a jump model (Eq. A10 of Lipari and Szabo), but in the simple theory we are dealing with here, all of the internal jump rates are assumed to be much faster than the overall tumbling rate, so that the second term of Eq. (28) can be neglected. In this limit, the order parameter S^2 is just the the jump model spectral density (Eq. 18) divided by a slow motion model in which the jump rates are much slower than overall tumbling (Eq. 16). [This definition is slightly different than that proposed by Lipari and Szabo (their Eq. 48), but has the advantage of keeping S^2 as a dimensionless quantity; our notation matches that of Olejniczak et al.[22].]

In essence then, for fast internal motions, our jump model is simply a means of computing an effective order parameter. A number of workers have adopted the strategy of determining this parameter in an empirical fashion, so that all methyls, for example, might be assigned a single, averaged value of S^2. This is often done under the guise of assigning effective correlation times, $\bar{\tau}_c$, to groups with internal mobility[9,24,26]. For typical biomolecules, where $\omega\tau_c \gg 1$ and the J(0) terms dominate the cross-relaxation rates, this is equivalent to assuming an order parameter with $S^2 = \bar{\tau}_c/\tau_c$. Still others ignore angular motions but compute averaged distances based upon a centroid distance or upon an r^{-3} weighted average of the distances in individualconformers[27].

In order to understand the relationships among these various motional models, we examine the situation of a methyl group cross relaxing with a single proton. The relaxation rates can be calculated by *(1)* the fast three-jump model of Eq. (19); *(2)* a r^{-6} average of three static conformers corresponding to the three methyl rotamers; this would be appropriate if the methyl rotation were slow relative to overall tumbling, as in

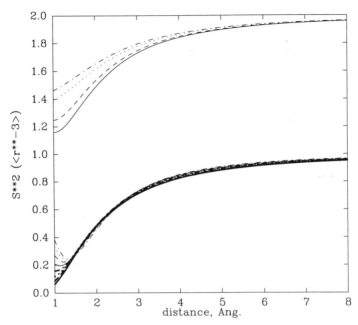

Figure 1. Effective order parameters, S^2, for a single proton interacting with a methyl group. *(bottom)* Values for $\phi = 30°$ and $\theta = 10, 20, 30, 40, 50, 60, 70,$ and $80°$ are given by heavy solid, heavy dashed, heavy dotted, heavy dot-dash, light solid, light dashed, light dotted and light dot-dash lines, respectively; *(top)* Values for $\theta = 40°$ and $\phi = 30, 50, 70$ and $90°$ are given by solid, dashed, dotted and dot-dash lines, respectively; values for S^2 have been shifted upward by one unit.

Eq. (16); *(3)* the average r^{-3} model that ignores angular motion; and *(4)* a static calculation using the distance from the unique proton to the centroid of the methyl protons. An effective S^2 parameter is then calculated as a function of position of the single proton relative to the methyl group for each of the models *(2)-(4)* by dividing the result for model "1" by that for models "2", "3" or "4", as explained above. In a sense, the "correct" comparison is between models *(1)* and *(2)*, i.e., between a static model and one with internal motion. Models *(3)* and *4)* come into play when it is recognized that the *radial* averaging for fast motion has an r^{-3} dependence or when a pseudo-atom is used to represent the methyl position.

For our calculations, we placed the methyl group to have proton coordinates at (0.89, -0.514, 0.0), (-0.89, -0.514, 0.0) and (0.0, 1.028, 0.0), so that its centroid is at the origin, and the z axis is along the axis of rotation. The polar coordinates r, θ and ϕ then locate the single proton with respect to the methyl group. In Figs. 1-3 S^2 is plotted as a function of r for various values of θ and ϕ for various models.

The results give some insight into the approximations inherent in simple motional models, using the jump model as a somewhat improved reference.

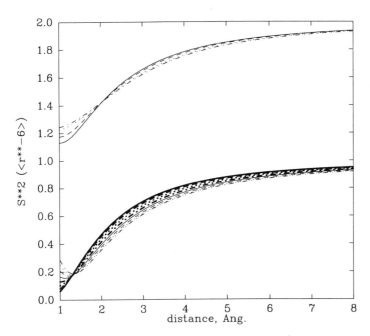

Figure 2. Same as Fig. 1, but using an r^{-3} average rather than r^{-6} for the rigid calculation.

It is clear that the motion of the methyl group becomes increasingly unimportant as the "reporter" proton moves further and further away since the distances and angles in Eq. (18) all become the same; this is indicated by S^2 values that approach unity at long distances. Hence, assuming a fixed (empirical) value for S^2 can never be rigorously correct. Nevertheless, if the cross peaks of interest are all in the 2.5-4.0 Å range, then using an average order parameter that is characteristic of this range may be an acceptable approximation, and will certainly be superior to ignoring internal motion altogether. For methyl groups, a value of around 0.6 for S^2 would be roughly appropriate, and values near to this have indeed been used[9]. It should be remembered, however, that such an approximation introduces errors of at least 20% or so, compared to the jump model, which itself is an imperfect description of methyl motion.

Figs. 2 and 3 illustrate the relationship of two other simple models to the jump model. In the first case (Fig. 2) the proper radial behavior is assumed, but angular effects of methyl rotation are ignored. This offers a slight improvement over the results of Fig. 1, since the order parameters are closer to unity, but the differences are not great. Fig. 3 shows results if a pseudo-atom is used to represent the location of the methyl protons. This has the difficulty that the effective order parameters are strong functions of angles as well as distances, so that simple empirical corrections are less likely to be appropriate. As emphasized by Olejniczak[18], the jump models

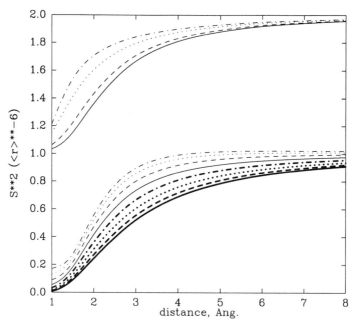

Figure 3. Same as Fig. 1, but using the centroid of the methyl protons for the rigid calculation.

themselves are so straightforward to use that there seems little point in pursuing simpler theories.

CONCLUSIONS

It is clear that a quantitative analysis of NOESY intensities requires close attention to effects of internal motion. The jump models described here are by no means the only alternative for such a description, but they have the advantage of being straightforward to implement and interpret, and of having a minimal number of adjustable parameters. We have discussed only simple cases in which all of the sites in the model are equally probable, and where the jump rates are either very fast or very slow relative to rotational diffusion. Extensions of these ideas are generally straightforward, and have been discussed in several contexts[12,15,28]. The mathematics of jumps with intermediate rates can be worked out, but requires a knowledge of the rate constants, i.e., of extra parameters whose values are not well understood. Models that invoke restricted (internal) diffusion are also available (see London[15] for a review.) It may also prove necessary to drop the assumption of isotropic rotational diffusion in favor of more complex motions[29]; this again will require more detailed knowledge of molecular motions, and perhaps additional adjustable parameters. Probably the next most complex model beyond those discussed here would use both terms in

Eq. (28), rather than just the first term, as we have done; this would require just one additional parameter, τ_e. Studies along these lines are in progress; an analysis of ^{13}C relaxation in a zinc-finger peptide suggests that internal motions are indeed fast enough that τ can be neglected relative to τ_c in Eq. (28) (A. Palmer, M. Rance and P. E. Wright, personal communication).

It is certainly feasible to imagine gaining improved understanding of the effects of internal motions by examining theoretical simulations, especially since these can now approach nanosecond time scales even when solvent effects are included. A number of studies along these lines have been reported[21,22,30,31], but it is not clear how to extract accurate models for the hundreds to thousands of cross peaks that need to be analyzed in a large-scale refinement. Fortunately, the distance dependence of NOESY intensities is sufficiently sharp that even sizeable errors in motional corrections lead to only small uncertainties in distances; turned around, this relationship implies that it will only rarely be possible to deduce motional characteristics from NOE data, since uncoupling distance and motional effects will be very difficult.

We have not dealt here with tests of the accuracy of the jump models themselves, other than to note that the physical assumptions are plausible. Elsewhere, we report detailed comparisons of NOESY buildup curves for plastocyanin, using both fixed- and non-fixed-distance pairs of protons. Those results suggest that the current level of approximations are sufficiently good to serve as a basis for a refinement scheme. We also expect further feedback between theory and experiment to lead to refined models for cross-relaxation effects.

ACKNOWLEDGEMENTS

We thank Peter Wright and Chris Lepre for many interesting discussions, and the National Institutes of Health (GM38794) for financial support.

REFERENCES

1. K. Wüthrich, *Science* **243**, 45-50 (1989).
2. R. Kaptein, R. Boelens,, R.M. Scheek, and W.F. van Gunsteren, *Biochemistry* **27**, 5389-5395 (1988).
3. G.M. Clore, and A.M. Gronenborn, *Crit. Rev. Biochem. Mol. Biol.* **24**, 479-564 (1989).
4. G.P. Gippert, P.F. Yip, P.E. Wright, and D.A. Case, *Biochem. Pharm.* **40**, 15-22 (1990).
5. A.T. Brünger, J. Kuriyan, and M. Karplus, *Science* **235**, 458-460 (1987).
6. P. Yip, and D.A. Case, *J. Magn. Reson.* **83**, 643-648 (1989).
7. R. Boelens, T.M.G. Koning, and R. Kaptein, *J. Molec. Struct.* **173**, 299-311 (1988).

8. B.A. Borgias, M. Gochin, D.J. Kerwood, and T.L. James, *Prog. NMR Spect.* **22**, 83-100 (1990).

9. J.D. Baleja, R.T. Pon, and B.D. Sykes, *Biochemistry* **29**, 4828-4839 (1990).

10. W. Nerdal, D.R. Hare, and B.R. Reid, *Biochemistry* **28**, 10008-10021 (1989).

11. E.P. Nakonowicz, R.P. Meadows, and D.G. Gorenstein, *Biochemistry* **29**, 4193-4204 (1990).

12. R.J. Wittebort and A. Szabo, *J. Chem. Phys.* **69**, 1722-1736 (1978).

13. J. Tropp, *J. Chem. Phys.* **72**, 6035-6043 (1980).

14. J.W. Keepers and T.L. James, *J. Am. Chem. Soc.* **104**, 929-939 (1982).

15. R.E. London, *Meth. Enzym.* **176**, 358-375 (1989).

16. D. Wallach, *J. Chem. Phys.* **47**, 5258-5268 (1967).

17. D.E. Woessner, *J. Chem. Phys.* **42**, 1855-1859 (1965).

18. E.T. Olejniczak, *J. Magn. Reson.* **81**, 392-394 (1989).

19. E.R. Andrew, W.S. Hinshaw, and M.G. Hutchins, *J. Magn. Reson.* **15**, 196 (1974).

20. D.A. Torchia, 1984, *Annu. Rev. Biophys. Bioeng.* **13**, 125-144 (1984).

21. D.M. LeMaster, L.E. Kay, A.T. Brünger, and J.H. Prestegard, J. H., *FEBS Lett.* **236**, 71-76 (1988).

22. E.T. Olejniczak, C.M. Dobson, M. Karplus, and R.M. Levy, *J. Am. Chem. Soc.* **106**, 1923-1930 (1984).

23. G. Lipari, and A. Szabo, *J. Am. Chem. Soc.* **104**, 4546-4559 (1982).

24. J.W. Keepers, and T.L. James, *J. Magn. Reson.* **57**, 404-426 (1984).

25. M. Madrid, and O. Jardetzky, *Biochim. Biophys. Acta* **953**, 61-69 (1988).

26. G.M. Clore and A.M. Gronenborn, *FEBS Lett.* **172**, 219-225 (1984).

27. H. Kessler, C. Griesinger, J. Lautz, A. Müller, W.F. van Gunsteren, and H.J.C. Berendsen, *J. Am. Chem. Soc.* **110**, 3393-3396 (1988).

28. S.B. Landy and B.D.N. Rao, *J. Magn. Reson.* **81**, 371-377 (1989).

29. A.J. Duben, and W.C. Hutton, *J. Am. Chem. Soc.* **112**, 5917-5924 (1990).

30. S.W. Fesik, T.J. O'Donnell, R.T. Gampe, and E.T. Olejniczak, *J. Am. Chem. Soc.* **108**, 3165-3170 (1986).

31. D. Genest, *Biopolymers* **28**, 1903-1911 (1989).

REFINEMENT OF THREE-DIMENSIONAL PROTEIN AND DNA STRUCTURES IN SOLUTION FROM NMR DATA[†]

Thomas L. James*, Miriam Gochin, Deborah J. Kerwood,
David A. Pearlman, Uli Schmitz, and Paul D. Thomas

Department of Pharmaceutical Chemistry
University of California
San Francisco, CA 94143 U.S.A.

INTRODUCTION

Two-dimensional NMR, in particular two-dimensional nuclear Overhauser effect (2D NOE) spectra, when used in conjunction with distance geometry and energy refinement calculations can be used to determine the high-resolution structure of DNA fragments and small proteins. To understand functional interactions of proteins and nucleic acids, it is important to know their solution structures to high-resolution. Problems addressed with DNA structure and with protein structure studies are often of a different nature. In general, we are interested in fairly subtle structural changes in the DNA helix which are sequence-dependent and, consequently, guide protein, mutagen or drug recognition. These subtle variations demand detailed knowledge of the structure and, therefore, accurate internuclear distance and perhaps torsion angle constraints. But one can define a protein tertiary structure with moderate accuracy using distance geometry or restrained molecular dynamics calculations without accurately determining interproton distances; a qualitative assessment of 2D NOE intensities is often all that is needed. However, in proteins possessing less common structural features, it may be especially valuable to have additional structural constraints and more accurate constraints for use with the computational techniques. And, even more

[†]This work was supported by the National Institutes of Health via grants GM39247, RR01695, and CA27343.

*Author to whom correspondence should be addressed.

Computational Aspects of the Study of Biological Macromolecules by Nuclear Magnetic Resonance Spectroscopy, Edited by J.C. Hoch *et al.*, Plenum Press, New York, 1991

331

importantly, we will want better defined structures at ligand binding sites (with and without ligand bound).

To obtain the best possible structures, one should utilize as many structural constraints as possible and determine these constraints as accurately as possible. (Certainly we wish to avoid systematically biased structural constraints.) There are different methods of analyzing 2D NOE spectra for internuclear distance and structural information. We have examined the use of the commonly employed two-spin or isolated spin pair approximation (ISPA) to obtain interproton distances from 2D NOE spectra and found, while the approximation leads to sizeable systematic as well as random distance errors[1,2], good protein structures can still be obtained, albeit with occasional local structural distortions when the distances are assumed to be more accurate than is warranted (*vide infra*). Use of a complete relaxation matrix approach (CORMA) to ascertain interproton distances from 2D NOE peak intensities enables more accurate distance determinations as well as a greater number of constraints to be derived[1-3]; this, consequently, offers the opportunity of determining protein solution structure with greater accuracy and resolution and is essential for determination of DNA helix structures. The most effective techniques employ iterative refinement against experimental 2D NOE spectra by calculating theoretical spectra for the molecular structures during refinement[1,3-5], in concert with molecular mechanics, molecular dynamics or distance geometry calculations. For example, the program MARDIGRAS (matrix analysis of relaxation for discerning geometry of an aqueous structure) seems to be rather robust and displays little dependence on the initial model while yielding reliable distances with computational efficiency; the resulting distances can be used with restrained molecular dynamics or distance geometry calculations.

The distance information from the 2D NOE analysis has been augmented by limited torsion angle information derived from coupling constants obtained from double-quantum-filtered COSY (2QF-COSY) and exclusive COSY (ECOSY) spectra. Broad lines prevented direct analysis of coupling constants, so we used simulation of 2QF-COSY cross peaks using the programs SPHINX and LINSHA[6]; to extract vicinal coupling constants and subsequently torsion angle constraints[7]. Some DNA fragments have been recently investigated by NMR to determine structure in solution; these will be used to illustrate the experimental implementation of the methodology.

DISTANCE CONSTRAINTS

Relationship of Interproton Distances and 2D NOE Peak Intensities

The effect of cross relaxation between two neighboring protons during the mixing time period τ_m of the 2D NOE experiment is to transfer magnetization between them. This results in cross-peak intensities in the spectrum that are approximately inversely proportional to the sixth power of the distance between the two neighboring protons. However, the neighboring protons belong to an array of all protons in the molecular structure. So the cross relaxation between the two is part of a coupled relaxation network. To be rigorous, the whole relaxation network should be considered.

The modification of the magnetization due to cross relaxation during the mixing time period τ_m of the 2D NOE experiment is expressed by[8]:

$$\mathbf{M}(\tau_m) = \mathbf{a}(\tau_m)\mathbf{M}(0) = e^{-\mathbf{R}\tau_m}\mathbf{M}(0). \tag{1}$$

In Eq. 1, $\mathbf{M}$ is the magnetization vector describing the deviation from thermal equilibrium ($\mathbf{M} = \mathbf{M}_z - \mathbf{M}_0$, and $\mathbf{R}$ is the matrix describing the complete dipole-dipole relaxation network, where diagonal and off-diagonal elements are

$$R_{ii} = 2(n_i - 1)(W_1^{ii} + W_2^{ii}) + \sum_{j \neq i} n_j(W_0^{ij} + 2W_1^{ij} + W_2^{ij}) + R_{1i};$$
$$R_{ij} = n_i(W_2^{ij} - W_0^{ij}). \tag{2}$$

Here n_i is the number of equivalent spins in a group such as a methyl rotor, and the zero, single, and double quantum transition probabilities W_n^{ij} are given (for isotropic random reorientation of the molecule) by:

$$W_0^{ij} = \frac{q\tau_c}{r_{ij}^6}; \quad W_1^{ij} = 1.5\frac{q\tau_c}{r_{ij}^6}\frac{1}{1 + (\omega\tau_c)^2}; \quad W_2^{ij} = 6\frac{q\tau_c}{r_{ij}^6}\frac{1}{1 + 4(\omega\tau^c)^2} \tag{3}$$

where $q = 0.1\gamma^4\hbar^2$. The term R_{1i} represents external sources of relaxation such as paramagnetic impurities and is generally ignored.

One problem to consider is the nature of the molecular motion. It has been shown that the error introduced by the assumption of *effective* isotropic motion is small, i.e., $\leq 10\%$[2]. This means that the simple mathematical form of the spectral density employed in Eq. 3 is satisfactory — but one may need to utilize different *effective* correlation times for different proton pairs in a protein.

$\mathbf{a}$ is the matrix of mixing coefficients which are proportional to the 2D NOE intensities. So this matrix of mixing coefficients is what we wish to evaluate. The exponential dependence of the mixing coefficients on the cross-relaxation rates complicates the calculation of intensities (or the distances). We note that the expression for the above rate matrix is actually still an approximation in that it neglects cross-correlation terms between separate pairwise and higher order interactions[9,10], and the expressions given above also do not account for second-order effects due to strong scalar coupling[11]. However, the magnitude of error due to neglect of these effects is small.

We have compared different methods of analyzing 2D NOE spectra for internuclear distance and structural information[1]. The method entailing the Isolated Spin Pair Approximation (ISPA) is the easiest and consequently the most commonly employed. The exponential expression of Eq. 1 can be expanded and the series truncated after the linear term for short ($\tau_m \rightarrow$ 0) mixing times. In some studies, NOE build-up curves are obtained to assess whether or not the short mixing time condition is achieved. The practical application of ISPA usually goes one step further to eliminate the dependence on correlation time by scaling all the distances with respect to a known reference distance which is assumed to have the same correlation time as the proton-proton pair of interest. The distance r_{ij} between nuclei i and j is usually determined by comparing the cross-peak intensity a_{ij} with that of a reference cross peak (a_{ref}) which originates from two protons whose internuclear distance r_{ref} is known (correlation time of ij pair and reference pair assumed to be equal) Then distances are calculated according to: $r_{ij} = r_{ref} \left(\frac{a_{ref}}{a_{ij}}\right)^{1/6}$ Clearly, the chief advantage of using ISPA lies in its simplicity. However, the major source of error in this approximation lies in neglecting multispin relaxation effects[1,2,5]. Estimates of the inherent error associated with ISPA vary widely throughout the literature. Some studies use the estimated distances only qualitatively, while others assume that the extreme (i.e. sixth power) dependence of intensity on distance allows distances to be specified more precisely. We have recently shown that, for mixing times generally accepted as being sufficiently short (i.e., 50 to 100 ms) and not including internal motions, ISPA can result in systematic errors of 45-80% in distances over 3.5 Å, the range which is most important in defining molecular structure[3].

An alternative method of estimating interproton distances from 2D NOE intensities involves consideration of all dipole-dipole interactions. Relaxation rates can be determined by solving the matrix equation

$$\mathbf{R} = \frac{-ln\left(\frac{\mathbf{a}(\tau_m)}{\mathbf{a}(0)}\right)}{\tau_m} \qquad (4)$$

Figure 1. Schematic diagram of Matrix Analysis of Relaxation for DIscerning GeometRy of an Aqueous Structure (MARDIGRAS). A model structure is used to generate a theoretical 2D NOE spectrum (using CORMA). Wherever possible, experimental intensities are substituted into the theoretical spectrum to yield a full spectrum (hybrid intensity matrix) suitable for direct solution of the rate matrix $\mathbf{R}$ and subsequently the distance matrix. Iterative steps require that (a) cross-relaxation rates R_{ij} between protons i and j with known interproton distance are not allowed to change; (b) diagonal and off-diagonal elements are required to be consistent; and (c) cross-relaxation rates corresponding to resolved and observed cross peaks only are allowed to change. This process incorporates all the effects of network relaxation and multiple spin effects.

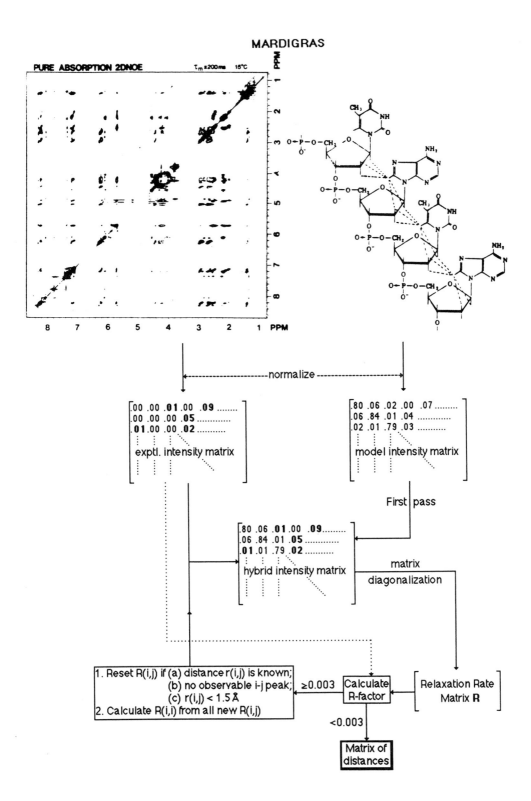

MARDIGRAS

PURE ABSORPTION 2DNOE τ_m =200ms   15°C

normalize

$$\begin{bmatrix} .00 & .00 & .01 & .00 & .09 & \cdots \\ .00 & .00 & .00 & .05 & & \cdots \\ .01 & .00 & .00 & .02 & & \cdots \\ \vdots & \vdots & \vdots & & & \end{bmatrix}$$

exptl. intensity matrix

$$\begin{bmatrix} .80 & .06 & .02 & .00 & .07 & \cdots \\ .06 & .84 & .01 & .04 & & \cdots \\ .02 & .01 & .79 & .03 & & \cdots \\ \vdots & \vdots & \vdots & & & \end{bmatrix}$$

model intensity matrix

First pass

$$\begin{bmatrix} .80 & .06 & .01 & .00 & .09 & \cdots \\ .06 & .84 & .01 & .05 & & \cdots \\ .01 & .01 & .79 & .02 & & \cdots \\ \vdots & \vdots & \vdots & & & \end{bmatrix}$$

hybrid intensity matrix

matrix
diagonalization

1. Reset R(i,j) if (a) distance r(i,j) is known;
   (b) no observable i-j peak;
   (c) r(i,j) < 1.5 Å
2. Calculate R(i,i) from all new R(i,j)

≥0.003   Calculate R-factor   Relaxation Rate Matrix **R**

<0.003

Matrix of distances

which is another form of matrix Eq. 1 above. Eq. 4 can be solved analytically without a series expansion, using linear algebra to solve the eigenvalue problem with $\mathbf{R} = \chi\lambda\chi^{-1}$, where χ is the matrix of characteristic eigenvectors for the matrix and λ is the diagonal matrix of eigenvalues. Because λ is diagonal, its exponential (or logarithmic) expansion collapses. This method explicitly accounts for multispin effects. Having calculated $\mathbf{R}$, interproton distances can then be extracted from cross-relaxation rates using Eqs. 2 and 3.

$\mathbf{a}(\tau_m)$, however, will be incomplete due to experimental limitations of resolution and spectral noise; a direct calculation of $\mathbf{R}$ therefore will not yield accurate, unbiased distances[1]. Consequently, we explored a more time-consuming approach of iterative least-squares refinement of structure based on the 2D NOE intensities; this entailed development of the program COMATOSE[1]. We have found COMATOSE to work reasonably well for DNA fragments[12], but it is relatively slow and subject to finding local minima. Another potentially useful iterative method for calculation of distances was termed iterative relaxation matrix analysis (IRMA)[4,13]. IRMA iteratively calculates distances with the aid of r-MD (or potentially DG), as it is feasible to supplement experimental intensities with intensities calculated for a model structure if those intensities would otherwise be missing in $\mathbf{a}(\tau_m)$. We have subsequently shown that an accurate set of distances can be determined by iteratively recalculating $\mathbf{R}$, starting with the hybrid intensity matrix and iteratively inserting experimental intensities, without relying at each cycle on more computationally-expensive techniques such as DG or r-MD[3,14]; the procedure, termed MARDIGRAS for Matrix Analysis of Relaxation for DIscerning the GeometRy of an Aqueous Structure, is illustrated in Fig. 1 and works as follows. Based on a starting structure, a 2D NOE relaxation matrix $\mathbf{R}$ is calculated; diagonalization of $\mathbf{R}$, using the program CORMA[1], yields a 2D NOE intensity matrix. Those elements for which experimental data are available are then replaced in the intensity matrix. The experimental values are scaled by the ratio of the sum of calculated intensities to the sum of observed intensities. The substituted NOE intensity matrix is back-transformed to a new relaxation matrix, from which a new set of distances can be derived. Certain elements of the new relaxation matrix are reset, namely cross-peak rates corresponding to fixed distances and cross-peak rates corresponding to distances for which no observed NOE is available. Diagonal rate constants (R_{ii}) are replaced by appropriate sums based on the calculated and constrained cross-relaxation rates. A new cycle of intensity matrix calculation, substitution and back-calculation is then repeated until the calculated and experimentally observed NOE intensities converge.

The MARDIGRAS refinement process is more time-consuming than simpler methods for estimating distances from 2D NOE intensities such as ISPA or the DIRECT method. However, it does not suffer from the systematic errors associated with such methods. No information is needed from the diagonal peak intensities (which are typically overlapping and difficult to measure), and overlapping or tiny cross peaks pose no problem (other than loss of potential additional information). Only well-resolved peaks need be considered. MARDIGRAS is capable of calculating distances from NOE intensities while taking into account spin diffusion and the effects of relaxation in a network of spins. It uses a model structure to generate a matrix of calculated intensities for aiding the distance calculation, but is relatively insensitive to the choice of model[3,14]. Also, the method is relatively fast. Depending on the number of cycles performed the time for calculating the distances from a set of 2D NOE intensities for a small protein (BPTI) is only about 35-50 minutes on a Sun Sparcstation 1, or about 35-60 seconds on a Cray Y-MP. These times show that this method, while substantially more involved than ISPA, is workable in essentially any computing environment, and represents a relatively small investment in computer time in relation to the complete structure determination process. The resulting distances can be used with restrained molecular dynamics or distance geometry calculations.

Use of the complete relaxation matrix methodology utilized by MARDIGRAS permits longer mixing times to be employed, with consequently larger intensities for weak cross peaks and the possibility of measuring more distances and longer distances. But spin diffusion effects ultimately put a limit on the extension to longer mixing times and larger cross-peak intensities[3].

Effect of Approximations on Protein Structure Determination

We can consider what can be gained by using a complete relaxation matrix analysis of protein 2D NOE data compared with ISPA. It is already known that the number of distance constraints is more valuable than the accuracy of the distances. How much more will we gain by using the 2D NOE data more quantitatively? This can be addressed by creating 2D NOE spectra from a hypothetical "true" protein structure, specifically, from the x-ray structure of bovine pancreatic trypsin inhibitor (the 5PTI structure in the Brookhaven Protein Data Bank) to which protons are added; random noise is added to the calculated intensities to simulate realistic signal intensities (the initial random noise level of ± 0.0025 is conservative, being somewhat worse than we actually obtain experimentally in our lab). We also consider peak overlap to be substantial, so the peaks to be used are from a list of experimentally resolved BPTI peaks. Three methods of obtaining distance constraints from 2D NOE peak intensities have been examined *vide infra*: one entails conservative use of ISPA with large error margins, one assumes ISPA to be a fairly good approximation, and one utilizes MARDIGRAS.

Table 1. Root-mean-square deviation (RMSD) between the "average structure"
for distance geometry structures and the "true" structure, the 5 pti crystal
structure of BPTI. The input distances to the distance geometry
program VEMBED were calculated using different methods of
analysis of simulated 2D NOE spectra[a], including random noise and peak
overlap effects, from the "true" structure.

Structure Set (Å)	Backbone Atoms (Å)	Sidechain Atoms (Å)	All Atoms (Å)
control	0.86	1.70	1.42
conservative ISPA	1.13	2.03	1.73
restrictive ISPA	1.13	1.94	1.64
MARDIGRAS	0.86	1.75	1.47

[a]Simulated assuming an isotropic correlation time of 5 nsec.

The distance geometry algorithm VEMBED was used to generate a family of structures for each resulting distance set. As seen in Table 1, the
quality of the average structure from each family was quite good. Basically, although the distances determined using ISPA are not too accurate, a
sufficiently large number of them together with the requirement that they
be self-consistent will constrain the structure adequately. The root-mean-
square deviation (RMSD) of the average structure from the true structure
was improved about 2-5% using the more restrictive rather than the more
conservative ISPA approach. Use of MARDIGRAS in a conservative fashion,
i.e., with a poor initial model, 5PTI with proton coordinates randomized by
an RMS displacement of 3, resulted in improvement in the RMSD by 8-15%.
A comparable or even better result was obtained using another poor initial
model — BPTI as an extended peptide chain with MARDIGRAS (Table 1).
With a better initial model (the 4PTI structure), MARDIGRAS obtained
even more accurate distances. These comparisons do not incorporate another advantage of MARDIGRAS, i.e., MARDIGRAS permits analysis of
2D NOE data at longer mixing times, yielding additional distances. There
is an additional *caveat* regarding use of the more restrictive ISPA distances,
which resulted in a few systematically incorrect structural features in local
regions of the protein, producing distortions of 2-3 Å.

Choice of Figure of Merit

We might consider what criteria should be used to assess structures during
refinement and for comparison between structures obtained using different
refinement strategies. First, it should be noted that 2D NOE peak intensities
are so sensitive to interproton distances that it is very easy to create plausible

model structures which yield theoretical spectra at strong variance with the experimental spectra. Various numerical indices, which could be used as a figure of merit to evaluate the quality of structures, can be defined. Lower numerical values are indicative of an improved fit of the theoretical with the experimental 2D NOE spectra. Among the indices that might be used are the average difference index ADI, the standard deviation σ, and the residual indices (or R factors) R_1 and R_2. We have in the past used a difference index as a means of comparing between closely related model structures[15,16].

$$ADI = \frac{\sum_i |I_o^i - I_c^i|}{N} \qquad \sigma = \left[\frac{\sum_i (I_o^i - I_c^i)^2}{N} \right]^{1/2} \qquad (5)$$

where the subscripts denote calculated (c) and observed (o) intensities, and the summation is over all N observed and calculated 2D NOE intensities. For ADI and σ, the size of a peak is directly proportional to its influence on the value of the index. So big peaks will dominate the evaluation. But big peaks have better signal-to-noise.

One can use a residual index, in keeping with the practice of crystallographers:

$$R_1 = \frac{\sum_i |I_o^i - I_c^i|}{\sum_i I_o^i} \qquad R_2 = \left[\frac{\sum_i (I_o^i - I_c^i)^2}{\sum_i (I_o^i)^2} \right]^{1/2} \qquad (6)$$

R_1 and R_2 give equal weighting to all peaks, including those with lower S/N[17]. For NMR, however, other functions may be more descriptive. We favor sixth-root residual indices which scale approximately with atomic distances:

$$R_1^x = \frac{\sum_i \left| a_o^{1/6}(i) - a_c^{1/6}(i) \right|}{\sum_i a_o^{1/6}(i)}$$

$$R_2^x = \left[\frac{\sum_i \left[a_o^{1/6}(i) - a_6^{1/6}(i) \right]^2}{\sum_i \left[a_o^{1/6}(i) \right]} \right]^{1/2} \qquad (7)$$

These equations attempt to relate the intensities, assuming here approximate dependence on r^{-6}, to the coordinate space of the model. Because of this extreme distance dependence, errors in the shortest, often least structurally interesting distances tend to dominate in the "crystallographic" R factors. The sixth-root scaling allows longer-range interactions (e.g., ~5 Å) to be considered as well, though they are still not weighted as heavily as the larger cross peaks.

For the calculations of protein structure utilizing the ISPA- and MARDI-GRAS-derived distances described above, it was found that comparison between experimental and calculated 2D NOE spectra for the distance geometry structures correlate with the RMSD, generally validating this as a method of structure evaluation. Of the different residual indices, R_1^x ranked structures within a given set most consistently with RMSD rankings, though the correlation was not significant for some sets.

DNA STRUCTURE FROM MARDIGRAS-DERIVED DISTANCES AND RESTRAINED MOLECULAR DYNAMICS

Examination of the molecular mechanics results for [d(TATATATATA)]$_2$ revealed that the methyl group on the thymine played a role in stabilizing the structure. So we began an investigation of this observation by studying [d(ATATATAUAT)]$_2$, which has one methyl missing in each of the complementary strands. Quantitative assessment of the 2D NOE cross-peak intensities was made using MARDIGRAS in conjunction with either A-DNA as starting structure or B-DNA as starting structure. As shown in Fig. 2, both starting structures yielded essentially the same values, thus establishing a set of upper and lower bound interproton distance constraints. These experimental structural constraints were used with restrained molecular dynamics calculations to determine the solution structure of the decamer. As with the MARDIGRAS distance calculations, the r-MD calculations were run using the two starting models standard A-DNA and standard B-DNA. The RMS difference between these two DNA models is 5.04 Å. The all-atom version of the program AMBER, Assisted Model Building with Energy Refinement[18-20], was used for the molecular dynamics calculations, with modifications to include a pseudo-energy term for the NOE-derived distance constraints:

$$
\begin{aligned}
E_{total} &= \sum_{bonds} K_r(r - r_{eq})^2 + \sum_{angles} K_\zeta(\zeta - \zeta_{eq})^2 \\
&+ \sum_{dihedrals} \frac{V_n}{2}[1 + \cos(n\theta - \gamma)] + \sum_{i<j}\left(\frac{A_{ij}}{r_{ij}^{12}} - \frac{B_{ij}}{r_{ij}^6} + \frac{q_iq_i}{\varepsilon r_{ij}}\right) \\
&+ \sum_{H-bonds}\left(\frac{C_{ij}}{r_{ij}^{12}} - \frac{D_{ij}}{r_{ij}^{10}}\right) \\
&+ \sum_{NOE}\left\{\begin{array}{ll} K_{NOE}(r - r_{min})^2 & \text{when } r < r_{min} \\ 0 & \text{when } r_{min} \leq r \leq r_{max} \\ K_{NOE}(r - r_{max})^2 & \text{when } r_{max} \leq r \end{array}\right.
\end{aligned}
\tag{8}
$$

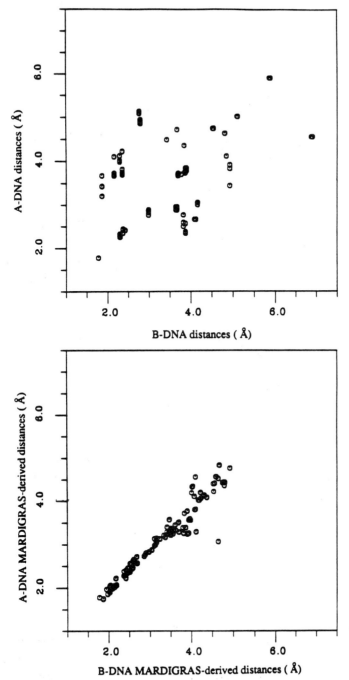

Figure 2. Comparison of the proton-proton distances of the two starting models, standard A- and B-DNA, of [d(ATATATAUAT)]₂ before (top) and after (bottom) refinement with MARDIGRAS.

where r_{min} and r_{max} are the minimum and maximum distances calculated from experimental 2D NOE intensities using MARDIGRAS. Force constant K_{NOE} values range from 1 to 10 kcal/mol-(Å)^2 and can be determined from estimated error in 2D NOE intensities, where peak overlap and signal-to-noise are taken into account. Each molecular dynamics run consisted of 15000 steps using a time step of 0.0015 ps. The temperature was initially 100° K and was gradually increased to 300° K during the first 7.5 ps. The NOE-pseudo-energy term was applied only after 3.0 ps, being linearly increased from 0.1 to 10.0 kcal/mol-(Å)^2 during the next 3.0 ps, and maintained at that level for the remainder of the run. The coordinates of the structures generated during the last 10.5 picoseconds of the 22.5 ps MD run were averaged, and the resulting average structure was subjected to restrained energy minimization utilizing AMBER. The RMS difference between the two final structures (from A- or B-DNA starting structures) is 0.9 Å, implying convergence of the two molecular dynamics runs. Detailed analysis of the structural features suggests that while the decamer is in the B-family of DNA structures, many torsion angle and helical parameters alternate from purine to pyrimidine, with kinks occurring at the U-A steps.

TORSION ANGLE INFORMATION FROM SCALAR COUPLING CONSTANTS

Most 2D NMR experiments are based on scalar coupling. But few are useful for obtaining scalar coupling constants quantitatively. Two which offer the possibility are the double-quantum-filtered COSY (2QF-COSY) and exclusive COSY (ECOSY) experiments. With a sufficient number of coupling constants, the conformation of the sugar rings in nucleic acid fragments may be defined. Linewidths in [d(GGTATACC)]$_2$ were sufficiently narrow that a vicinal coupling constant and sugar puckering analysis could be carried out[21]. Subsequent developments have yielded a superior means of assessing sugar pucker, which may enable determination of coupling constants even when the linewidth exceeds the value of the coupling constants[6,7,22].

Basically, the coupling constants can be extracted from ECOSY or 2QF-COSY spectra via simulation of those spectra utilizing the SPHINX-LINSHA program[6]. Torsion angle constraints in the sugar ring may consequently be derived by use of Karplus-type relationships correlating measured coupling constants to dihedral angles. In more detail, the sugar conformation is analyzed as follows. Individual endocyclic torsion angles ν_i in a particular sugar conformation can be related to the pseudorotation angle P and the amplitude of pucker θ_m, which describe sugar pucker[23]:

$$\nu_i = \theta_m \cos(P + 2(i-2)(360°)/5), i = 0 - 4. \tag{9}$$

The five individual H-H torsion angles $\phi_j (j = 1, 5)$, have been empirically correlated with P and θ_m (usually $\sim 35°$) using 178 crystal structures[24]. The five H-H torsion angles are related to the respective vicinal proton coupling constants J_j by the modified Karplus equation[25]:

$$J_j = a \cos^2 \phi_j + b \cos \phi_j + \sum \Delta \chi_k \left[d + e \cos^2 (\xi_k \phi_j + f |\Delta \chi_k|) \right] \qquad (10)$$

where constants $a, b \ldots f$ have been parameterized using 178 x-ray structures, $\Delta \chi_j$ is the difference in Huggins' electronegativity between substituent S_k and H, and ξ_k is $+1$ or -1, depending on orientation of S_k relative to the geminal protons. In principle, the five vicinal proton coupling constants J_j (1'-2', 1'-2'', 2'-3', 2''-3', 3'-4') can be obtained from 2D NMR spectra via spectral simulation using SPHINX and LINSHA[6,7,22]. Sometimes it is found that a single conformation with characteristic P cannot account for all coupling constants measured for a particular sugar. There may be rapid exchange between n sugar conformations with

$$J_j(obs) = \sum_n f_n J_j(n) \qquad (11)$$

where f_n is the fraction of conformer n. Altona and colleagues assume a two-state model with minor N-conformer ($P = 9°$, i.e., 3'-endo) and major S-conformer (typically $P = 171°$, i.e., 2'-endo). The program PSEUROT can be used to estimate P, θ_m and the fraction of each of the two conformers from the five coupling constants[26].

With our initial studies of [d(GTATATAC)]$_2$, it was not possible to directly extract vicinal proton coupling constant values from any spectrum including ECOSY or 2QF-COSY due to the large linewidths. But comparison of quantitative 2QF-COSY spectral simulations, using SPHINX and LINSHA, with experimental spectra enabled elucidation of coupling constants[7]. The power of the SPHINX method lies in its ability to extract coupling constant information from peak intensity patterns without the necessity for observable resolution of scalar splitting; the cross-peak patterns are very sensitive to the parameters involved: vicinal and geminal coupling constants, linewidth, lineshape, filter functions applied, acquisition times, digital resolution and phase. As the experimental parameters are known, the coupling constants can be estimated by comparison of the experimental and simulated cross peaks of 2QF-COSY spectra. The major difficulty with this approach is in establishing the correct linewidth to be employed[7]. However, the choice of linewidth can usually be constrained such that limitations on sugar pucker can be made. Sometimes, this can lead to a single pucker model which will enable simultaneous fitting of all pertinent 2QF-COSY cross peaks, but sometimes there are two possible solutions for the set of sugar coupling constants[7]. This is illustrated in Fig. 3. In the case of $T2$, simulated cross

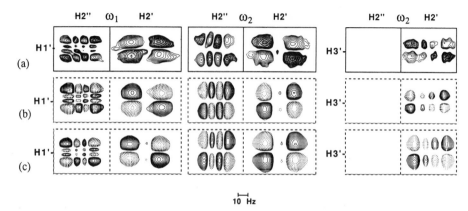

Figure 3. Comparison of experimental 2QF-COSY cross peaks (solid boxes) and best fit simulations (dashed boxes) for T2 in [d(GTATATAC]$_2$. The bottom row of simulated cross peaks corresponds to a three-state model with: $P = 162°(52\%)$, $P = 126°(33\%)$, and $P = 9°(15\%)$, all with $\theta_m = 35°$. The middle row of simulated cross peaks corresponds to a two-state model with: $P = 144°(85\%)$ and $P = 9°(15\%)$, both with $\theta_m = 35°$. Actual chemical shifts and individual linewidths were used. Chemical shift positions are indicated by the proton labels H1', H2', H2'' and H3'.

peaks for a variety of conformational mixtures were computed using a variety of pseudorotation phase angles for the major conformer. The best fit to experimental cross peaks was obtained for a mixture with a major S-conformer (85%) with $P = 144°$ and $\theta_m = 35°$ and a minor N-conformer with $P = 9°$. However, molecular dynamics simulations with oligomers suggest low conformational barriers for the deoxyribose ring, so a three-state model might be more realistic. Employing two different conformers in the S-range of pseudorotation, with $P = 162°(52\%)$ and $P = 126°(33\%)$, along with the minor N-conformer with $P = 9°(15\%)$ all with $\theta_m = 35°$, also resulted in a conformational mixture that provided good simulated peaks.

DNA STRUCTURE FROM NMR-DERIVED DISTANCE AND TORSION ANGLE CONSTRAINTS WITH r-MD

Our early work on [d(AT)$_5$]$_2$ and [d(GGTATACC)]$_2$ suggested that, of the standard DNA structures, a wrinkled D (wD) structure, a member of the B family, best accounted for the experimental 2D NOE spectra for d(TATA) as well as molecular mechanics calculations[15,16]. The experimental data and calculations suggest alternating structural features at TA and AT steps. A question arises as to whether these structural features are characteristic of alternating d(TATA) or whether alternating purine-pyrimidine sequences containing dA-dT base pairs would manifest these alternating structural features. Consequently, the structure of [d(GT)$_4$·d(CA)$_4$] has been investigated by restrained molecular dynamics simulations. In this case, constraints

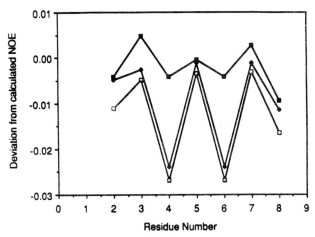

Figure 4. Comparison of the error between calculated and experimental H1'-H4' for three model structures of [d(GT)$_4$·d(CA)4]: B-DNA (- □ -), wD-DNA (- ◆ -) and B-DNA with sugar puckers altered to the values in the last column of Table 2 (- ■ -).

included distances derived from MARDIGRAS analysis of 2D NOE intensities and sugar ring torsion angles derived from SPHINX-LINSHA analysis of 2DF-COSY spectra. The force field utilized in the r-MD calculations included a penalty function of the form

$$\sum_{J-coupling} K_J(\theta_i - \theta_{i_{exp}})^2 \tag{12}$$

to constrain dihedral angles θ_i. A force constant of 25 kcal/mol-Å^2 was utilized.

Both 2D NOE and 2QF-COSY data are consistent with some alternating structural features in [d(GT)$_4$·(CA)$_4$]. One example is shown in Fig. 4. The $H1' - H4'$ distance sensitively reflects sugar pucker and consequently backbone torsion angle δ. CORMA calculations on standard B- and wD-DNA models yield $H1' - H4'$ cross-peak intensities which can be compared with experimental intensities; Fig. 4 exhibits oscillations in the difference between experimental and calculated NOE intensities using B-DNA or wD-DNA, but much reduced oscillations using the sugar puckers listed in the last column of Table 2. These sugar puckers were deduced by fitting the SPHINX-simulated 2QF-COSY spectra to experimental spectra.

As the data clearly demonstrated an alternating character, r-MD runs were made utilizing either wrinkled B-DNA or wrinkled D-DNA as starting structures with the sugar ring torsion angle constraints and MARDIGRAS-derived distance constraints. Within limits, some structural characteristics of [d(GT)$_4$·d(CA)$_4$] could be elucidated[27,28]. Certainly, it is possible to ascertain overall structure, and easily distinguish between different DNA families, i.e., A- or B-DNA. However, one cannot reliably determine

Table 2. Sugar puckers and amplitudes for B-type DNA duplexes

	B	wB	wD	$d(AC)_4d(GT)_4$
Puckers				
purine	171°	178°	178°	180°±10°
pyrimidine	171°	161°	152°	144°±6°
Amplitudes				
purine	35°	43°	44°	35°
pyrimidine	35°	53°	44°	35°

long-range structural features such as bending. Within a family, the 2D NOE approach can also be used to identify deviations from canonical structure, such as wrinkling of alternating sequence DNA, and possibly even the type of wrinkling that is occurring, i.e., distinguish between wB- and wD-DNA.

SUMMARY

It is possible to define a good protein tertiary structure in solution utilizing distances calculated from 2D NOE cross-peak intensities with the unjustified isolated spin pair (or two-spin) approximation *if* the accuracy of those distances is not overestimated. However, for little or no additional effort, considerably more accurate distances may be calculated using the program MARDIGRAS which entails complete relaxation matrix analysis. The use of this approach also enables 2D NOE data obtained at longer mixing times to be analyzed, yielding a greater number of distances. More accurate distances and more distances independently lead to better defined structures. While this is beneficial for analysis of protein solution structure, it is essential for elucidating the sequence-specific structural details of nucleic acid helices. Additional structural constraints are gathered by augmenting the accurate MARDIGRAS-derived distances with torsion angle values estimated from vicinal coupling constants determined by simulating 2QF-COSY and ECOSY spectra. The use of these structural constraints in restrained molecular dynamics calculations has yielded structures for a couple DNA oligonucleotide helices.

ACKNOWLEDGEMENTS

This work was supported by National Institutes of Health grants GM 39247, CA 27343, and RR 01695. We also gratefully acknowledge use of the Cray-YMP supercomputer which was supported by a grant from the Pittsburgh Supercomputing Center through the NIH Division of Research Resources

346

cooperative agreement U41RR04154 and a grant from the National Science Foundation Cooperative Agreement ASC-8500650.

REFERENCES

1. B. A. Borgias and T. L. James, *J. Magn. Reson.* **79**, 493-512 (1988).
2. J. W. Keepers and T. L. James, *J. Magn. Reson.* **57**, 404-426 (1984).
3. B. A. Borgias and T. L. James, *J. Magn. Reson.* **87**, 475-487 (1990).
4. R. Boelens, T. M. G. Koning, and R. Kaptein, *J. Mol. Struc.* **173**, 299-311 (1988).
5. B. A. Borgias and T. L. James, <u>in</u> "Methods in Enzymology, Nuclear Magnetic Resonance, Part A: Spectral Techniques and Dynamics," N. J. Oppenheimer and T. L. James, ed., vol. 176, pp. 169-183, Academic Press, New York (1989).
6. H. Widmer and K. Wüthrich, *J. Magn. Reson.* **74**, 316-336 (1987).
7. U. Schmitz, G. Zon, and T. L. James, *Biochemistry* **29**, 2357-2368 (1990).
8. S. Macura, and R. R. Ernst, *Mol. Phys.* **41**, 95-117 (1980).
9. T. E. Bull, *J. Magn. Reson.* **72**, 397-413 (1987).
10. L. Werbelow and D. M. Grant, *Adv. Magn. Reson.* **9**, 189 (1978).
11. L. E. Kay, T. A. Holak, B. A. Johnson, I. M. Armitage, and J. H. Prestegard, *J. Am. Chem. Soc.* **108**, 4242-4244 (1986).
12. T. L. James, D. J. Kerwood, M. Gochin, P. A. Mills, B. A. Borgias, and V. J. Basus, 1991, <u>in</u> "Fourth Cyprus Conference on New Methods in Drug Research, Proceedings," Makriannis, A., ed., in press, Pergamon Press, Oxford.
13. R. Boelens, T. M. G. Koning, G. A. van der Marel, J. H. van Boom, R. Kaptein, *J. Magn. Reson.* **82**, 290-308 (1989).
14. B. A. Borgias, M. Gochin, D. J. Kerwood, and T. L. James, <u>in</u> "Progress in Nuclear Magnetic Resonance Spectroscopy," J. W. Emsley, J. Feeney, and L. H. Sutcliffe, eds., vol. 22, pp. 83-100, Pergamon Press, Oxford (1990).
15. E.-I. Suzuki, N. Pattabiraman, G. Zon, and T. L. James, *Biochemistry* **25**, 6854-6865 (1986).
16. N. Zhou, A. M. Bianucci, N. Pattabiraman, and T. L. James, *Biochemistry* **26**, 7905-7913 (1987).
17. W. C. Hamilton, "Statistics in Physical Science," Ronald Press, New York (1964).
18. U. C. Singh, P. K. Weiner, J. Caldwell, and P. A. Kollman, "AMBER 3.0," University of California, San Francisco (1986).
19. P. K. Weiner, and P. A. Kollman, *J. Comp, Chem.* **2**, 287-303 (1981).
20. S. J. Weiner, P. A. Kollman, D. T. Nguyen, and D. A. Case, *J. Comp. Chem.* **7**, 230-252 (1986).
21. N. Zhou, S. Manogaran, G. Zon, and T. L. James, *Biochemistry* **27**, 6013-6020 (1988).
22. B. Celda, H. Widmer, W. Leupin, W. J. Chazin, W. A. Denny, and K. Wüthrich, *Biochemistry* **28**, 1462-1470 (1989).
23. C. Altona and M. Sundaralingam, *J. Am. Chem. Soc.* **94**, 8205-8212 (1972).
24. C. A. G. Haasnoot, F. A. A. M. de Leeuw, H. P. M. de Leeuw, and C. Altona, *Org. Magn., Reson.* **15**, 43-52 (1981).
25. L. J. Rinkel, and C. Altona, *J. Biomol. Struct. Dyn.* **4**, 621-649 (1987).
26. F. A. A. M. de Leeuw and C. Altona, *J. Comp. Chem.* **4**, 428-437 (1983).
27. M. Gochin, G. Zon, and T. L. James, *Biochemistry* **29**, 11161-11171 (1990).
28. M. Gochin, and T. L. James, Biochemistry, **29**, 11172-11180 (1990).

HOW TO DEAL WITH SPIN-DIFFUSION AND INTERNAL MOBILITY IN BIOMOLECULES. A RELAXATION MATRIX APPROACH

R. Kaptein, T.M.G. Koning, and R. Boelens

Bijvoet Center for Biomolecular Research
NMR Spectroscopy, Padualaan 8, 3584 CH Utrecht
The Netherlands

ABSTRACT

Errors in proton-proton distances derived from NOEs are mainly due to the effects of spin-diffusion and local mobility. Both can be treated in the frame-work of relaxation matrix theory. The iterative relaxation matrix approach (IRMA), by which errors due to spin-diffusion can be corrected, is reviewed. Symmetrical exchange processes in biomolecules such as aromatic ring flips and methyl group rotation can be easily treated. To account for fast local motions on a picosecond time-scale a method is proposed, in which generalized order parameters S^2 are calculated from a free molecular dynamics run of the molecule in H_2O. The S^2 values are then used to correct NOE derived distances, which in turn are used for the final structure refinement.

INTRODUCTION

The primary source of information for the structure determination of bio-molecules in solution is the nuclear Overhauser effect (NOE). Structures are thus far based on proton-proton distances (or distance ranges) obtained from NOEs measured as cross-peak intensities in 2D NOE spectra. It is therefore important to examine and possibly eliminate the sources of error involved in this distance determination. The most important errors are those due to indirect magnetization transfer or spin-diffusion and due to local mobility. It is well known that for a rigid biomolecule of known structure the NOE intensities including the spin-diffusion effect can be calculated to a very good approximation by solving the coupled relaxation equations (Bloch or Solomon equations). For molecules of unknown structure this cannot be done in a simple way. However, we have devised an iterative relaxation matrix approach (IRMA) by which biomolecular structures can

Computational Aspects of the Study of Biological Macromolecules by Nuclear Magnetic Resonance Spectroscopy, Edited by J.C. Hoch *et al.*, Plenum Press, New York, 1991

be determined with full account of the spin diffusion contribution to the NOEs[1,2].

In fact the relaxation matrix is also a good vehicle to treat some forms of local mobility, the second source of error in distance determinations from NOEs. Symmetrical exchange processes, such as aromatic ring flips of tyrosine and phenylalanine and methyl group rotations can be dealt with in a straightforward way[3]. Other forms of local mobility are probably best studied by measuring ^{15}N and ^{13}C relaxation times T_1 and T_2 in conjunction with the proton-proton cross-relaxation rates. However, since these data are rarely available we have suggested another approach, whereby at least for the fast (picosecond) motions information is obtained from molecular dynamics (MD) simulations and subsequently used in structure refinement. In the following we shall review the relaxation matrix theory, discuss the IRMA method and show how local mobility effects can be treated within the framework of IRMA.

RELAXATION MATRIX THEORY

Multispin relaxation in a biomolecule in solution can be approximately described by the generalized Bloch equations. The evolution of the longitudinal magnetization components is described by the following set of differential equations[4]:

$$\frac{d\Delta \vec{M}(t)}{dt} = -\mathbf{R}\Delta \vec{M}(t). \tag{1}$$

The vector $\Delta M(t)$ comprises the deviations from thermal equilibrium for all spins i:

$$\Delta M(t)_i = [M_z(t) - M_0]_i \tag{2}$$

where M_0 is the equilibrium magnetization. The cross-relaxation matrix $\mathbf{R}$ is given by:

$$\mathbf{R} = \begin{bmatrix} \rho_1 & \sigma_{12} & \cdot & \cdot & \cdot & \sigma_{1n} \\ \sigma_{21} & \rho_2 & \cdot & \cdot & \cdot & \sigma_{2n} \\ \cdot & \cdot & \cdot & \cdot & \cdot & \cdot \\ \cdot & \cdot & \cdot & \cdot & \cdot & \cdot \\ \cdot & \cdot & \cdot & \cdot & \cdot & \cdot \\ \sigma_{n1} & \sigma_{n2} & \cdot & \cdot & \cdot & \rho_n \end{bmatrix} \tag{3}$$

where ρ_i is the diagonal relaxation rate for proton i according to[4,5]:

$$\rho_i = K \sum_{\substack{j=1 \\ j \neq i}}^{N} \left(\frac{1}{r_{ij}^6}\right) [6J_2(\omega) + 3J_1(\omega) + J_0(\omega)] + R_{leak}. \tag{4}$$

The first term describes the dipolar relaxation rate, while all other sources of relaxation are taken together in the second term, R_{leak}. The off-diagonal elements of **R** contain the cross-relaxation rates, σ_{ij}, between protons i and j given by:

$$\sigma_{ij} = K \left(\frac{1}{r_{ij}^6} \right) [6J_2(\omega) - J_0(\omega)]. \tag{5}$$

In Eqs. 4 and 5 r_{ij} is the distance between the proton i and j and $K = 0.1\gamma^4\hbar^2(\mu_0/4\pi)^2$. For isotropic tumbling of the molecule with a correlation time τ_c the spectral densities $J_n(\omega)$ take the simple form:

$$J_n(\omega) = \frac{\tau_c}{1 + n^2\omega^2\tau_c^2} \tag{6}$$

where ω is the Larmor frequency of the protons. The solution of Eq. 1 is:

$$\Delta \vec{M}(t) = \left[e^{-\mathbf{R}t} \right] \Delta \vec{M}_0. \tag{7}$$

Thus, if only spin j is perturbed initially the magnetization of spin i is given by:

$$\Delta M_i(t) = \left[e^{-\mathbf{R}t} \right]_{ij} \Delta M_{j0}. \tag{8}$$

Macura and Ernst[6] have shown that analogous to Eq. 8 the integrated intensities in a 2D NOE spectrum recorded with a mixing time τ_m are given by:

$$I_{ij}(\tau_m) = \left[e^{-\mathbf{R}\tau_m} \right]_{ij} M_{j0} = A_{ij}(\tau_m) M_{j0}. \tag{9}$$

This defines the exponential matrix of normalized NOE intensities

$$\mathbf{A} = \exp\left[-\mathbf{R}\tau_m \right]. \tag{10}$$

Therefore, the matrix **A** will be referred to as the "NOE" matrix.

Formally, the matrix equations can be solved as

$$\mathbf{A} = \mathbf{X} \exp\left[-\Lambda\tau_m \right] \mathbf{X}^{-1} \tag{11}$$

where Λ is the diagonal eigenvalue matrix obtained after diagonalization of **R**

$$\mathbf{X}^{-1}\mathbf{R}\mathbf{X} = \Lambda. \tag{12}$$

Therefore, given a molecular model the NOE matrix can be calculated from the relaxation matrix for each mixing time τ_m of a 2D NOE experiment and the buildup of NOE intensities in a time series can be obtained.

The reverse procedure is also possible[7,8]. When the complete NOE matrix **R** is known, the relaxation matrix **A** can be obtained after diagonalization of the NOE matrix

$$\mathbf{X}^{-1}\mathbf{A}\mathbf{X} = \mathbf{D} = \exp\left[-\Lambda\tau_m \right] \tag{13}$$

351

$$R = -X \frac{|\ln D|}{\tau_m} X^{-1}. \qquad (14)$$

Thus, theoretically the matrix R can be obtained from one 2D NOE experiment taken at a suitably chosen mixing time or more accurately by averaging over a series of τ_m values. In practice for biomolecules this cannot be done directly because the NOE matrix is not completely known due to incomplete assignments and peak overlap. However, the diagonalization of Eq. 13 can still be carried out for suitable combinations of experimental and theoretical NOE matrices. This forms the basis of the IRMA procedure as is discussed below.

ITERATIVE RELAXATION MATRIX APPROACH (IRMA)

A flow diagram of the procedure is shown in Fig. 1. It starts with an initial structural model, possibly obtained from a set of approximate distance constraints. For this starting structure and a choice for the motional behavior of the molecule the relaxation matrix is set up and the NOE matrix is calculated. The off-diagonal elements of this matrix for which experimental NOEs are available are then replaced by the measured ones. The combined NOE matrix is diagonalized to obtain a relaxation matrix with cross-relaxation elements now including the effect of spin diffusion. After averaging a set of matrices for a series of mixing times this then leads to an improved set of distance constraints, from which a better structural model can be obtained using distance geometry or restrained molecular dynamics methods. The procedure can be repeated in subsequent cycles until all experimental NOEs are satisfactorily explained.

The IRMA method was tested on theoretical[1] and experimental[2] data sets of a DNA octamer and, more recently, on the protein crambin[9]. The conclusions are that the method has very good convergence properties, with respect to both distance and structure determinations. For the latter, of course, good NOE data sets must be available. The main advantage of the method is that one is not limited to 2D NOE spectra recorded with short mixing times and hence one can work under conditions of optimal sensitivity. By accounting for the spin-diffusion effect longer distances can be derived with better precision than hitherto possible.

SYMMETRIC EXCHANGE PROCESSES: AROMATIC RING FLIP AND METHYL GROUP ROTATION

Often single lines are observed for the ortho and meta-protons of tyrosine and phenylalanine residues in a protein indicating that the aromatic rings

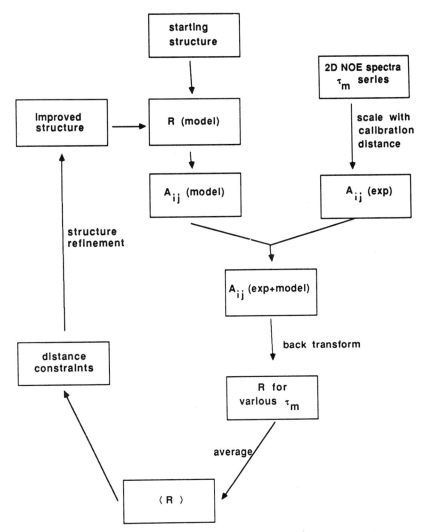

Figure 1. Flow diagram for the iterative relaxation matrix approach (IRMA) for structure determination of biomolecules from proton-proton NOEs. Details of the procedure are discussed in the text.

undergo 180° flips on a time scale fast compared to the chemical shift differences. This situation can be easily described by an extension of the relaxation matrix treatment. One way is to add a kinetic matrix $\mathbf{K}$ to the relaxation matrix, that describes the exchange process, and diagonalize $\mathbf{R} + \mathbf{K}$ instead of $\mathbf{R}$ in Eq. 12. In fact, all exchange rates from slow to fast could be treated in this way. However, when exchange is fast it is more simple to average the pertinent elements of $\mathbf{R}$ in the way suggested by Landy and Rao[10]. Considering a system with N multiple exchanging sites, the probability for each conformation is p_i and its NOE matrix $\mathbf{A}_i$ is determined by the cross-relaxation rate matrix $\mathbf{R}_i$. Provided that the exchange is faster than the cross-relaxation rate the following relations hold:

$$\sum_{i=1}^{N} p_i = 1, \quad \sum_{i=1}^{N} \mathbf{A}_i = \mathbf{A}, \quad \text{and} \quad \mathbf{A}_i = p_i \mathbf{A}.$$

The set of differential equations now has the form:

$$\frac{d\mathbf{A}}{dt} = \frac{d\sum_{i=1}^{N} \mathbf{A}_i}{dt} = -\left(\sum_{i=1}^{N} p_i \mathbf{R}_i\right) \sum_{i=1}^{N} \mathbf{A}_i. \tag{15}$$

The summations run from 1 to N, the number of exchanging sites. For each conformation involved in the exchange process the relaxation matrix $\mathbf{R}_i$ is calculated and the probability p_i of the occurrence of this conformation serves as a weighting factor. The solution of Eq. 15 is:

$$A = \sum_{i=1}^{N} \mathbf{A}_i = \mathbf{X} \exp\left(-\Lambda^* \tau_m\right) \mathbf{X}^{-1} \tag{16}$$

with

$$\Lambda^* = \mathbf{X}^{-1} \left(\sum_{i=1}^{N} p_i \mathbf{R}_i\right) \mathbf{X}. \tag{17}$$

This method corresponds with $\langle r^{-6} \rangle$ averaging over the different allowed conformations. It should be noted that in the case of symmetrical exchange it is not necessary to calculate separate relaxation matrices. The average relaxation matrix can also be obtained by averaging the values of the cross-relaxation rates in the matrix belonging to one conformation. For instance, for two proton exchange the calculation then runs as follows: in each row and column i the values at the positions $j, j+1$, belonging to two exchanging protons are averaged. This method saves computing time, since only one relaxation matrix has to be calculated.

Model calculations have been performed to assess the effect of 180° ring flips on the NOE intensities of proton around aromatic rings. In Fig. 2

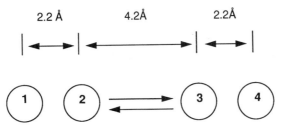

Figure 2. Model system for the aromatic ring flip. Exchange of spins 2 and 3 mimics the 180° flip of tyrosine or phenylalanine rings.

the exchange of protons 2 and 3 simulate the ring flip. The NOE build-up between protons $(1,2) + (1,3)$ (for fast exchange) and $(1,4)$ was calculated in two ways: i) with the kinetic matrix added to the relaxation matrix (rate constant 1000 s^{-1}) and ii) by averaging the corresponding matrix elements, both for a fixed and a flipping ring. The results are very similar as is shown in Fig. 3. It can be concluded that the ring flip does not affect the $(1,2) + (1,3)$ NOE intensity to any appreciable extent. However, Fig. 3b shows the $1,4$ NOE is greatly affected. In the case of exchange of protons 2 and 3 the buildup corresponds to a virtual distance of about 4-5 Å, whereas the actual distance is 8.6 Å. Thus, sometimes the relay effect due to aromatic ring flips may lead to erroneous distances between protons near these rings if the effect is not recognized.

For a rotating methyl group usually the correlation time for the rotation is much shorter than that for overall molecular tumbling. Therefore the spectral density functions changes and $\langle r^{-6} \rangle$ averaging is not correct. Assuming a three-site jump model the spectral density functions take the form[11]:

$$J_{ij}(\omega) = \frac{1}{5} \frac{\tau_c}{1 + n^2\omega^2\tau_c^2} \sum_{m=-2}^{2} \left| \frac{1}{3} \sum_{i=1}^{3} \frac{Y_m^2 \left(\theta_{mol}^i \phi_{mol}^i \right)}{r_{ij}^3} \right|^2 \tag{18}$$

where θ_{mol}^i and ϕ_{mol}^i are the polar angles in the molecular frame of the internuclear vector of length r_{ij} between a proton j and each of the methyl protons. Using these spectral density functions in Eq. 5 the cross-relaxation rates and the NOE intensities can be calculated; this method corresponds to $\langle r^{-3} \rangle$ averaging.

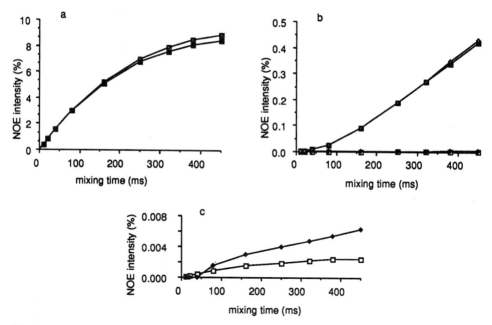

Figure 3. NOE intensities for the aromatic ring flip model calculated with a correlation time of 1 ns. a) The buildup curve for the cross peak between proton 1 and the two proton $(2+3)$ at the aromatic ring. b) The buildup for the cross peak between 1 and 4. c) Expanded region for the cross peak between 1 and 4. NOEs were calculated with two spin approach ▣; static approach ◆; kinetic matrix addition ▢; weight average method $\langle r^{-6} \rangle$ ◇.

The intramethyl relaxation depends on the correlation time for the fast motion τ_1 and is described by the spectral density functions[12]:

$$J^n(\omega) = \frac{1}{4\pi r_{ij}^6} \left[\frac{1}{4} \left(3\cos^2 \Delta - 1 \right) \frac{\tau_c}{1 + n^2 \omega^2 \tau_c^2} \right.$$
$$\left. + \frac{3}{4} \left(\sin^2 2\Delta \sin^4 \Delta \right) \frac{\tau_{c1}}{1 + n^2 \omega^2 \tau_{c1}^2} \right] \qquad (19)$$

where Δ is the angle between the rotation axis and the interproton vector with

$$\frac{1}{\tau_{c1}} = \frac{1}{\tau_c} + \frac{1}{\tau_1}.$$

These spectral density functions can be used to calculate the extra diagonal relaxation contribution of the methyl spin interaction, using the following equation:

$$\frac{1}{T_1} = \frac{2\pi}{5} \gamma^4 \hbar^2 \left[3J_1(\omega) + 12J_2(\omega) \right]. \qquad (20)$$

Model calculations simulating a rotating methyl group were performed similar to the aromatic ring flip as discussed above[3]. Here also a relay effect

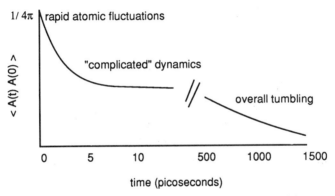

Figure 4. Idealized behavior of the rotational correlation function $C(t)$, defined in Eq. 21.

may occur for protons near methyl groups, although the effect is not as strong as for a flipping aromatic ring. An interesting conclusion from these calculations was that for the pseudoatom correction for a rotating methyl group a value of 0.3 Å should be taken and not 1.0 Å as is often done. This is due to the $\langle r^{-3} \rangle$ averaging, that should occur for NOEs involving protons to fast rotating methyl groups. This gives less weight to the shorter distances than the $\langle r^{-6} \rangle$ averaging on which the correction of 1.0 Å is based and consequently the virtual NOE distance is closer to the geometric mean position of the methyl protons.

FAST (PICOSECOND) LOCAL MOBILITY

For the interpretation of NOEs in terms of distances usually the assumption of a single rotational correlation time corresponding to a rigid molecule approximation is made. We have recently investigated the effect of fast internal motions of the interproton vectors in the context of the relaxation matrix approach for structure determination of biomolecules[13]. Our approach has been to simulate the dynamics of a biomolecule by a long free MD run including the solvent and use the dynamic information to interpret the spectral densities in terms of distances that are then used in a final structure refinement.

Fast internal motions can be conveniently described by the "model free" approach Lipari and Szabo[14]. Here the system is described in terms of correlation times for the overall rotation τ_0 and internal motions (τ_p) and a generalized order parameter S^2. It varies between 0 and 1, the former corresponding to completely isotropic motion and the latter to the absence of local motion (rigid model). The generalized order parameter can be determined from the time correlation function of the interproton vector. The time correlation function is described by:

$$C(t) = \left\langle \frac{A(t)A(0)}{r_{ij}^3(t)r_{ij}^3(0)} \right\rangle \qquad (21)$$

where $A(t) = Y_n^2\left(\Phi_{lab}(t)\right)$ is the second order spherical harmonics and the $\Phi_{lab}(t)$ specifies the orientation of the interproton vector relative to an external magnetic field. The general form of such a time correlation function is shown in Fig. 4. After a rapid initial decay which is due to the fast internal motion a plateau value is reached in a time τ_p in the order of a few picoseconds. On a much longer time scale (nanosecond range) the function decays to zero due to overall motions of the molecule. It is the plateau value of the normalized correlation function to which the generalized order parameter is related in the following way:

$$C(t) = \left\langle \frac{A(t)A(0)}{r_{ij}^3(t)r_{ij}^3(0)} \right\rangle = S^2 \left\langle r_{ij}^{-6} \right\rangle \quad \text{for} \quad t > \tau_p. \qquad (22)$$

The time correlation function can now be rewritten in another form:

$$C(t) = \frac{1}{4\pi} \left[S^2 \left\langle r_{ij}^{-6} \right\rangle + (1 - S^2) \left\langle r_{ij}^{-6} \right\rangle \exp(-t/\tau_p) \right] \qquad (23)$$

where the first part describes the plateau and the second part describes the decay with the correlation time τ_p to the plateau value $S^2 \left\langle r_{ij}^{-6} \right\rangle$. When in addition overall isotropic motion with a correlation time τ_0 occurs the correlation function has the form:

$$C(t) = \frac{1}{4\pi} \left[S^2 \left\langle r_{ij}^{-6} \right\rangle \exp(-t/\tau_0) + \left(1 - S^2\right) \left\langle r_{ij}^{-6} \right\rangle \exp(-t/\tau_c) \right] \qquad (24)$$

where $\tau_c^{-1} = \tau_0^{-1} + \tau_p^{-1}$.

The spectral density functions, being the cosine Fourier transforms of the time correlation functions, are now:

$$J_{ij}^n(\omega) = \frac{1}{4\pi} \left[S^2 \left\langle r_{ij}^{-6} \right\rangle \frac{\tau_0}{1 + (n\omega\tau_0)^2} + (1 - S^2) \left\langle r_{ij}^{-6} \right\rangle \frac{\tau_c}{1 + (n\omega\tau_c)^2} \right]. \qquad (25)$$

If the internal motions are in the motional narrowing limit ($\omega\tau_c \ll 1$) the spectral density functions are:

$$J_{ij}^n(\omega) = \frac{1}{4\pi} \left[S^2 \left\langle r_{ij}^{-6} \right\rangle \frac{\tau_0}{1 + (n\omega\tau_0)^2} + (1 - S^2) \left\langle r_{ij}^{-6} \right\rangle \tau_c \right]. \qquad (26)$$

Furthermore, when the internal motions are very fast compared to the overall motion ($\tau_p \ll \tau_0$) the second term of Eq. 26 can be neglected and the spectral density functions are uniformly reduced to:

$$J_{ij}^n(\omega) = \frac{S^2 \left\langle r_{ij}^{-6} \right\rangle}{4\pi} \left[\frac{\tau_0}{1 + (n\omega\tau_0)^2} \right]. \qquad (27)$$

358

Eq. 27 can be used in Eqs. 4 and 5 to calculate the relaxation rates and the NOE intensities.

The effect of fast local motions was investigated for the same DNA octamer, that was used previously to test the IRMA procedure[13]. From a 180 ps MD trajectory of the octamer in the presence of water and counter ions the generalized order parameters S^2 were obtained for all proton pairs for which NOEs were measured. For most proton-proton vectors a good plateau value (see Fig. 4) was indeed observed. The S^2 values varied between 0.9 for intrabase proton pairs to 0.6 for some internucleotide pairs. Although the corresponding distance corrections were only of the order of 10% at most, a structure refinement based on the S^2 values resulted in a much better agreement between calculated and experimental NOEs.

ACKNOWLEDGEMENT

This work was supported by the Netherlands Foundation for Chemical Research (SON) with financial aid from the Netherlands Organization for Scientific Research (NWO).

REFERENCES

1. R. Boelens, T. M. G. Koning, and R. Kaptein, *J. Mol. Struc.* **173**, 299 (1988).
2. R. Boelens, T. M. G. Koning, G. A. van der Marel, J. H. van Boom, and R. Kaptein, *J. Magn. Res.* **82**, 290 (1989).
3. T. M. G. Koning, R. Boelens, and R. Kaptein, *J.Magn. Res.*, **90**, 111 (1990).
4. I. Solomon, *Phys. Rev.* **99**, 559 (1955).
5. D. Neuhaus and M. Williamson, "The Nuclear Overhauser Effect in Structural and Cornformational Analysis", VCH Publishers, Inc., New York (1989).
6. S. Macura and R. R. Ernst, *Mol. Phys.* **41**, 95 (1980).
7. C. L. Perrin and R. K. Gipe, *J. Am. Chem. Soc.* **106**, 4036 (1984).
8. E. T. Olejniczak, R. T. Jr. Gaupe, and S. W. Fesic, *J. Magn. Reson.* **67**, 28 (1986).
9. J. A. C. Rullmann, A. M. J. J. Bonvin, R. Boelens, and R. Kaptein, to be published.
10. S. B. Landy and N. Rao, *J. Magn. Reson.* **81**, 371 (1989).
11. J. Tropp, *J. Chem. Phys.* **72**, 6035 (1980).
12. D. E. Woessner, *J. Chem. Phys.* **36**, 1(1962).
13. T. M. G. Koning, R. Boelens, G. A. van der Marel, J. H. van Boom, and R. Kaptein, *Biochemistry*, **30**, 3787 (1991).
14. G. Lipari and A. Szabo, *J. Am. Chem. Soc.* **104**, 4546 (1982).

INTERACTIVE COMPUTER GRAPHICS IN THE ASSIGNMENT OF PROTEIN 2D AND 3D NMR SPECTRA

Per J. Kraulis

Department of Molecular Biology, BMC
Uppsala University, Box 590
S-751 24 Uppsala, Sweden

It has long been clear that it should be possible to computerize the assignment of protein 2D NMR spectra to a large degree. Work in this area has mainly focused on various automatic procedures[1-7].

We have chosen another approach. The aim was to design a computer program that allows the user to assign spectra in basically the same way as with paper plots, but interactively on a computer graphics screen. The program ANSIG[8] allows contour computation, viewing of several spectra at once, semi-automatic peak-picking, peak editing, connection of peaks to indicate various relationships, and assignment of peaks. The program checks extensively for input errors, keeps tables reflecting the current status of the assignment, updates assignments for other affected cross peaks, and saves all changes to file. In effect, the problem of the computerized bookkeeping of assignments in 2D NMR spectra has been solved by these features in the ANSIG program.

The reason for choosing this approach is that no automatic assignment procedure can reasonably be expected to work perfectly. Inspection and validation of the results by the user, and intervention in some cases, will be required. There should be a considerable advantage in integrating automatic assignment procedures tightly with an interactive graphics program which already provides basic assignment capabilities.

An important feature of ANSIG is the ability to check systematically for possible assignments of long-range NOEs once the sequence-specific assignments have been made, thereby making this task simpler and safer. NOE intensities are currently mapped to distance intervals via intensity classes, and output to file as distance restraints.

The implementation of ANSIG attempts to maximize portability. About 80% of the source code is completely device-independent, the rest being

Computational Aspects of the Study of Biological Macromolecules by Nuclear Magnetic Resonance Spectroscopy, Edited by J.C. Hoch *et al.*, Plenum Press, New York, 1991

361

easily exchangeable modules of procedures that interact directly with the graphics device and operating system. Implementations exist for the Evans & Sutherland PS300 (VAX/VMS), Silicon Graphics IRIS-4D, and PostScript (plot files only).

New features currently being tested include a general NOE output file format, improved integration of cross peaks, and spectrum data matrix compression. The integration of various automatic assignment procedures[1-7] and an algorithm for identifying corresponding cross peaks in different spectra are being studied.

The arguments for computerized viewing and bookkeeping of assignments are even more important for protein 3D NMR work[9]. Design and implementation of such a 3D assignment program has been initiated in collaboration with Prof. R. R. Ernst (ETH, Zürich) and co-workers.

REFERENCES

1. K. P. Neidig, H. Bodenmueller, and H. R. Kalbitzer, *Bioch. Bioph. Res. Com.* **125**, 1143-1150 (1984).
2. M. Billeter, V. J. Basus, and I. D. Kuntz, *J. Magn. Res.* **76**, 400-415 (1988).
3. C. Cieslar, G. M. Clore, and A. M. Gronenborn, *J. Magn. Res.* **80**, 119-127 (1988).
4. P. L. Weber, J. A. Malikayil, and L. Mueller, *J. Magn. Res.* **82**, 419-426 (1989).
5. C. D. Eads and I. D. Kuntz, *J. Magn. Res.* **82**, 467-482 (1989).
6. G. J. Kleywegt, R. M. J. N. Lamerichs, R. Boelens, and R. Kaptein, *J. Magn. Res.* **85**, 186-197 (1989).
7. F. J. M. van de Ven, *J. Magn. Res.* **86**, 633-644 (1990).
8. P. J. Kraulis, *J. Magn. Res.* **84**, 627-633 (1990).
9. H. Oschkinat, C. Griesinger, P. J. Kraulis, O. W. Sørensen, R. R. Ernst, A. M. Gronenborn, G. M. Clore, *Nature* (London), **332**, 374-376 (1988).

DETERMINATION OF LARGE PROTEIN STRUCTURES FROM NMR DATA: DEFINITION OF THE SOLUTION STRUCTURE OF THE TRP REPRESSOR[†]

Russ B. Altman, Cheryl H. Arrowsmith, Ruth Pachter, and Oleg Jardetzky*

Stanford Magnetic Resonance Laboratory
Stanford University
Stanford, CA 94305-5055, USA

ABSTRACT

We have determined the solution structure of the *E. Coli* trp repressor (a 25 kD dimer) from NMR data. This is the largest protein structure thus far determined by NMR. The determination of the solution structure of larger proteins (MW ±15 kD) by NMR requires a different methodology, both experimental and computational, than that developed for peptides and small proteins. This structure determination illustrates a general paradigm for this purpose. We have found it necessary to first, use isotopic spectral editing and second, to develop new strategies for sequential assignment of the resonances. Third, since the protein contains more than one peptide chain it was necessary to distinguish between intra-chain and inter-chain contacts by spectroscopy of isotopically labeled hybrids. Fourth, in order to efficiently search a very large conformational space with a relatively small data set, we have used a hierarchical method of data analysis: the heuristic refinement method coded in PROTEAN has proven both accurate and computationally efficient in solving the structure of the trp repressor and placing the bound L-tryptophan ligand. Our NMR structure has the same general topology, and same binding site for the L-tryptophan molecule as the previously reported crystal structures. There are, however, some differences in the backbone trace. In addition, the solution structure shows significant structural uncertainty in the DNA binding region of the molecule.

INTRODUCTION

The tryptophan repressor is a 25 kD dimer (107 amino acid residues per monomer) which regulates the *de novo* synthesis of tryptophan in *E. Coli.* In the presence of exogenous tryptophan, the repressor binds to operator DNA,

[†]This work was supported by NIH grants RR02300 and GM33385 and RBA is supported under NIH GM07365.
*Correspondence Author.

Computational Aspects of the Study of Biological Macromolecules by Nuclear Magnetic Resonance Spectroscopy, Edited by J.C. Hoch *et al.,* Plenum Press, New York, 1991

363

inhibiting the production of the enzymes in the tryptophan biosynthetic pathway. Crystal structures of the repressor (in the presence and absence of L-tryptophan) as well as of a DNA-repressor complex have been reported[1-4]. To further characterize the specific interactions involved in DNA and ligand binding, we are studying the trp repressor in solution using NMR (nuclear magnetic resonance spectroscopy). The crystallographic studies show that there is some structural variability, especially in the DNA-binding region, that may be functionally important. One may expect this variability to be more pronounced in the solution state. Therefore, the structural and dynamic information available from NMR data is an important complement to the crystal studies.

NMR information is generally obtained from observations of the nuclear Overhauser effect (NOE) which occurs between protons that are roughly within 6 Å of one another. Distance constraints derived from NOEs, along with covalent bond and bond angle information, can be used to calculate a three dimensional solution structure[4,5]. This protein is particularly challenging for NMR studies. The repressor exists in solution as a 25 kD dimer and is therefore larger than any structure previously determined by conventional NMR methods. Larger molecules have slower tumbling times, and this contributes to the broadening and overlap of spectral lines. In addition, the abundance of α-helical secondary structure, which is known from CD measurements[6], concentrates many of the α-carbon signals into a relatively small and densely populated region of the spectrum[7]. Finally, it is necessary to find ways to disambiguate intrasubunit NOEs from intersubunit NOEs. To overcome these assignment problems, we have prepared selectively deuterated repressor analogs as discussed below and reported elsewhere[8].

In addition, the protein presents a difficult computational problem. With 214 amino acids and approximately 3,500 atoms, there is a very large conformational space to search. With a preliminary set of long range distance constraints, the potential for multiple minima using optimization procedures is high. A strategy in which alternative conformations are maintained until they are shown to be incompatible with the input data is more conservative. The PROTEAN system has been designed specifically for this purpose, i.e., the structure calculation of large proteins with relatively sparse data sets[4].

In this report, we describe how we have used the PROTEAN system to calculate the dimer structure using 165 short and medium range and 39 long range NMR derived connectivities per monomer. The long range NOEs involve only 17 amino acids. We have avoided the combinatorial explosion of conformational possibilities by building the structure in steps, using NMR, covalent, bond angle, and dihedral angle constraints. We have first used the subset of short and medium range backbone NOEs to calculate the location of five helices within the polypeptide sequence. We have then used a cluster of long range NOEs between three helices to construct a core segment of the protein. The symmetry of the dimer was solved by docking two core segments with only covalent and steric constraints. Finally, the remaining segments of the protein were added using a second cluster of long range NOEs. The L-tryptophan ligand, which provides a number of NOE constraints, was included in the structure calculation and, in fact, allowed us to better define the relative location of the DNA binding segments.

METHODS AND RESULTS

Sequence Specific Assignment of NMR Spectra

The detailed strategy employed to make sequence and monomer-specific assignment of the proton resonances and interresidue NOEs in the trp repressor are reported elsewhere[9]. The results are summarized in Fig. 1. Over 90% of the amide backbone proton resonances and 70% of the alpha-proton and sidechain proton resonances were assigned based on NOESY spectra of a series of selectively deuterated repressors. The identification of NOEs between residues as intramonomeric or intermonomeric was made as follows. Backbone NOEs between residues within four amino acids of one another in the primary sequence were assumed to be intramonomeric. For long range NOEs, heterodimers were prepared from two different selectively deuterated proteins. In the heterodimer, NOE cross peaks due to intermonomeric interactions are absent or severely attenuated if one of the residues is deuterated in one monomer and protonated in the other monomer. Fig. 1 summarizes the short, medium and long range NOEs available for the structure determination.

Structure Calculation

The trp repressor dimer contains a total of 214 amino acids (107 in each monomer). The first 14 amino acids of each monomer do not have any significant short range sequential NOEs and therefore were not included in the structure calculation. It has been shown experimentally that at least the first 7 residues of this section of the protein are in rapid motion on the NMR time-scale[10].

Secondary Structure

The short range NOE data suggest five regions of the primary sequence which may contain α-helical secondary structure (24–42, 43–65, 66–76, 77–93, 94–105). These regions give rise to the characteristic NOE cross-peaks

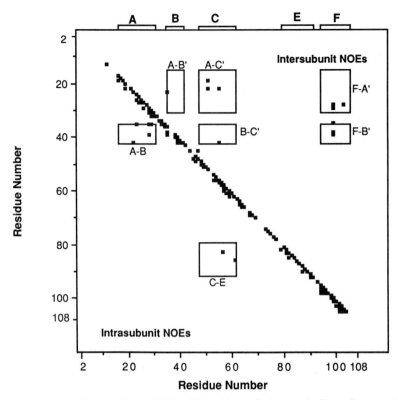

Figure 1. Summary of interresidue NOEs. Helical boundaries are indicated across the top. Points close to the diagonal correspond to residues involved in short and medium range NOEs. Off-diagonal points are between residues with long range NOEs. Intramonomeric NOEs are indicated below the diagonal and intermonomeric NOEs above the diagonal. The boxes encompass regions of the diagram in which inter-helical NOEs appear.

between protons NH^i-NH^{i+2}, α^i-NH^{i+3}, α^i-NH^{i+1}, $\beta-NH^{i+1}$ and $\alpha-NH^{i+4}$, typical of helical sections of polypeptides. In order to establish the location of the secondary structures as implied solely by the NMR data, we calculated structures of each of these segments of polypeptide using only the short and medium range intrasubunit NOE measurements reported by Arrowsmith et al.[9] and summarized in Fig. 1. The calculations were performed using the double iterated Kalman Filter methodology described in Altman and Jardetzky[4] and Pachter et al.[11] and deposited in the Quantum Chemistry Program Exchange[12,13]. The input to this program consists of distance and angle constraints in the form of a mean value and variance and calculates a mean structure as well as the variance and covariance of all cartesian coordinates. In addition to the normal bond lengths (variance = 0.1) and bond angles (variance = 0.1), we provided dihedral angle constraints corresponding to the ranges of allowed values for ϕ and ψ backbone angles ($\phi = -90$ mean, variance = 999; $\psi = 60$, variance = 8,000) as well as the experimental NOE constraints (with mean values varying from 3.1 Å, to 4.6 Å as a function of NOE strength and proton type, and variance varying from 0.5 to 1.1 Å^2). Each calculation was performed using a pseudo-atom representation for all sidechains[14], with the appropriate corrections to the NOE distances.

These calculations showed helical structures within each of the segments to which we will refer as A, B, C, E and F to conform to the nomenclature used for the crystal structure. The boundaries of the helices within segments A, B, C, E and F are similar to, but not the same as, the boundaries of the crystallographic helices. The RMS deviation of these helical segments from ideal helices (generated using dihedral angles of $\phi = -57°$, $\psi = -47°$) are in the range of 1.7–2.5 Å. The segment from amino acids 66–76 did not form (using only short and medium range NOEs) a structure that was clearly a helix (RMS from ideal = 3.7). It should be noted that this segment exists as a somewhat distorted helix in the crystal structure[1].

Tertiary Structure

To calculate the full tertiary structure, we introduced long range NOE constraints between amino acids not adjacent in the primary or secondary structure. These correspond to the off-diagonal points in Fig. 1. Preserving the helicity of the segments from the secondary structure calculation, we added a matrix of α-carbon distances for each of the helical segments. These distances were assigned a sizeable variance of 0.25 in order to allow bending and other distortions resulting from tertiary interactions. We also used long range NOE constraints between the polypeptide segments and the L-tryptophan ligand. To avoid conformational combinatorics, the model was constructed in steps and full pseudo-atomic representations were used for only those sidechains involved in long range NOEs. We will refer to the segments of the second monomer of the repressor dimer as A', B', C', D', E' and F'.

Construction of A-B Segment

The structure of the segment A-B (containing amino acids 16–44) was calculated using 36 short and medium range NOEs from the backbone and one long range NOE (Ala29-His35) confirmed to be intra-molecular by the selectively deuterated heterodimer experiments. We used the double iterated Kalman filter and a pseudo-atom representation for all sidechains in segments A and B. The structure of the A-B segment satisfied the input constraints with an average error of 0.3 SD. The average variance in Cartesian coordinates of the backbone was 1.6 Å^2. There is a short turn between helices A and B involving amino acids 32, 33 and 34. The axes of the two helices are at an angle of approximately 60°.

Addition of C' to A-B Segment

There are NOEs between the A-B segment and the C (or C') segment (Phe22-Val55, Trp19-Leu51). In order to determine which of the C segments gave rise to these constraints, we calculated the structure of the A-B-C and A-B-C' segments using the double iterated Kalman filter. For this calculation, only sidechains involved in NOEs were used (again, at the pseudo-atomic level). The constraints used were identical (four long range NOEs) except that a covalent connection between segments B and C was introduced in the former calculation, but not in the latter. The double iterated Kalman filter was not able to find structures satisfying the covalent constraint between B and C while also satisfying the NOE constraints (the average and maximal constraint errors were 1.6 and 21.9 SD, respectively), whereas it was able to satisfy all the constraints (the average and maximal constraint errors were 0.5 and 8.3 SD) when B and C were not forced to be covalently connected. This by itself is strong evidence that C' and not C is the segment in close contact with segment A. The fact is conclusively demonstrated by experiments which showed the disappearance of the NOEs between amino acids Phe22 and Val55 and between Trp19 and Leu51 in the heterodimer species[9]. The A-B-C' structure satisfies the constraints with an average error of 0.4 SD. The introduction of segment C' reduced the average variance in the positions to 0.4 Å^2. The structure shows the helix within segment C' to be cradled within the V-shaped A-B structure.

Calculation of A-B-C/A'-B'-C' Dimer Core

In order to establish the position of the A'B'C segments relative to the ABC' structure, we used the PROTEAN solid level modelling geometry system. This system is described in detail in Brinkley et al.[15] It allows us to dock rigid molecules using van der Waals and covalent connections in addition to NOE constraints. We fixed the A-B-C' backbone structure (as previously calculated) at its average position and introduced a second unit with identical backbone structure. The constraints between the two units were the covalent connections between B and C (as well as B' and C'), two long range NOEs (Phe22-Met42 and Pro45-Glu47) as well as the steric bulk of the two moieties. The geometry system systematically sampled possible locations for the A'-B'-C unit relative to A-B-C' (varying position by 2 Å and orientation angle by 20°). All structures that satisfied the covalent and steric constraints were retained. Primarily because of the relatively large

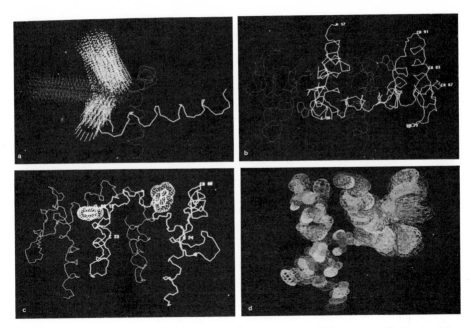

Figure 2. a) The structure of the dimer core after docking A'B'C to ABC'. The backbone of segment C' is shown as the horizontal coil on the right. The backbones of segments A and B are shown as the overlapping vertical coils. The accessible volumes of segments A' and B' (as calculated by the PROTEAN geometry modelling system) are shown as the lighter cylinders (A' on top) and for C is shown as the darker horizontal cylinder to the left. These volumes were summarized and used as starting points for further refinement of the dimer core. b) The dimer structure of the *trp* repressor looking down the central axis of symmetry. One monomer is lighter, the other is darker and the two L-tryptophan ligands are the smaller medium-intensity structures, one on top of the darker monomer and the other to the bottom left of the lighter monomer. c) The dimer structure is shown rotated horizontally 90 degrees relative to Fig. 2b. The crystallographic DNA binding surface runs along the top of the molecule in this view. d) The half-dimer ABC'D'D'F' structure is shown with uncertainty ellipsoids (drawn at 2 standard deviations) for the backbone atoms. Segment D' is on the far right and has noticeably larger uncertainty than segments A, B, and C' (smaller ellipsoids on left).

volume of the A-B-C' structure, only 17 satisfactory positions were found at this sampling interval. All 17 positions are spatially close, as shown in Fig. 2a. We used these 17 positions to calculate the average coordinates and variance for the atoms of the A'-B'-C segment and refined the A-B-C'-A'-B'-C ensemble with the double iterated Kalman filter, now allowing all atoms to move in amounts commensurate with their variance. For this refinement, no explicit constraint on the symmetry of the dimer was included, and only sidechains involved in NOEs were represented. We provided equivalent NOE constraints for each monomer and refined the dimer, allowing independent motion of all atoms within the constraints.

The ABC-A'B'C' dimer satisfies all experimental constraints with an average error of 0.4 SD. The atoms have an average variance of 0.4 Å². The structure is symmetric with a 180 degree axis of symmetry located near the N-terminal end of segments C and C′ and oriented roughly perpendicular to the axes of the helices within C and C′. The helices within segments C and C′ are at an angle of ∼ 125°.

Addition of D′-E′-F′ and Tryptophan to Core Dimer

Having calculated the structure of the core of the dimer containing both A-B-C segments, we introduced the remainder of one monomer. Segments E′, F′ and a single tryptophan molecule can each be unambiguously placed relative to the dimer:

1. Segment F′ has five NOEs to the A-B segment (or A′-B′). From hetero-dimer experiments the NOEs between amino acids Ala29-Trp99, Asn28-Trp99 and Leu34-Trp99 were proven to be intermonomer NOEs, thereby demonstrating contact between F′ and A-B. The remaining two NOEs between F′ and the A/B (or A′/B′) segment (Val23-Leu103, Leu38-Trp99) involve amino acids adjacent to these three unambiguous NOEs and are also on the same face of the helices. Examination of the structure at this point revealed that they were not consistent with F′-A′ contacts.

2. An NOE between Ile57 and Ile82 was shown to be intramonomeric[9] thereby demonstrating the proximity of E′ to C′. A second NOE between Leu61 and Ser86 was also assigned to E′ and C′ because of proximity in the primary sequence and location on the same face of the helix as amino acids 57 and 82.

3. Finally, the tryptophan ligand has NOEs to three residues in segments C/C′ and E/E′ (Thr81, Ile57 and Val58). Since these amino acids are spatially adjacent (as implied by the E′-C′ NOEs between 57 and 82), these tryptophan NOEs can be reliably assigned to E and C of the same subunit. Since the trp repressor is known to bind two trypto-phan molecules[16], we can conclude that one tryptophan has NOEs to E-C and the other has NOEs to E′-C′. There are also two other NOEs to the L-tryptophan ligand from residues 43 and 44 at the junction be-tween segments B and C (or B′ and C′). The known NOE from 42→55 indicates that the middle of helix C′ (which contains Val55′) is close to the B-C turn (which contains Met42) of the opposite subunit; therefore the NOEs to residues 43 and 44 were assigned to the B-C segment of the opposite subunit from the other ligand NOEs. The assignment of these NOEs to B′ and C′ results in incompatible and unsatisfiable distance constraints when combined with the NOEs between L-tryptophan and the E′-C′ segment.

We used the PROTEAN solid level geometry system to separately dock the individually calculated structures of segments F′, E′ as well as the tryp-tophan molecule to the ABC-A′B′C′ core dimer. The average positions of the segments and tryptophan molecule were taken as rigid fragments and distance constraints to the core dimer were derived from the NOEs. A list of positions (at sampling of 2 Å in position and 20°in orientation) was generated for each of these structures. These lists were further reduced by application of the distance constraints between E' and F' (a single covalent bond) and of the NOE constraints between the tryptophan and segments E' an F'. We calculated the mean coordinates and variances for each of these segments and the tryptophan molecule as a starting point for refinement by the double iterated Kalman filter. We also introduced the atoms of segment D' with covalent constraints between C' and E' and several short range sequential NOE constraints. (No long range NOE constraints were available to more accurately position segment D' relative to the other segments.)

Calculation of the Full Dimer with Tryptophan Molecules.

Having positioned A'-B'-C'-D'-E'-F' with respect to A-B-C and a single tryptophan molecule, we generated the positions of D-E-F and the second L-tryptophan ligand by applying the symmetry transformation relating A-B-C and A'-B'-C'. This yielded a full representation of the dimer and the positions of two tryptophan ligands. We refined the dimer structure using an equivalent set of NOE constraints for each subunit. Sidechains involved in NOEs were represented using pseudo-atoms. Other sidechains were represented by a single large super-atom to approximate their steric bulk and prevent close backbone contacts. We used no explicit constraint forcing the monomers to be identical. The resulting dimer along with the tryptophan ligands is shown in Fig. 2b and c.

COMPARISON WITH CRYSTAL STRUCTURE

The dimer structure shown in Fig. 2b and c matches the ortho-rhombic crystal structure to 5.8 Å RMS. The atoms of the crystal structure are an average of 5.9 SD from the NMR structure. This large deviation is due predominantly to the positional uncertainty of segment D. Specifically, RMS deviations (Å) from the crystal positions are 2.0 for the A-B-C' unit, 2.9 Å for the A-B-C–A'-B'-C' dimer coil, and 4.1 Å for the A-B-C-E-F–A'-B'-C'-E'-F' structure. Detailed comparison between the solution and crystal structures awaits our refinement of the solution structure. However, it is clear that the position and orientation of the L-tryptophan is remarkably similar in both structures. It comes in close contact with helices C and E of one subunit and the B-C turn of the opposite subunit. In addition, there is significant uncertainty in the orientation of segments D and D'. Fig. 2d shows the uncertainty distributions for alpha carbons of segments D, E, and F. It is clear that segment D has significantly more uncertainty than segments E and F. This is consistent with crystallographic results showing shifts in the orientation of these segments upon binding L-tryptophan, upon binding DNA, and in different crystal space groups[2,3,17].

DISCUSSION

The use of a small set of long range NOEs has allowed us to construct a relatively detailed solution structure of this protein dimer in solution. We believe that there are two reasons we have been able to calculate this large (by NMR standards) structure with a moderate data set. First, we have benefitted from the abundance of regular α-helical secondary structure. Although this makes the assignment problem more difficult, it helps in structure calculations because these secondary structures can be modelled as rigid fragments for the purposes of the initial docking calculations. Second, we have not attempted to calculate the entire structure from an initial random starting point. Instead, we have systematically calculated the local secondary structure for a number of segments, and then used the NOE and known covalent connectivities to construct first a core for a single monomer, then a dimer core, followed by a complete dimer. Thoughtful sequencing of docking and refinement calculations allowed us to avoid direct calculation in the space of an extremely large conformational search space. The double iterated Kalman filter is well suited to this strategy because it can use estimates of positional error from previous calculations as starting points for new calculations. Thus, both the direction and magnitude of positional uncertainties can be preserved and refined throughout all computations.

It is interesting to note that the symmetry of the dimer core could be clearly defined based on only four long range NOEs, the covalent connections, and the van der Waals envelope of each core monomer. Test calculations showed that any two of these three types of constraints were inadequate to position the monomers accurately. The importance of van der Waals constraints in limiting conformational search has been stressed many times before[18] and is underscored in our calculation. These constraints allowed us to position the atoms with an average uncertainty of only 3 Å.

Construction of the D-E-F segment was considerably facilitated by the use of the tryptophan molecule. Since we had NOE constraints between the tryptophan and three different segments of the protein, the incorporation of this molecule added significant constraining power to our docking calculations with the geometry system as well as subsequent refinements. Preliminary analysis of sidechain packing also shows general agreement in sidechain conformations in the solution and crystal structures. At this stage, it is clear that the general topology in solution is the same as in all crystal environments studied to date, but there are differences in the average conformation of individual segments.

The structure presented here has been calculated from NMR data completely independently from the crystal structures. It is important to note that no "starting structure" generated from known homologies or other sources of data were used for these calculations. The only assumption implicit in our structure is that the dimer is symmetric, and so positions of segments of one monomer can be transformed into equivalent positions for the second monomer once the symmetry relationship is established. In fact, we have only used this assumption for initial placement of segments, and have allowed subsequent refinements of the two monomers to proceed independently. Any asymmetry of the dimeric structure is small in the context of the calculated uncertainties.

ACKNOWLEDGEMENTS

The authors would like to thank Dr. Jerzy Czaplicki for his help in manual model building of this structure. He provided insight into the appropriate strategy for assembling the structure. We would also like to thank Stardent Computers Inc. for use of the TITAN minisupercomputer for the calculations.

Note Added in Proof

The NOE between Pro45 and Glu47 has subsequently been found to have been incorrectly assigned. However, the docking of the A-B-C' to A'-B'-C core without this NOE shows only a slight increase in uncertainty, which does not affect the results of subsequent refinement by the Kalman filter.

REFERENCES

1. R. Schevitz, Z. Otwinowski, A. Joachimiak, C. L. Lawson, and P. B. Sigler, *Nature*, **317**, 782 (1985).
2. R.-G. Zhang, A. Joachimiak, C. L. Lawson, R. W. Schevitz, Z. Otwinowski, and P. B. Sigler, *Nature* **327**, 591 (1987).
3. Z. Otwinowski, R. W. Schevitz, R.-G. Zhang, C. L. Lawson, A. Joachimiak, R. Q. Marmorstein, B. F. Luisi, and P. B. Sigler, *Nature* **335**, 321 (1988).
4. R. B. Altman and O. Jardetzky, in "Methods in Enzymology" 177, N. J. Oppenheimer and T. L. James, eds., Academic Press, New York, p. 218 (1989).
5. K. Wüthrich, "NMR of Proteins and Nucleic Acids", John Wiley and Sons, Inc., New York (1986).
6. A. N. Lane and O. Jardetzky, *Euro. J. Biochem.*, **164**, 389 (1987).
7. L. Szilágyi and O. Jardetzky, *J. Magn. Res.* **83**, 441 (1989).
8. C. H. Arrowsmith, L. Treat-Clemons, L. Szilágyi, R. Pachter, and O. Jardetzky, in "Die Makromolekulare Chemie, Macromolecular Symposia" 34, H. Höcher, and B. Sedláček, eds.), Hüthig and Wepf Verlag, Heidelberg (1990).

9. C. H. Arrowsmith, R. Pachter, R. B. Altman, S. Iyer, and O. Jardetzky, *Biochemistry* **29**, 6332 (1990).

10. C. H. Arrowsmith, J. Carey, L. Treat-Clemons, and O. Jardetzky, *Biochemistry* **28**, 3875 (1989).

11. R. Pachter, R. B. Altman, and O. Jardetzky, *J. Mag. Res.*, **89**, 578 (1990).

12. E. A. Carrara, J. F. Brinkley, C. C. Cornelius, R. B. Altman, J. Brugge, R. Pachter, B. Buchanan, and O. Jardetzky, "PROTEAN — PART I", QCPE *Bulletin 10:4, Program 596*, Indiana University (1990).

13. R. B. Altman, R. Pachter, E. A. Carrara, and O. Jardetzky, "PROTEAN — PART II", QCPE *Bulletin 10:4, Program 596*, Indiana University (1990).

14. K. Wüthrich, M. Billeter, and W. Braun, *Mol. Biol.*, **169**, 949 (1983).

15. J. F. Brinkley, R. B. Altman, B. S. Duncan, B. Buchanan and O. Jardetzky, *J. Chemical Info. and Comput. Sci.* **28**, 4, 194 (1988).

16. A. Joachimiak, R. L. Kelley, R. P. Gunsalus, C. Yanofsky, and P. B. Sigler, *Proc. Natl. Acad. Sci*, USA, 80, 668 (1983).

17. C. L. Lawson, R.-G. Zhang, R. W. Schevitz, Z. Otwinowski, A. Joachimiak, and P. B. Sigler, *Proteins* **3**, 18 (1988).

18. A. M. Gronenborn and G. M. Clore, *Biochemistry*, **28**, 5978 (1989).

INTERPRETATION OF NMR DATA IN TERMS OF PROTEIN STRUCTURE
Summary of a Round Table Discussion[¶]

Oleg Jardetzky*

Panel Members: G. Marius Clore[†], Dennis Hare[‡], and Andrew Torda[§]

*Stanford Magnetic Resonance Laboratory
Stanford University
Stanford, CA 94305-5055
[†]Laboratory of Chemical Physics
NIDDKD
National Institutes of Health
Bethesda, MD 20892
[‡]Hare Research, Inc.
14810 216 Ave. NE.
Woodville, WA 98072
[§]Department of Physical Chemistry
University of Groningen
9747 AG Groningen

The closing round table of the meeting was focussed on the fundamental issues of calculating protein structures from NMR data, with special emphasis on the as yet unsolved problems of NMR protein structure determination[1].

These problems were first brought to the attention of the conference in the opening lecture by Richard Ernst, who summarized them as in Fig. 1, which I have modified only slightly by including refinement as a distinct step.

The diagram reminds us of the fundamental difference of structure determination from x-ray diffraction data on one hand and NMR data on the other: in x-ray crystallography the structure can in principle be calculated

---

[¶]No papers were presented in this session and, to stimulate wide participation, all individual comments were limited to 3 minutes.

Computational Aspects of the Study of Biological Macromolecules by Nuclear Magnetic Resonance Spectroscopy, Edited by J.C. Hoch *et al.*, Plenum Press, New York, 1991

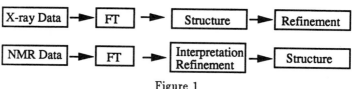

Figure 1

by Fourier transformation from the diffraction pattern (provided the phase problem is solved). Perfect data would, by straightforward calculation, give an exact structure. No such simple mathematical transformation of data to structure exists in NMR. The problem of structural interpretation of NMR data is not simply a data-fitting problem, as it is in crystallography, but involves several steps that require a subjective judgment on the nature of the data. The final result depends not only on the quality of the data, but also on the quality of these judgments.

Although this problem has been known for a long time[1,2] it has been seriously neglected in the recent literature. In the burst of enthusiasm for having enough data to build plausible structures, all too often plots of elegant 2DFT experiments were immediately followed by colorful pictures of 3D protein structures, leaving the audience with the impression that the gray box which converted data to structure was straightforward, unproblematic and well-understood by those "in the know." Anyone questioning the working of the box was clearly not "in the know" and deserved to be silenced. Focussing attention right at the beginning of the conference on the fact that the structural interpretation of NMR data is far from unproblematic was a welcome departure from the more prevalent avoidance of the subject. It was the first step to showing that appreciation of the fundamental problems is by now far more widespread than would appear from the literature.

The leitmotiv for the discussion was the question: "How does one establish that a proposed NMR structure is accurate?"

The question has no single, simple answer. Quite often one finds claims of accuracy supported by measures of the precision with which the final structure has been calculated. This confusion of accuracy and precision is unfortunate. The two are clearly distinct (Figs. 2 and 3).

Precision of the structure is usually measured by the variance (or RMSD) obtained in repeated runs of an optimization (refinement) procedure. Yet it must be borne in mind that optimization procedures with the strongest built-in biases are apt to show the greatest reproducibility — the greatest precision — of the final result. No optimization procedure can by itself guarantee the accuracy of the result.

The answer to the question of accuracy, i.e., the faithfulness to the structure that exists in reality, depends on careful scrutiny of the subjective

Figure 2. Accuracy

judgments required at several steps of the interpretation. The questions
that need to be asked are, in logical order:

(1) How accurate is the assignment of spectral lines and cross peaks in 2D
(or multi-D) spectra?

(2) Is the number of interresidue NOEs sufficient to yield a structure?

(3) How reliable is the quantitation of an NOE? Is it best to distinguish only
between the coarse categories of "strong," "medium," and "weak," or can
reliance be placed on more precise intensity measurements?

(4) How does one best translate observed NOE intensities into a set of
distance constraints? By assigning distance ranges to the coarse categories
of "strong," "medium," and "weak?" Calibrating the distances precisely
against a "known" standard? Or fitting calculated relaxation parameters to
measured NOE intensities?

One limit on the precision with which distance constraints can be specified is given by the precision with which the NOE intensity can be measured.
The other limit will be set by the decisions on questions of dynamics:

(5) Is the molecule to be treated as rigid (relating NOE intensity to distance
by a single correlation time) or flexible (relating NOE intensity to distance
using different conversion factors for different NOEs)?

(6) Do the observed NOEs reflect a single structure or a conformational
equilibrium between two or more structures?

(7) Do all of the observed NOEs reflect the same structure (or the same
conformational equilibrium), or do different subsets of the observed NOE
set reflect different structures in equilibrium?

(8) What is the best choice of the method of computation — i.e., the best
choice of the assumptions on which a molecular model is fitted to the set of
constraints derived from experiment?

Figure 3. Precision

For clarity of presentation, the discussion is summarized in the logical, rather than in the chronological, order in which the foregoing questions were raised.

(1) Assignment: There was general agreement that for the time being assignment remains a largely subjective procedure, although attempts are continuing to automate the process. Several pattern-recognition algorithms have been devised and successfully tested on relatively small peptides (up to 10-12 residues). However, as the molecular weight and line overlap increase and the uncertainties in the chemical shift are taken into account, one is faced with a combinatorial explosion which continues to defy a successful resolution. This specifically also applies to the classification of NOEs as intra- or interresidue, a nontrivial problem where line overlap and/or spin diffusion are extensive. The intuition and prior knowledge of related (or even unrelated) molecular systems on which spectroscopists depend to reach the correct decision have thus far not been summarized in a set of simple, formal, generally applicable rules required for effective programming.

(2) The basic question: "How do we know in the first place when we have enough constraints to determine a structure?" was raised by Dennis Hare early in the discussion, but found few respondents. A partial answer was offered by Marius Clore in his definition of successive "generations" of structures on a scale of the total number of NOEs observed (1st generation 200, 4th generation 2,000). However, this answer is as deceptive as it is simple. No one could argue that more and better data would not, generally speaking, result in a more accurate structure. Yet the total number of constraints is not a valid measure of their adequacy. NOE constraints differ widely in their constraining power: intraresidue constraints have a rather low con-

straining power compared to interresidue constraints between side chains far apart in the peptide sequence. Among the latter, those that connect domains with a well-defined secondary structure are more constraining than those connecting random coil regions. Constraints along the backbone based on stereospecific assignments are among the most stringent, and so are constraints that uniquely define a dihedral angle. Clusters of constraints from neighboring groups often carry information that is largely redundant. This was clearly shown some years ago in the calculations of Lichtarge et al.[3]. Thus, there exists a hierarchy of constraints, and it is a sufficient number of *relevant, strategically placed* (severe) constraints that makes it possible to determine a structure. The relationship of this number to the total number of observed constraints is likely to depend on the specific protein structure. For the present we lack any theoretical — or even statistical — basis for predicting this relationship or the number itself a priori. As Nobuhiro Gō pointed out, we are dealing with a severely underdetermined problem, and it is the "lucky" distribution of the observed constraints that allows us to obtain structures with a bare minimum of data.

The difficulty in answering the question: are the data adequate? appears only in borderline cases: when the data are grossly inadequate, no method of structure calculation will converge and thereby signal that the data are insufficient. Since, however, one is dealing with an underdetermined problem in any case, there can be instances where a structure — or a family of structures — can be calculated, but the result will be incorrect. It might be noted that this difficulty is much more severe when using methods of structure calculation which rely heavily on global optimization — such as distance geometry[4,5] and restrained molecular dynamics[6,7]. In the heuristic refinement method[8,9], which relies on hierarchical model building and retains information on the uncertainty of all atomic positions throughout the calculation, the inadequacy of the data is automatically reflected in the larger uncertainty limits calculated for each coordinate. Thus the question: are the data adequate? is answered by examining the uncertainty factors — and the answer may well be different for different parts of the structure. Because of the reliance on a systematic sampling step, the accuracy of the resulting structure is not likely to be in question — only the precision with which it can be specified. For this reason the development and application of methods that allow an objective assessment of the uncertainty of a structure proposed on the basis of a given set of constraints is of utmost importance. Eric Zuiderweg underscored that the introduction of new types of constraints — such as heteronuclear coupling constants — would help to narrow down the choices between different possible structures. He also pointed out that in NMR we have a much larger arsenal of potential experiments to sort out ambiguities than exists in x-ray diffraction. For the present, however, the decision that the data in hand are sufficient to give an accurate structure remains in most cases fundamentally an act of faith.

(3) Quantitation of NOE intensities is a separate and distinct step that has to precede quantitation of the distance constraints derived from them.

The tendency to discuss the two steps in the same breath, as if only one step were involved, is rather confusing. As pointed out by Andrew Torda and Martin Billeter, attempting to quantify NOEs accurately — i.e., to obtain an accurate integral of cross peak intensity as a first step — is not the same as over-interpreting them by specifying an overly-precise distance calculated from the relationship $\sigma = \frac{k}{r^6}\tau$ in the second step. The reason that the separability of the two steps is often neglected is that the impetus for more stringent quantitation of NOE intensities has come from some of the groups that have adopted an iterative relaxation matrix approach to structure refinement. The problems of obtaining the best fit of calculated to observed intensities by adjusting the structure have come to overshadow the problems of obtaining an accurate number for the experimentally measured intensity in the first place. This is very unfortunate. It is necessary to know what the data are before drawing conclusions as to what they mean (in structural terms). The precision with which a cross peak intensity (and hence the cross-relaxation rate) can be specified is by no means well-defined and depends on a number of experimental variables. Marius Clore aptly summarized the views of several members of the audience, including the present writer: "you are on much more solid and safer ground if you interpret your NOEs in as approximate and conservative a manner as possible ... the quantification of NOEs is really problematic. In 2D when there is overlap, it is undoubtedly difficult to integrate. In 3D it may be impossible — you have to take into account what the transfer efficiencies are to integrate cross peaks accurately. These can vary and are not always easily accessible ..." Additional problems occur if the data are derived from deuterated analogs. I think it is fair to say that an accurate experimental measurement of cross peak intensities and an accurate calculation of cross-relaxation rates from their dependence on the mixing time requires more stringent control of instrumental variables and much more labor than is commonly invested in spectroscopic structure determinations today. Without very extensive and stringent controls, a more precise specification of experimental cross peak intensities than that given by the widely used classification "strong," "medium," and "weak" is probably not warranted. It also must be emphasized that a reliable quantitative measure of the precision with which NOE intensities can be measured remains to be defined (no one has proposed to run a statistically significant number of NOESY experiments for each structure determination).

The relative lack of precision with which the cross-relaxation rate can be experimentally measured must be regarded as the major factor limiting the ultimate precision of the calculated structure. The relative imprecision (or uncertainty) in the experimental value of the cross-relaxation rate is also one of the main factors limiting the usefulness of the relaxation matrix approach to NMR structure determination. The benefits of fitting a precise set of calculated intensities to an inherently imprecise set of experimental data are illusory. Only if the accuracy and precision of the experimental NOE intensities could be defined in a manner beyond question and only if the dynamics of the protein molecule were understood (and properly taken into account) could the power of the relaxation matrix approach receive a fair test.

(4) Translation of observed NOE intensities into a set of distance constraints can conservatively be described as a hotly contested issue. The views recorded in the literature range from the skepticism about trying to be very precise, voiced most strongly by Jeff Hoch and Marius Clore at this meeting, to attempts at very precise calibration of internuclear distances (to ±0.1-0.2 Å) by reference to a known distance and the intensity of its associated NOE[10-12]. Such calibrations can be considered to be warranted if — and only if — (1) the experimental value of σ is precisely determined, (2) there is no spin diffusion so that the two-spin approximation is valid and (3) the spectral density function of the pair of atoms can be specified precisely.

As pointed out by Marius Clore at this meeting and by us on many occasions before[1,2,13,14], these conditions are rarely satisfied. The lack of a simple unique relationship between observed NOE intensity and internuclear distance in proteins has been amply demonstrated by the work of James[15,16], Lane[17], Madrid et al.[18,19], and Clore[20]. The imprecision caused by spin diffusion and the imprecision resulting from specific assumptions about internal motions or lack of them can be quantitatively evaluated and are found to introduce distance errors of the order of a factor of 2-3 (Jardetzky and Lane[14], and others cited above). Tom James stressed the need for more accurate distances and the potential of relaxation matrix analysis for providing them. Marius Clore put it succinctly: "Obviously more accurate distances are highly desirable. The question is whether they *are* more accurate. Let us say you measure an NOE of 30%, but let us say the true NOE is actually 60% and it has been attenuated two-fold (by instrumental variables, etc.). You may still be able to fit it at 30% perfectly well, but you are still in error. That is why one should be cautious about it ... If you look empirically, you never observe anything beyond five Å. So if you say an NOE is weak and I just say it's less than five Å, whether it is direct or indirect, no error has been made." At the present state of the art of spectroscopy and of our knowledge, there is no escape from the conclusion that only rather imprecise distance constraints can be deduced from the observed NOEs.

Several different formulations of the allowed error limits have been adopted: upper and lower bounds[7,11,21], approximate ranges[20], approximate probability distributions derived from solutions of Bloch equations on proteins of known structure[8,18,19]. No rigorous or extensive comparisons of the relative magnitude of the errors induced by the different formulations have been made, but preliminary indications are that all conservative error limits used in different methods of structure computation lead to approximately the same results[13]. The final word on the best way of deriving distance constraints from NOE data is probably yet to be spoken, but at the present state of the art a generous error margin appears strongly indicated.

(5), (6), and (7) The problem the NMR spectroscopist faces is therefore no less than: How to convert a set of imprecise constraints into a precise set of coordinates? This problem is compounded by inadequate knowledge — and consideration — of protein dynamics. A large (generally an infinite) number of interproton distances can be calculated for a given proton pair from a given measured NOE, depending on the assumptions about the correlation time. Many of these can be defended as "reasonable." Clearly, it would be better to actually measure the correlation time for different parts of the molecule, but this is rarely done. Needless to say, structures published to date, with few exceptions[22], have been calculated on the simplest assumption that the protein is rigid, i.e., only one correlation time and only one conformation need to be taken into account, and all NOE distance constraints can be treated as if they were satisfied simultaneously. As a first approximation for the purpose of roughly defining the topology, this assumption, though, strictly speaking, incorrect, is useful, and will likely remain so. If, on the other hand, one attempts to refine structures to a precision comparable to the precision of x-ray structures, the generally unwarranted use of this assumption is a likely source of serious errors.

There are two principal sources of such errors: (1) assuming the wrong correlation time for a local internuclear vector (e.g., the "overall tumbling time" of the molecule in the presence of internal motion) can introduce an inaccuracy of a factor of 2-3 in the distance estimated from a given NOE intensity[14]. (2) Assuming that a motionally averaged distance calculated from an NMR parameter can be used to calculate an "average" structure can lead to meaningless structures[1], unless the states being averaged are known and the non-linear nature of the averaging is correctly taken into account on the basis of an independently verifiable model (the "averaging problem"). Both of these problems are well known[1,2,11,14], but they have no general and easy solution. The solution depends on information not provided by the set of NMR experiments conventionally used for structure determination.

The impetus for the discussion of the topics dealing with protein dynamics — questions (5), (6) and (7) on our initial list — was given by the elegant attempts of Andrew Torda to calculate correctly averaged internuclear distances from molecular dynamics simulations[23]. While one could argue that in principle this approach could eliminate both of the cited sources of error, it is too early to assess whether it can do so within the practical — and rather severe — limitations of molecular dynamics simulations themselves. Averaging between alternative conformers, as we know it from kinetic experiments[24], occurs on time scales of micro- to milliseconds, well beyond the reach of MD calculations. Evaluation of spectral density functions for very rapid local motions (pico- to nanosecond time scale) is on a sounder footing, but comparison with experimental data, which is still almost entirely lacking, will be essential to assess its accuracy. Here one faces the additional problem that relaxation theory for calculating spectral density functions for complex non-rigid structures is approximate at best[2,25]. Given the cumulation of uncertainties in the interpretation of experimental data on protein dynamics, the question as to whether we can achieve a significant gain in precision for specifying protein solution structure over the "imaginary rigid protein which doesn't have motions" (and for which a calculated high degree of precision is illusory), remains open.

Opinions on the best approach to this difficult problem remained divided throughout the session. Some of the participants defended the viewpoint that the rigid body approximation should be considered adequate (Occam's razor — fit the simplest plausible model to the data), unless there is specific evidence for significant internal motion or for the averaging of multiple conformational states. Such evidence would have to come from experiments — using NMR relaxation or exchange parameters or other methods and types of experiments — not on the current list of essential steps for NMR structure determination. Others took the position that existing general evidence for internal motions in proteins is overwhelming[24,26] and needs to be taken into account in any event (Occam's razor properly aimed).

Andrew Torda emphasized that using a time-averaging approach would also mean using a different calibration of the NOEs in terms of distances, since the $1/r^6$ dependence of NOE intensity would weight any distribution of motionally averaged distance toward shorter distances. This would apply to both the averaging of oscillations around a single minimum and to the averaging of the position of the minimum in different conformations. The principal difficulty of taking dynamics correctly into account is that there exists no clear evidence for the shape of the local potential well for local oscillations — and rarely a clear experimental proof for the number and nature of multiple conformers. Absent such evidence, multiple hypotheses can be advanced to decompose in a "more or less realistic" manner the observed NOE data which represent a weighted average of all accessible states. Specific comments on factors to be taken into account were not lacking: How long a simulation would be necessary to "realistically" represent internal motions in proteins? How much difference would inclusion of solvent make? How does one take correlated motions into account? On the other hand, almost entirely lacking at present are experimental criteria by which contributions, amplitudes and time scales of internal motions could be reliably and quantitatively assessed.

The judgment as to what is "more or less realistic" in describing the dynamics of a given protein remains in large measure subjective. This being the case, the discussion of protein dynamics and its implications for NMR protein structure determination can be fairly summarized by saying that: (1) experimental studies of protein dynamics and the development of new methods for such studies have a very high priority and (2) until protein dynamics are better understood, judgment on the attainable precision of NMR structures has to remain reserved, regardless of the choice of method of computation.

(8) Methods of structure calculation from NMR data have been at the center of attention in the field for some time, and yet the published reports and discussion have mostly centered on technical details of implementation and on results rather than on fundamental questions which still remain open. Three basic points are worth making at the outset:

(1) The fundamental differences between structure calculation from x-ray and NMR data respectively should always be borne in mind: as stressed by David Cowburn, the quality, the nature of the experimental data in the two cases is fundamentally different. In x-ray diffraction the geometric distribution of electron density can be directly calculated by Fourier Transformation from the diffraction pattern, so that the primary data directly yield a molecular shape — a severe spatial constraint; the electron density calculated from diffraction data provides directly a volume constraint for each atom — and with good data the constraints are severe.

In contrast, NMR provides only linear (distance) constraints between pairs of atoms and only for those at short distances from each other. This, even together with the covalent constraints known from the primary structure, restricts the relative position of different atoms to a highly variable degree — and generally amounts to a much looser set of constraints.

Since the NMR data input into a structure calculation is a set of approximate, short-range distance constraints in distance space derived from experimental data on the basis of subjective judgments discussed under points

384

(1)-(7) above, it is to be expected that the conversion of the data into a 3D geometry will require much more extensive sampling of the conformational space accessible to the polypeptide chain than is needed in crystallography — *before* any refinement is attempted. As a consequence, the potential for introducing bias, error and misinterpretation is much greater in NMR than in crystallography. This point was repeatedly underscored by several discussants. Nobuhiro Gō asked a searching question: Is NMR information good enough to allow a meaningful second (refinement) step? The literature to-date seems to have given this question an affirmative answer, but much of the discussion at this session signalled greater caution. Dennis Hare pointed out that bias can creep into the selection of starting structures, into sampling algorithms, and into spectral simulation. Jim Prestegard noted that error functions are quite intolerant, and Andrew Torda illustrated four different classes of penalty functions, each of a different analytical form, hence containing a somewhat different analytical bias. A clear quantitative understanding of these factors is yet to be achieved.

(2) All existing methods of structure calculation have two clearly distinguishable (though not always clearly separated) components: (a) the sampling of conformational space and (b) optimization.

Distance Geometry relies on random sampling in distance space for the former and minimization of a global (least squares type) target function for the latter. Restrained Molecular Dynamics relies on random sampling of starting structures and simulated annealing for conformational sampling and energy-minimization for global refinement. Heuristic Refinement relies on coarse systematic sampling for the former and on the Kalman filter, which uses no global target function, for (limited) local optimization.

It is worth underscoring that the accuracy of the final structure depends entirely on the adequacy with which the first step — i.e., conformational sampling — is carried out. No optimization procedure, especially no procedure using a global target function, can by itself guarantee accuracy because of its vulnerability to built-in bias and to local minima. As again pointed out by Dennis Hare, very little is known about the relative bias of different optimization procedures. We know that some implementations of distance geometry are biased and fail to adequately sample conformational space[27], but very few comparative studies have been reported. As in the case of evaluating accuracy, objective numerical evaluation of either sampling bias or target function bias is almost entirely lacking.

(3) Despite remarkable successes in calculating apparently very precise structures from NMR data and structures in good agreement with known crystal structures, there are, as Dennis Hare pointed out, no objective measures for the goodness of fit for NMR structures to the "real" structures: "only assertions, no numbers." The fact that precision cannot serve as a measure of

accuracy has already been pointed out above. The lack of a suitable metric for objectively measuring accuracy poses a problem that remains unsolved.

Most of the ensuing discussion centered on the issue of the adequacy and accuracy of sampling. Here again opinions remained divided. Those who have published several structures felt that "sampling is not a problem" and the existing methods are adequate. On the other hand, the published report by Dennis Hare and coworkers leaves no doubt that sampling bias can be introduced in the process of coding the distance geometry algorithm. Although the view has been expressed that this is a problem only for the specific implementation studied by Metzler et al.[27], but not for other implementations, convincing objective proof for this statement is lacking. Lack of an objective measurement of the adequacy of sampling in nearly all published NMR structure determination remains the central difficulty in settling the issue of accuracy of the proposed structures. Resort to strong personal convictions, which are clearly held by the proponents of different methods of calculation, are, unfortunately, not a substitute for objective verification. An encouraging preliminary comparative study of the *lac* repressor structure calculated by distance geometry, restrained molecular dynamics and heuristic refinement procedures[13] seems to indicate that to a first approximation all existing methods of calculation lead to similar, but not identical results. The differences observed thus far remain beyond the limits generally held desirable by the standards of crystallography, but this may change as more data and "cleaner," less biased, algorithms are introduced.

An escape from the dilemma was proposed by Marius Clore — instead of worrying about the absolute accuracy of a proposed protein structure, we should ask whether the information contained in it is adequate to answer specific questions about its biological function. It might be said that rather imprecise knowledge of the structure is often sufficient for this purpose, but inaccurate structures could lead to wrong conclusions.

In summary, the discussion brought out most of the aspects of NMR structure determination that remain problematic. The remarkable success of NMR in providing the essential experimental data to construct solution structures of proteins is a clear fulfillment of a prophecy made 25 years ago[28]. If the data set contains evidence of extensive secondary structure and a substantial number of NOE constraints between residues far apart in the primary sequence, the accuracy of the proposed model can be ensured to first order by conventional manual or computer model building as well as by several computational techniques. The product of such model building does, in the language of crystallography, constitute a "low resolution" structure. What remains uncertain is the degree of precision to which NMR structures can be refined without over-interpreting the data — or even sacrificing accuracy for the appearance of high precision.

There is general agreement that inadequate data sets may be a problem for an individual investigator, but are not a general problem at the present state of the art in the field. Quite the contrary, with the best spectrometers and experimental techniques one is overwhelmed by the wealth of data. The same applies to the accuracy of spectral assignments. The generally unresolved problems of NMR data interpretation, which can in all cases severely limit the precision and can in individual cases also limit the accuracy of the proposed molecular model, can be narrowed down to a few major sources of uncertainty:

(1) subjectivity in the quantitation of the observed NOEs (here over-interpretation may be more dangerous than imprecision),

(2) lack of quantitative experimental information on protein dynamics (which provides a temptation to use the rigid molecule model when it is not appropriate),

(3) lack of an objective quantitative measure of the adequacy of conformation space sampling in different computational regimens (this is a potential source of serious inaccuracies), and

(4) lack of a quantitative measure of bias introduced by different refinement procedures.

Because of the inescapable subjective judgments required in an NMR structure determination, the inherent precision of an NMR structure cannot be evaluated, as has been attempted, by taking the RMS difference between repeated runs of a single computational procedure. Rather, it is better measured by the RMS difference between the results of different computational procedures. Another important point was made by Marius Clore: one should not look merely at global numbers, but evaluate the precision of the structure residue by residue.

What needs to be done to put NMR methods of macromolecular structure determination on a secure footing and to define the limits of the information they provide is fairly clear:

(1) The reproducibility of NOE intensities, their dependence on instrumental parameters and the solution conditions must be better documented, so that the confidence limits for the experimental measurement of cross-relaxation rates can be accurately defined.

(2) For each protein (or nucleic acid) structure determination, the dynamics of the molecule must be evaluated by experimental studies (NH exchange rates, relaxation measurements, etc.) before a precise final structure is proposed. (3) The adequacy of conformational space sampling must be quantitatively established for all methods of structure calculation. Although validation studies using artificial NMR data generated from known crystal structures and recalculating the initial structures have been reported for all of the main algorithms in use, studies such as those by Metzler et al.[27] are

more informative. Extensive comparisons of grid search and random sampling procedures, defining confidence limits for each procedure, are needed. (4) Comparisons of structures produced by different refinement procedures are among the most urgent needs in the field. They will serve both to evaluate the bias built into each algorithm, the magnitude of the correction needed to correct for the bias, and the limits of precision to which a (relatively) unbiased structure can be specified from NMR data (and the known covalent and van-der-Waals chemistry of the molecule).

We can hope that as new methods and experimental measurements are brought to bear on the problem of protein dynamics, a better understanding of the quantitative relationship between observed NOE intensity and the distance(s) implied by it will result. Computational methods for evaluating the adequacy of sampling and bias of data fitting and optimization procedures, though imperfect, are generally available[29]. They just have not been applied extensively in this field. A series of careful comparative studies, comparing structures calculated by different computational methods from the same set of data, should give us a better understanding of the magnitude of errors introduced by different methods of analysis.

Procedures for producing accurate, though not necessarily precise, structures are available[8]. The essential ingredient of such procedures is a systematic sampling (grid search) step. Exhaustive random sampling could be a good approximation, if it were shown that it is truly random. Unfortunately, for most of the existing algorithms the claim that the search of the entire available conformational space is random remains unproven at best. Caution is required, because in the one case that has been properly studied[27] non-randomness in the coding of the algorithm was found. It is unfortunate that a rigorous systematic grid search at the atomic level rapidly leads to a combinatorial explosion for molecules larger than a decapeptide: it remains the only method by which the accuracy of an NMR structure can be rigorously established. More efficient implementations of approximate grid searches, relying on fewer simplifications than existing methods do, may therefore be well worth the effort.

The precision — and to some extent the accuracy — with which a protein structure can be determined from NMR data will remain an open question until the factors which influence it as discussed above are more clearly understood and more quantitatively and thoroughly evaluated for each of the available methods of structure determination. The question as to whether NMR can yield structures that could be judged "high resolution" by crystallographic standards does not yet have a clear answer. It may be of interest to note, however, that — as Tom James and others have pointed out — a major contribution of NMR may have been to raise the question in the minds of at least some crystallographers, whether the crystallographic standards for high resolution are not in some measure artefactual. For purposes of understanding protein function — and given that motion is ubiquitous — the issue of high precision may prove not to be very relevant. Resolving the issues raised in this round table remains relevant in any case, however, if one is to avoid introducing inaccuracies in the quest for precision.

REFERENCES

1. O. Jardetzky, *Biochim. Biophys. Acta* **621**, 227 (1980).
2. O. Jardetzky and G. C. K. Roberts, "NMR in Molecular Biology", Academic Press: New York (1981).
3. O. Lichtarge, C. W. Cornelius, B. G. Buchanan, and O. Jardetzky, *Proteins* **2**, 340 (1987).
4. G. M. Crippen, "Distance Geometry and Conformational Calculations", Wiley: Chichester, England (1981).
5. I. D. Kuntz, J. F. Thomason, and C. M. Oshiro, (1989) in "Methods in Enzymology", 177, N. J. Oppenheimer and T. L. James, eds., p. 159, Academic Press: New York.
6. G. M. Clore, A. M. Gronenborn, A. T. Brünger, and M. Karplus, *J. Mol. Biol.* **186**, 435 (1985).
7. R. M. Scheek, W. F. van Gunsteren, and R. Kaptein (1989), in "Methods in Enzymology", 177, N. J. Oppenheimer and T. L. James, eds., p. 204, Academic Press: New York.
8. R. B. Altman and O. Jardetzky (1989), in "Methods in Enzymology," 177, N. J. Oppenheimer and T. L. James, eds., p. 218, Academic Press: New York (1989).
9. R. B. Altman, C. H. Arrowsmith, R. Pachter, and O. Jardetzky, in this volume.
10. G. M. Clore and A. M. Gronenborn, *J. Magn. Res.* **61**, 158 (1985).
11. K. Wüthrich, "NMR of Proteins and Nucleic Acids", Wiley: New York (1986).
12. W. Braun, *Quart. Rev. Biophys.* **19**, 115 (1987).
13. R. B. Altman, R. Pachter, and O. Jardetzky, (1989) in "Protein Structure and Engineering", O. Jardetzky, ed., p. 79, Plenum Press: New York.
14. O. Jardetzky and A. N. Lane, (1988) in "Physics of NMR Spectroscopy in Biology and Medicine", B. Maraviglia, ed., p. 267, North-Holland Physics Publishing: Amsterdam.
15. B. A. Borgias, and T. J. James, *J. Magn. Res.* **79**, 493 (1988).
16. J. W. Keepers, and T. James, *J. Magn. Res.* **57**, 404 (1984).
17. A. N. Lane, *J. Magn. Res.* **78**, 425 (1988).
18. M. Madrid and O. Jardetzky, *Biochim. Biophys. Acta* **953**, 61 (1988).
19. M. Madrid, J. E. Mace, and O. Jardetzky, *J. Magn. Res.* **83**, 267 (1989).
20. A. M. Gronenborn and G. M. Clore, *Biochemistry* **28**, 5978 (1989).
21. A. D. Kline, W. Braun, and K. Wüthrich, *J. Mol. Biol.* **189**, 377 (1986).
22. Y. Kim and J. H. Prestegard, *Biochemistry* **28**, 8792 (1989).
23. A. Torda, R. M. Scheek, and W. M. van Gunsteren in this volume.
24. R. O. Fox, P. A. Evans, and C. M. Dobson, *Nature* **320**, 192 (1986).
25. O. Jardetzky, *Accts. Chem. Res.* **14**, 291 (1981).
26. J. A. McCammon and S. C. Harvey, "Dynamics of Proteins and Nucleic Acids", Cambridge University Press: Cambridge, England (1987).
27. W. J. Metzler, D. R. Hare, and A. Pardi, *Biochemistry* **28**, 7045 (1989).
28. O. Jardetzky in "Proc. Int. Conf. Magnetic Resonance", Tokyo, Japan, N-3-14, 1 (1965).
29. W. H. Press, B. P. Flannery, S. A. Teukolsky, and W. T. Vetterling, "Numerical Recipes", Cambridge University Press: Cambridge, England (1989).

FAST CALCULATION OF THE RELAXATION MATRIX

Mark J. Forster

National Institute for Biological Standards and Control
Blanche Lane, South Mimms
Hertfordshire, EN6 3QG
United Kingdom

ABSTRACT

The relaxation matrix for a macromolecule is a 2D array holding the set of auto and cross relaxation rates. This is used in the numerical simulation of NOEs for given candidate structures. Structure refinement routines that compare experimental and simulated NOEs require many recalculations of the matrix and algorithm efficiency should be an important consideration. Three algorithms are described and their relative efficiencies compared by computing relaxation matrices for model macromolecules. An interaction distance cutoff of 8 Angstroms greatly reduces the calculation time for large systems and introduces only a small truncation error. Further performance improvements, at the same level of accuracy, were obtained by using a 3D array/linked list data structure to avoid calculating the full set of $N(N-1)/2$ distances.

INTRODUCTION

For a system of N spins, such as the set of protons in a biological macro-molecule, the set of cross- and auto-relaxation rates are arranged into an N by N array called the relaxation matrix. This is used in the numerical simulation of theoretical nuclear Overhauser effects for candidate structures. In a structure optimization scheme that compares experimental and simulated NOEs, many recalculations of this matrix will be necessary. It is therefore important to find efficient algorithms for computing this matrix. Three algorithms are described and their relative efficiencies compared by computing relaxation matrices for model structures including carbohydrates and oligonucleotides. Neglecting direct interactions for spins separated by more than 8 Angstroms greatly reduces the calculation time for large systems, without a major loss of accuracy. Further significant improvements in performance, at the same level of accuracy, are obtained by the use of

Computational Aspects of the Study of Biological Macromolecules by Nuclear Magnetic Resonance Spectroscopy, Edited by J.C. Hoch *et al.*, Plenum Press, New York, 1991

391

an 8 Angstrom cutoff in conjunction with a 3D array-linked list data structure. This is achieved by avoiding the computational overhead involved in calculating the full set of $N(N-1)/2$ distances.

The time development of nuclear Overhauser effects for a system of interacting spins are described by the Solomon equations[1].

$$\frac{d\eta_i}{dt} = \sum_{k=1}^{N} \Gamma_{ik} \eta_k. \tag{1}$$

Where the diagonal and off diagonal elements of the relaxation matrix Γ are the auto and cross-relaxation rates respectively. The theoretical NOEs for the candidate structure are found by specifying the initial conditions, then solving the Solomon equations using either numerical integration or eigenvalue-eigenvector methods.

The cross-relaxation rate expression is

$$\sigma_{ik} = \alpha_{HH} r_{ik}^{-6} \left[6J\left(\omega_i + \omega_k\right) + J\left(\omega_i - \omega_k\right) \right] \tag{2}$$

where r_{ik} is the $i-k$ spin-spin distance and α_{HH} is the dipolar interaction constant for two protons and is given by,

$$\alpha_{HH} = 0.1 \left(\mu_0/4\pi\right)^2 \gamma_i^2 \gamma_k^2 \left(h/2\pi\right)^2. \tag{3}$$

The value of this constant is $5.6965 \times 10^{10} \text{Å}^6 s^{-2}$. $J(\omega)$ is the spectral density function describing the motional model used for the $H-H$ interaction vector. The auto relaxation rate is a sum of pairwise dipole-dipole terms and a term for the 'external' rate (the total rate due to other relaxation mechanisms).

$$\rho_i = \sum_k \rho_{ik} + \rho_i^*. \tag{4}$$

The pairwise auto relaxation rate expression is

$$\rho_{ik} = \alpha_{HH} r_{ik}^{-6} \left[6J\left(\omega_i + \omega_k\right) + 3J\left(\omega_i\right) + J\left(\omega_i - \omega_k\right) \right]. \tag{5}$$

The simplest spectral density function expression is that for a rigid body that tumbles isotropically in solution, if τ is the reorientational correlation time

$$J(\omega) = \tau / \left(1 + \omega^2 \tau^2\right). \tag{6}$$

The spectral density function for a $H-H$ vector in a rigid body undergoing symmetric top tumbling depends upon two diffusion constants and upon the angle between the interaction vector and the top axis[2]. In both of these cases the $H-H$ vector has a fixed length and orientation with respect to the molecular axes. The Tropp model[3] is a mathematical approach for

392

describing spectral density functions when the $H - H$ vector rapidly jumps between a number of conformations thereby changing both its length and its orientation. The NOEMOL program[4] models NOEs using either isotropic or symmetric top diffusion models and models methyl group rotations using the Tropp model. The following three algorithms have been used to compute relaxation matrices for model structures.

ALGORITHM 1 *Calculate the relaxation matrix using the approach described above. Take all spin interactions into account (no distance cutoff used).*

ALGORITHM 2 *As Algorithm 1 but neglect interactions between spins separated by more than 8 Angstroms. The algorithm can be summarized as*

> *LOOP $i = 1,N$*
> *LOOP $j = 1,N$*
> *IF $(i > j)$ and $r^2(i, j) > 64$ then compute dipolar interaction*
> *NEXT j*
> *NEXT i.*

For a large system of spins calculating the full set of $N(N - 1)/2$ distances is a significant part of the overall computational burden.
Using this distance cutoff criterion typically affects calculated auto relaxation rates by around 0.5 per cent. This is considered small compared to the neglect of external relaxation, the error due to simplified motional models and the accuracy of typical experimental data.

ALGORITHM 3 *As algorithm 2 but avoid calculating the set of $N(N - 1)/2$ distances. Divide Cartesian space into a box of cubes of size 4 Angstroms. For each spin find the x, y, z indices of the cube in which it resides. Set up 3D array-linked list data structure to find which spins are in a given cube, for each spin, find the spins in the same cube and in the adjacent two layers. Apply the distance cutoff criterion and compute the dipolar interactions.*
The 3D array (of integers) holds the spin number of the next spin in the same cube, the list is terminated by a zero entry. This data structure allows us to quickly and simply find the spin present in a given cube, for large spins systems the number of distance calculations is greatly reduced and the relaxation matrix calculation is speeded up.

The following model structures were used:
1: N-Acetyl Heparin dodecasaccharide with the repeat unit

$$[4]\text{-}\alpha\text{-L-IdoA2SO}_3^- (1\text{-}4)\text{-}\alpha\text{-D-GlcNCOCH}_3\text{-}6SO_3\text{-}(1\text{-}]$$

this has 109 protons 6 methyl groups).

Table 1. Isotropic Diffusion Model

	Heparin	Trp Operator	30 bp DNA	50 bp DNA
Algorithm 1	36.4	330	589	300
Algorithm 2	6.1	51.9	91.9	177
Algorithm 3	5.2	28.4	37.3	21.8

Table 2. Symmetric Top Diffusion Model

	Heparin	Trp Operator	30 bp DNA	50 bp DNA
Algorithm 1	61.1	821	1699	3554
Algorithm 2	11.7	87.6	154	293
Algorithm 3	10.8	64.0	99.9	139

2: Trp operator 20 base pair B-DNA, $d(CGTACTAGTTAACTAGTACG)_2$ with 456 protons (12 methyl groups).
3: 30 base pair B-DNA, $d((CGTA)_7CG)_2$, having 678 protons (14 methyl groups).
4: 50 base pair B-DNA, $dC_{50}\text{-}dG_{50}$, 1104 protons (no methyl groups).

Relaxation matrices were calculated for each model structure using the three algorithms, the number of CPU seconds required on a SUN 3/160 are shown in the Tables 1 and 2. For all of the model structures algorithm 3 is the fastest method of calculating the relaxation matrix. Choosing the most efficient algorithm will have a major bearing on the efficiency of structure optimization routines based upon comparing experimental and theoretical NOEs.

REFERENCES

1. Solomon, I., *Phys. Rev.* **99**, 559 (1955).
2. Woessner, D.E., *J. Chem. Phys.* **37**, 647 (1962).
3. Tropp, J., *J. Chem. Phys.* **72**, 6035 (1980).
4. Forster, M., Jones, C., and Mulloy, *J. Mol. Graph.* **7**, 196 (1989).

NMR STRUCTURES OF PROTEINS USING STEREOSPECIFIC ASSIGNMENTS AND RELAXATION MATRIX REFINEMENT IN A HYBRID METHOD OF DISTANCE GEOMETRY AND SIMULATED ANNEALING

J. Habazettl[†], M. Nilges[‡], H. Oschkinat[†], A. T. Brünger[‡],
and T. A. Holak[†]

[†]Max-Planck-Institut für Biochemie
D-8033 Martinsried bei München, F.R.G.
[‡]Howard Hughes Medical Institute
and Department of Molecular Biophysics and Biochemistry
Yale University, New Haven, CT 06511, U.S.A.

ABSTRACT

The hybrid method combining the early stages of a distance geometry program with molecular dynamics/simulated annealing in the presence of NMR constraints was optimized to obtain structures consistent with the observed NMR data. Two novel methods of stereospecific assignments of the protons at the prochiral carbons are used in simulated annealing, the "floating" chirality assignment and a high-dimensional potential. These two methods were compared with stereospecific assignments obtained from the coupling constant data. There is good agreement between the three methods in predicting the same stereospecific assignments. As the high-dimensional potential uses more relaxed absolute distance constraints and also takes into account the relative distance constraint patterns, it reduces possible overinterpretation of the NOE data. The structures obtained from the hybrid method were further refined using the relaxation matrix approach. This approach employs the analytical form of the gradient of the calculated spectrum. Compared to the structures determined with the two-spin approximation, the fit to the NMR data improves significantly with only minimal r.m.s. shifts in the structure during simple conjugate gradient minimization. The R-factors, defined similarly to the crystallographic R-factors, are 0.51 for the structures calculated using the two-spin approximation and 0.26 for the refined structures. Large shifts of approx. 1 Å occur during a dynamics/simulated annealing calculation. The various stages of refinement and stereospecific assignments are tested on the NOE data for the small squash trypsin inhibitor, CMTI-I. In the case of CMTI-I, the last step of the refinement improved the agreement with the X-ray structure significantly.

INTRODUCTION

Over the past few years methods have been developed to determine the three-dimensional structure of proteins by NMR. The methods based on

Computational Aspects of the Study of Biological Macromolecules by Nuclear Magnetic Resonance Spectroscopy, Edited by J.C. Hoch *et al.,* Plenum Press, New York, 1991

395

molecular mechanics[1-4], molecular dynamics[2,5-10], and least-squares minimization of torsion angle space[1,11-15], have been used successfully in the determination of three-dimensional structures of proteins by NMR. We have recently shown that a substantial saving in computing time may be achieved with a hybrid approach that combines only early stages of a distance geometry program with energy minimization[4]. The substructures obtained from phase 2 of the distance geometry method (DISGEO)[16,17], which have approximately the correct polypeptide fold, were used as starting structures for a two-stage energy minimization. This hybrid method was subsequently improved by replacing the energy minimization stage by simulated annealing[18,19]. Simulated annealing as compared to energy minimization should be better suited to locating the global minimum of the potential energy of the protein.

In the present paper we describe the determination of the full three-dimensional structure of the squash seed trypsin inhibitor CMTI-I using the hybrid method of DISGEO and dynamical simulated annealing. The two novel methods of stereospecific assignments of protons at the prochiral centers, the "floating" chirality assignment[19,20] and assignment with the high-dimensional potential, are applied in the simulated annealing stage of the structure calculations to obtain stereospecific assignments at prochiral centers. Using these methods, the stereospecific assignments are obtained automatically during the structure calculation itself. We tested the methods on CMTI-I because, despite its small size, it possesses most of the features of a small size protein. Due to its compactness and rigidity, and to a very large number of assigned NOEs (324 NOEs for 29 residues) it is well suited to a detailed analysis of the quality of structures and for the effectiveness with which the stereospecific assignments can be obtained with the method used in our calculations.

Once the stereospecific assignments were determined, we refined the CMTI-I structures using the differences between the calculated and observed NOESY spectra as the driving force to change the conformation[21,22]. A formula for the gradient of NOE intensities calculated with the relaxation matrix approach has recently been derived by Yip and Case[23]. This method, interfaced with the refinement program X-PLOR[24], was used in our calculations on CMTI-I.

THE HYBRID DISTANCE GEOMETRY-ENERGY MINIMIZATION/DYNAMICAL SIMULATED ANNEALING METHOD

The molecular dynamics and mechanics calculations require suitable starting structures. They operate in real space and are susceptible to being trapped in local energy minima as opposed to the global minimum that fits all constraints. This is especially true for the molecular mechanics approach. Distance geometry algorithms provide a means of generating structures without the necessity of having a starting structure. However, the distance geometry method has a number of drawbacks which stem from the fact that the program does not incorporate a complete description of molecular energy terms. This is particularly evident in the poor stereochemistry of the local segments of the structures, especially in terms of non-bonded contacts[14]. Other disadvantages include: inefficient sampling of the conformational space around the global minimum[25] and long computational times necessary for the embedding of the complete structure[26]. For acyl carrier protein, a protein of 77 amino acids, the program required 6 h on a VAX 8600 to complete the structure[4]. We found, however, that substructures produced at the end of the first three phases of DISGEO (phases 0-2) are sufficiently close to the global minimum to allow efficient minimization using a pseudoenergy approach. These three phases of distance geometry on a VAX 8600 required only 20 min per structure for embedding of the main-chain subset of the atoms[4]. The simulated annealing stage took about 2 h on the VAX computer. Thus the savings in computing time for a protein of the size of ACP is twofold compared to the complete DISGEO run. For smaller peptides, like CMTI-I, the saving in CPU time is smaller. A simplified flowchart of the calculational protocol for the hybrid method is shown in Fig. 1. The two-stage minimization was carried out with the AMBER program[27] and simulated annealing with the X-PLOR program.

Transformation of Peak Intensities into Distance Constraints.

The intensities of the NOE cross peaks were determined from volume integrals. The volume integrals consisted of the sum over all data points within a rectangular base plane. For partially overlapping peaks, we use a procedure that calculates volume integrals by combining column and row line integrals that have base lines chosen interactively[28]. The NOEs used in the calculations were derived from a single NOESY spectrum at a mixing time of 100 ms. All distances were calibrated using a single proportionality constant. For the 2D NOE spectra in water the largest resolved $d_{\alpha N}$ cross peaks in the spectrum were used as the calibration with a distance of 2.3 Å. This calibration was also used for the D_2O spectra after an appropriate scaling between the two spectra.

The input distance constraints for DISGEO consisted of two tables. The principal table contained the upper bounds only. A second table with the upper and lower limits was used for the non-observed NOEs (see below). The pseudoatom representation of Wüthrich et al. was used[29]. Appropriate corrections were added to the upper bounds with the exception that no pseudoatom correction of 1 Å was used for the methylene protons with

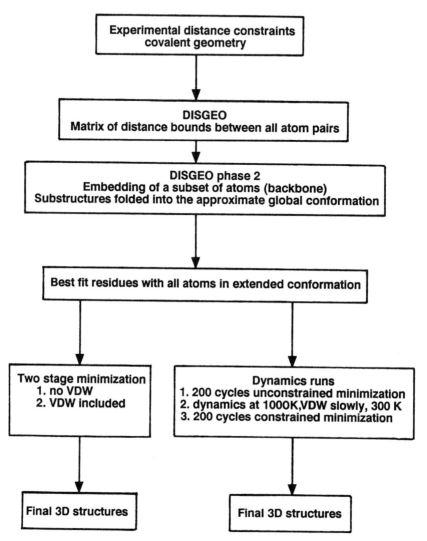

Figure 1. Diagram of the overall strategy used to compute 3D structures of proteins from the NMR data.

two separate signals. Also, no correction was applied to methyls on valine and leucine with two separate signals. Instead the highest possible upper bound was used as a distance to the pseudoatom center[19]. For any distance constraints involving methyl protons a correction of 1 Å was added to the upper distance bounds for each methyl group. Although all protons were explicitly defined in the dynamical simulated annealing calculations, in some cases additional terms were added to the upper bounds. For any experimental distance constraints involving methyl protons, aromatic protons, and methylene protons for which only one signal was observed, an appropriate correction was added to the upper distance bounds and the distance was referred to the average positions of these protons[29]. When the difference in chemical shifts between the 2 methylene protons or 2 methyl groups on valine or leucine was large enough that NOEs involving each group could be resolved, they were assigned arbitrarily to H_1 and H_2, or Me_1 and Me_2. In the case when only an NOE to one of the partners from the resolved pair was observed, the distance constraint to the other partner was set to 4.2 Å (the estimated maximum cutoff distance), with the lower bound -0.6 Å and the upper bound +1.8 Å and 3.0 Å for methylene protons and methyls, respectively[19].

80 non-NOE distance constraints were also present in the calculations[30]. These constraints were introduced after a few initial calculations to exclude any close contacts (ca. 2.0 Å) that contradicted the lack of an observable very strong NOE in the NOESY spectra. They were kept throughout all of the subsequent calculations, although at the later stages of calculations they seem to be of a rather limited usefulness in restriction of the allowed conformational space for the structures, as the RMS difference between the experimental and calculated constraints was always less than 0.01 Å in most of the structures.

Structure Calculations.

All minimization and dynamical simulated annealing calculations were carried out with the program X-PLOR[6,24] on a CONVEX C1-XP2 computer. Distance geometry calculations were carried out with the program DISGEO[16,31]. The basic protocol used for the calculations has been presented in our previous publications[19,30]. The protocol consists of 5 stages. In stage 1, the substructure coordinates are obtained from the distance geometry program DISGEO[31]. In the second stage all of the atoms are added to a subset of atoms already present in substructures. Steps three and four consist of simulated annealing. The fifth stage involves 1000 cycles of constrained Powell minimization. These last four stages are carried out with the program X-PLOR[6,24].

The total target function E_{tot} comprises the following terms:

$$E_{tot} = E_{empirical} + E_{relaxation}$$

$E_{empirical}$ is the target function for maintaining correct bond lengths, angles, planes, and chirality. $E_{relaxation}$ contains a pseudoenergy target function for the distance and torsion constraints derived from NMR[19,30].

The "floating" chirality calculations involved lowering the force constants during the two SA stages for angles HA-CT-HA, CT-CT-CT, and HA-CT-CT, where HA is a proton and CT is a tetrahedral carbon with four explicit substituents, to 5.0, 5.0, and 30.0 kcal mol^{-1}rad^{-2}, respectively. All 1-4 non-bonded interactions involving $C^{\gamma,\delta}$ proton and carbon atoms of any amino acid were also removed. For the final fifth stage of the energy minimization all angle terms were set to 200 kcal mol^{-1} rad^{-2} and the 1-4 non-bonded interaction was reintroduced.

To demonstrate that only a single prochiral configuration is consistent with the distance constraints, the calculations were also performed reversing the initial assignments. The assignments (i.e., H_1 and H_2, or Me_1 and Me_2) were interchanged in the input NOE tables and the whole calculation procedure was repeated for some substructures. Also 10 final SA structures (out of 34) were subjected to the simulated annealing procedure in the presence of the interchanged NOE assignments. Only these configurations for which the assignments were inverted at the end of these calculations were considered to be stereo-assigned.

The calculations produced structures with the global folding of the polypeptide chain of CMTI-I uniquely defined. All final SA structures satisfied the experimental constraints, exhibited very small deviations from idealized covalent geometry, and had very good nonbonded contacts as shown by small values of F_{repel} and negative Lennard-Jones van der Waals energies (Table 1). The efficiency of the floating stereospecific assignment is shown in Table 2. A comparison to other methods of the assignments is discussed in section 5.

NMR ASSIGNMENTS OF SIDECHAIN CONFORMATIONS IN PROTEINS USING A HIGH-DIMENSIONAL POTENTIAL IN THE SIMULATED ANNEALING CALCULATIONS

The high-dimensional pseudoenergy function for the distance constraints of the prochiral protons has been described in our recent publication[32]. Here we apply the high-dimensional potential in the simulated annealing (SA) protocol to calculate structures of CMTI-I.

The high-dimensional potential function works by forcing a sidechain to rotate into the conformation that fits the pattern of all NOEs to the

Table 1. Average deviations from ideality of the structures

Structure	Bonds (Å)	Angles (deg.)	Impropers (deg.)
Starting structures	0.012 ± 0.001	2.13 ± 0.03	0.36 ± 0.037
Refined structures	0.008 ± 0.001	2.25 ± 0.02	1.77 ± 0.146
X-ray Structure	0.020	3.04	3.12
Minimized X-ray	0.016	2.38	0.53

Table 2. Comparison of the stereospecific assignments[a]

Residue	NOEs	High-dimensional potential	Floating assignment	Coupling constant
Val-2	12	(-217±58)	(10±53)	180±60
Cys-3	12	-152±35	-183±4	180±60
Leu-7	6	-162±20	(-89±31)	-[b]
Met-8	8	-162±7	(74±18)	-[b]
Glu-9	6	-65±13	-57±11	-60±60
Cys-10	10	66±32	45±6	60±60
Lys-12	8	-44±6	(51±7)	-[b]
	6	168±11	179±6	-[b]
Asp-13	14	(145±88)	-71±4	-60±60
Ser-14	6	(64±8)[d]	(60±6)[d]	60±60
Cys-16	6	-49±8	-72±19	-60±60
Cys-20	18	-69±9	-101±28	-60±60
Val-21	10	-73±11	-65±3	-60±60 and 60±60
Cys-22	14	171±6	-160±3	180±60
Leu-23	10	-41±33	-53±18	-60±60
	16	-175±10	173±5	-[b]
Glu-24	6	(38±99)	-159±13	-[c]
His-25	10	-75±8	-41±6	-60±60
Tyr-27	10	(3±85)	-64±5	-60±60
Cys-28	10	-47±12	-61±10	-60±60

[a]The second column gives the number of NOEs to the prochiral protons obtained from the NOESY spectra. The average χ^1 angles (deg) for the SA structures are entered in columns 3 to 4 together with standard deviations; for Lys-12 and Leu23 second entry refers to χ^2. 20 SA structures were calculated for each entry in columns of 3 and 4 with the slightly up-dated distance constraints table compared to that in (30). The ranges for χ^1 torsion angle (5th column) are estimated as described in Wagner et al.[14]. No stereospecific conformation could be determined for the χ anglesgiven in parentheses.

[b]No unambiguous assignment was possible.

[c]See text.

[d]The chiral assignment was obtained by fixing the χ^1 angle to the value given in column 5 c.f.[30].

401

given proton. If there are n NOEs to the H_1 and H_2 protons, the potential is n-dimensional and has two energy minima. The two possible minima are a priori equivalent. Since the assignment of the H_1 and H_2 protons is fixed in the starting structures, the energies of the two protons are not equal, the assignment with the lowest penalty function is preferred. In the n-dimensional target function, all n NOEs are correlated and only one pathway of relative patterns of NOEs will be selected provided that there are enough NOEs to determine a sidechain conformation. Once a single or restricted conformation of the sidechain is secured, the stereospecific assignment of the H_1 and H_2 protons in terms of their chemical shifts can be accomplished by comparing the distribution of NOEs for three possible rotamers of the β-methylene group using the method described by Hyberts et al.[33], or for γ,δ protons, by comparison with the floating stereospecific assignment[30].

The pseudoatom corrections are only used for protons with completely degenerate chemical shifts in our calculations. In all other cases, even when no preferred rotamer conformation is obtained from the calculations, the pseudoatom corrections are not necessary. Our input data is thus more accurate than that obtained from only intraresidue NOEs and J-coupling constant information, provided the same number of prochiral centers is determined by both methods.

COMPARISON OF VARIOUS METHODS OF STEREOSPECIFIC ASSIGNMENTS

The comparison of three different methods of the stereospecific assignments: the high-dimensional potential, floating stereospecific, and coupling constant method, is presented in Table 2. Our current coupling constant data is derived from the E. COSY spectrum and is thus more comprehensive than published previously[30]. The second entry in Table 2 gives the number of NOEs to the prochiral protons obtained from the NOESY spectra. This number was ca. 4 for residues for which no stereospecific assignments were obtained. Table 2 shows that there is good agreement between all three methods in predicting the same stereospecific assignments. The high-dimensional potential seems to produce the most conservative results. For example, Asp13, Glu24, and Tyr27 were not assigned stereospecifically by the high-dimensional method whereas they were by the floating chirality method.

The floating stereospecific method requires relatively tight boundaries for the distance constraints in order for the floating protons to flip yet remain in a particular conformation. Application of such tight constraints could produce structures for which not all possible conformations of a sidechain are sampled. This limitation can be corrected by repeating the calculations

402

in the presence of the more relaxed constraints with the fixed stereospecific assignments (after the prochiral centers were assigned stereospecifically). We decided, however, to refine the structures using the relaxation matrix approach (see below).

The structures calculated with the high-dimensional potential represent a model of a molecule that allows for additional flexibility in the structures, yet still producing structures with stereospecifically assigned sidechains. One of the advantages of the high-dimensional potential is that the input parameters provide an adjustable degree of structural variability. Two types of bounds are used in this potential. Restrictive bounds are used for the relative distance constraints within each NOE proton pair in the prochiral group. The absolute boundaries of the distance constraints, however, are more relaxed than those in the floating stereospecific assignments. Thus the high-dimensional potential should represent the experimental NOEs more realistically.

Once the stereospecific assignments were determined, we refined the CMTI-I structures using the differences between the calculated and observed NOESY spectra as the driving force to change the conformation. A formula for the gradient of NOE intensities calculated with the relaxation matrix approach has recently been derived by Yip and Case[23]. This method, interfaced with the refinement program X-PLOR[24], was used in our calculations on CMTI-I.

RELAXATION MATRIX REFINEMENT

The intensities of the NOE cross-relaxation were measured from NOESY spectra in H_2O. The spectra were recorded interleaved with mixing times of 50, 100, 150, 200, and 250 ms.

Empirical Part of Target Function.

The form of the empirical part of the target function is very similar to the target functions previously used to calculate the starting structures with the NOE derived distance constraints[30].

Experimental Part of Target Function.

We define the general form of the experimental part of the target function, $E_{relaxation}$, as:

$$E_{relaxation} = K_R \sum_{i=1}^{N} w_i \cdot dev(I_i^c, kI_i^o, \Delta_i, n)^m \qquad (1)$$

where K_R is the energy constant for the relaxation term, I_i^c and I_i^o are the calculated and observed intensities, respectively, Δ_i is an error estimate for I_i^o, w_i is a weight factor, k is the scale, and N is the number of cross peaks. $dev(a, b, \Delta, n)$ is defined as the absolute value of the difference between the nth powers of a and b where b has an error estimate Δ:

$$
dev(a, b, \Delta, n) \equiv
\begin{cases}
(b - \Delta)^n - a^n & \text{if } a^n \leq (b - \Delta)^n \\
0 & \text{if } (b - \Delta)^n < a^n < (b + \Delta)^n \\
a^n - (b + \Delta)^n & \text{if } a^n \geq (b + \Delta)^n.
\end{cases}
\tag{2}
$$

As part of the target function, we only use $n = 1$ and $m = 2$ in the present paper. The error estimates Δ_i are generally set very narrow (around 1% for large peaks). Their main purpose is to be able to include peaks which cannot be measured properly due to overlap. In this case, the error is set to a high value such that effectively only an upper bound is given for the peak size. As a measure of the fit of the refined structure to the NOE data we use a generalized R-factor

$$
R^n = \frac{\sum_{i=1}^{N} dev(I_i^c, kI_i^o, \Delta_i, n)}{\sum_{i=1}^{N} (kI_i^o)^n}.
\tag{3}
$$

For $n = 1$, and $\Delta_i = 0$ this is similar to the crystallographic R-factor. The value of the scale factor k is evaluated simply as

$$
k = \frac{\sum_{i=1}^{N'} I_i^c}{\sum_{i=1}^{N'} I_i^o}
\tag{4}
$$

where the sum runs over all N' peaks which are "well determined": For a peak to be included in the calculation of the scale, it has to be at least three times as big as the error estimate.

Increasing the Efficiency.

The major drawback of the use of the relaxation matrix in NMR refinement is the high computational cost. The diagonalization necessary to calculate intensities from a conformation has $\mathcal{O}(N^3)$ complexity, where N is the number of spins, and the evaluation of the gradient (equation 12 of Yip and Case[23]) has $\mathcal{O}(N^4)$ complexity for *every cross peak*. The complexity is reduced to $\mathcal{O}(N^3)$ by rewriting the equation for the gradient with respect to a coordinate μ due to a cross peak ij

$$
\Delta_\mu [\exp(-\mathbf{R}\tau)]^{(ij)} = \text{Trace}[(\Delta_\mu \mathbf{R}) \mathbf{L} \mathbf{F}^{(ij)} \mathbf{L}^T]
\tag{5}
$$

Table 3. Atomic r.m.s. differences (Å) between starting,
refined, and X-ray structures of CMTI-I

Structure	Heavy Backbone Atoms	All Heavy Atoms
Starting versus Starting	0.46± 0.08	1.18± 0.16
Refined versus Refined	0.40± 0.13	1.19± 0.13
Starting versus Refined	0.87± 0.07	1.54± 0.17
Starting versus X-ray	0.95± 0.05	1.77± 0.10
Refined versus X-ray	0.63± 0.08	1.54± 0.14
Pairwise Refined versus starting	0.85± 0.07	1.24± 0.05

where $\mathbf{R}$ is the relaxation matrix, $\mathbf{L}$ the matrix of eigenvectors, τ the mixing time, and $\mathbf{F}^{(ij)}$ is defined as

$$
\mathbf{F}_{ru}^{(ij)} \equiv
\begin{cases}
-\mathbf{L}_{ir}\mathbf{L}_{uj}^{T}\dfrac{\exp(-\lambda_r\tau)-\exp(-\lambda_u\tau)}{\tau(\lambda_r-\lambda_u)} & \text{if } r \neq u \\[2ex]
\mathbf{L}_{ir}\mathbf{L}_{rj}^{T}\exp(-\lambda_r\tau) & \text{else}
\end{cases}
\tag{6}
$$

where λ_r is the r'th eigenvalue of the relaxation matrix. This definition is similar to that of the matrix $\mathbf{M}^{(ij)}$ in Eq. 13 of Yip and Case[23]. As noted by Yip and Case, the contribution of the relaxation matrix to the gradient need not be calculated every step during the dynamics calculation. We found it is sufficient to recalculate the relaxation matrix whenever a hydrogen atom has moved by more than 0.075 Å. Increasing this number beyond 0.075 Å resulted in heating the system and large temperature fluctuations.

A large reduction in computation time was achieved by the introduction of a cutoff for the relaxation matrix calculation. For every cross peak, we include only atoms in the calculation whose distance is smaller than the cutoff from either hydrogen atom contributing to the cross peak.

Refined Structures.

The X-ray structure exhibits some large deviations with respect to the NMR starting structures calculated from the distance constraints[30]. The largest distance violations are for the side-chain and long range contacts. In contrast, the R-factors in the refined structures, which are a measure for the NOE violations, are quite low. The R-factors were ca. 0.51 for the starting structures and 0.26 for the refined structures. Thus the fit to the NMR data improves significantly. The improvement is also evident in the structures obtained from simple conjugate gradient minimization. This is accompanied with only minimal r.m.s. shifts. However, large shifts of approx. 1 Å occur during the dynamics/simulated annealing calculations (Table 3). Plots of

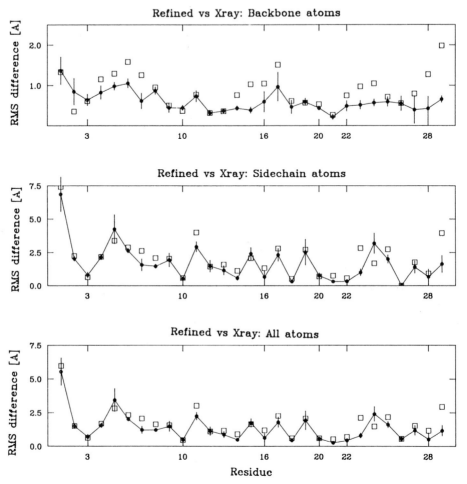

Figure 2. Atomic r.m.s. difference between the Refined (dots) and Starting (squares) structures on the one hand the X-ray structure of CMTI-I complex. The filled circles represent the average r.m.s. difference between the Refined structures and the X-ray structure at each residue and the squares the average r.m.s difference between the Starting structures and the X-ray structure. The bars represent the standard deviations in these values. The residue numbers are given for the 6 Cys residues.

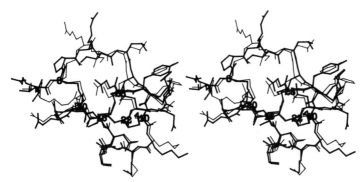

Figure 3. Best superposition (residues 2 to 29) of one refined n.m.r. structure (thick line) with the X-ray structure of CMTI-I (thin line). All atoms including the disulfide bridges, with the exception of hydrogen, are shown.

the atomic r.m.s. difference between the refined and X-ray structures are shown in Fig. 2, the r.m.s difference between the starting and X-ray structures are indicated as squares. Fig. 2 shows that compared to the starting structures, the backbone as well as the side-chains of the refined structures are closer to the X-ray structure. The best superposition of the refined structures with the X-ray structure shows an overall similarity of the two structures (Fig. 3). Furthermore the backbone conformation of the binding loop (residues 2 to 10) is identical in the two structures. Fig. 3 also shows that in many stereospecifically assigned side-chains, their positions in the refined structure are similar to those of the X-ray structure at least up to the C^γ atoms.

ACKNOWLEDGEMENTS

We thank David Case, James H. Prestegard, Thomas Simonson and Ping Yip for discussions. This work was supported by a research grant from the Bundesministerium für Forschung und Technologie (Grant no. 0318909A).

REFERENCES

1. K. Wüthrich, in "NMR of Proteins and Nucleic Acids," pp. 117-199, Wiley-Interscience, New York (1986).
2. R. Kaptein, E. R. P. Zuiderweg, R. M. Scheek, R. Boelens, and W. F. van Gunsteren, *J. Mol. Biol.* **182**, 179-182 (1985).
3. T. A. Holak, J. H. Prestegard, and J. D. Forman, *Biochemistry* **26**, 4652-4660 (1987).
4. T. A. Holak, S. K. Kearsley, Y. Kim, and J. H. Prestegard, *Biochemistry* **27**, 6135-6142 (1988).
5. R. Kaptein, R. Boelens, R. M. Scheek, and W. F. van Gunsteren, *Biochemistry* **27**, 5389-5395 (1988).

6. A. T. Brünger, G. M. Clore, A. M. Gronenborn, and M. Karplus, *Proc. Natl. Acad. Sci. USA* **83**, 3801-3805 (1986).

7. G. M. Clore and A. M. Gronenborn, *C.R.C. in Biochemistry and Mol. Biol.* **24**, 479-564 (1989).

8. M. Nilges, A. M. Gronenborn, A. T. Brünger, and G. M. Clore, *Protein Eng.* **2**, 27-38 (1988).

9. J. M. Moore, D. W. Case, W. J. Chazin, G. P. Gippert, T. F. Havel, R. Powls, and P. E. Wright, *Science* **240**, 314-317 (1988).

10. M. J. Tappin, A. Pastore, R. S. Norton, J. H. Freer, and I. D. Campbell, *Biochemistry* **27**, 1643-1647 (1988).

11. W. Braun and N. Gō, *J. Mol. Biol.* **186**, 611-626 (1985).

12. K. Wüthrich, *Science* **243**, 45-50 (1989).

13. W. Braun, *Quart. Rev. Biophys.* **19**, 1115-1157 (1987).

14. G. Wagner, W. Braun, T. F. Havel, T. Schaumann, N. Gō, K. Wüthrich, *J. Mol. Biol.* **196**, 611-639 (1987).

15. V. Saudek, R. J. P. Williams, and G. Ramponi, *FEBS Lett.* **242**, 225-232 (1989).

16. T. F. Havel, and K. Wüthrich, *J. Mol. Biol.* **182**, 281-294 (1985).

17. M. P. Williamson, T. F. Havel, and K. Wüthrich, *J. Mol. Biol.* **182**, 295-315 (1985).

18. T. A. Holak, M. Nilges, J. H. Prestegard, A. M. Gronenborn, and G. M. Clore, *Eur. J. Biochem.* **175**, 9-15 (1988b).

19. T. A. Holak, M. Nilges, and H. Oschkinat, *FEBS Letters* **242**, 218-224 (1989).

20. P. L. Weber, R. Morrison, and D. Hare, *J. Mol. Biol.* **204**, 483-487 (1988).

21. B. A. Borgias, M. Gochin, D. J. Kerwood, and T. L. James, *Progress in NMR Spectr.* **22**, 83-100 (1990).

22. R. Boelens, T. M. G. Koning, G. A. Van der Marel, J. H. van Boom, and R. Kaptein, R., *J. Magn. Reson.* **82**, 290-308 (1989).

23. P. Yip, and D. A. Case, *J. Magn. Reson.* **83**, 643-648 (1989).

24. A. T. Brünger, *J. Mol. Biol.* **203**, 803-816 (1988).

25. W. J. Metzler, D. R. Hare, and A. Pardi, *Biochemistry* **28**, 7045-7052 (1989).

26. T. F. Havel, I. D. Kuntz, and G. M. Crippen, *Bull. Mth. Biol.* **45**, 673-698 (1983).

27. V. C. Singh, and P. A. Kollman, *J. Comput. Chem.* **5**, 129-145 (1984).

28. T. A. Holak, J. N. Scarsdale, and J. H. Prestegard, *J. Magn. Reson.* **74**, 546-549 (1987).

29. K. Wüthrich, M. Billeter, and W. Braun, *J. Mol. Biol.* **169**, 949-961 (1983).

30. T. A. Holak, D. Gondol, J. Otlewski, and T. Wilusz, *J. Mol. Biol.* **210**, 635-648 (1989).

31. T. F. Havel, DISGEO, *Quantum Chemistry Exchange*, Program no. 507, Indiana University (1986).

32. J. Habazettl, C. Cieslar, H. Oschkinat, and T. A. Holak, *FEBS Letters* **268(1)**, 141-145 (1990).

33. S. G. Hyberts, W. Mäki, and G. Wagner, *Eur. J. Biochem.* **164**, 625-635 (1987).

A CRITIQUE OF THE INTERPRETATION OF NUCLEAR OVERHAUSER EFFECTS OF DUPLEX DNA

Jane M. Withka, S. Swaminathan, and Philip H. Bolton

Chemistry Department
Wesleyan University
Middletown, Connecticut 06457

ABSTRACT

The shape of a duplex DNA molecule can be modeled as a cylinder with symmetry axis z. The effective correlation time of an internuclear vector depends on the orientation of the vector to the symmetry axis as well as the internuclear distance and extent of internal motion. Calculations are presented which illustrate these principles. Since the effective correlation times can depend on the orientation of internuclear vectors so do NMR relaxation parameters such as the rate of buildup of nuclear Overhauser effects and the rates of spin diffusion. The approach is illustrated using a model nucleotide to show how the magnetization transfer can be affected by the orientation and spin diffusion rates.

INTRODUCTION

The determination of the conformations of molecules in solution has been one of the major areas of NMR over the past several years and is the major topic of this volume. Such studies on nucleic acids and proteins have relied, to a considerable extent, on the distance information present in proton-proton NOEs[1,2].

One of the key questions in this area of research is the accuracy of the distances determined from NOEs. A variety of approaches have been used to estimate the error range for the distances determined from NOEs. It has been widely assumed that the rate of buildup of NOEs can be directly related to the interproton distance through the familiar relationship

$$\sigma_{ij} = \left\{ 5.7 \times 10^4 / (r_{ij})^6 \right\} \left\{ \tau_c - 6\tau_c / \left(1 + 4\omega^2 \tau_c^2 \right) \right\} \tag{1}$$

with σ_{ij} the cross-relaxation rate between protons i and j in seconds, r_{ij} the distance between protons i and j in nM, ω the Larmor frequency and

Computational Aspects of the Study of Biological Macromolecules by Nuclear Magnetic Resonance Spectroscopy, Edited by J.C. Hoch *et al.,* Plenum Press, New York, 1991

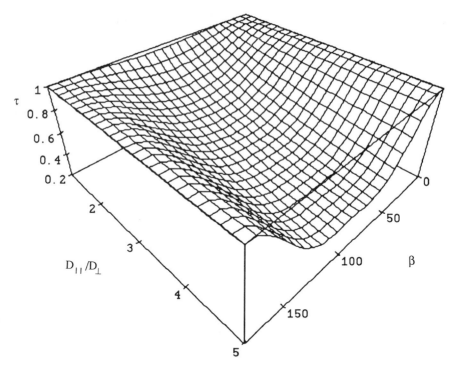

Figure 1. The three-dimensional plot shows the dependence of the effective correlation time on the angle, β, of a vector to the z axis of a cylinder with diffusion rates $D_\parallel$ about the symmetry axis and $D_\perp$ perpendicular to the symmetry axis in the absence of internal motion. All plots were generated using the program Mathematica[11].

τ_c the correlation time. As will be presented below orientation of internuclear vectors and the interplay between orientation and spin diffusion have significant effects on σ_{ij}.

A number of studies have shown that to accurately characterize the relaxation of nucleic acids, which is the case of primary interest here, that consideration of only pairs of spins is not sufficient and that all proton-proton interactions need be taken into account (12 and 13 are reviews). These approaches using complete relaxation matrix calculations take into account alternate relaxation pathways in addition to the pairwise interaction giving rise to a particular NOE. It is noted that even the complete relaxation matrix can be incomplete in certain cases[3]. That is, even with perfect data different spatial arrangements of nuclei can not be distinguished.

What all of these studies have had in common is the assumption that the relationship between NOE buildup rates and interproton distances is a scalar relationship. However, since double stranded DNA is an asymmetric molecule under certain circumstances the relationship between NOEs and distances is actually a vector relationship. That is, the NOE buildup rate

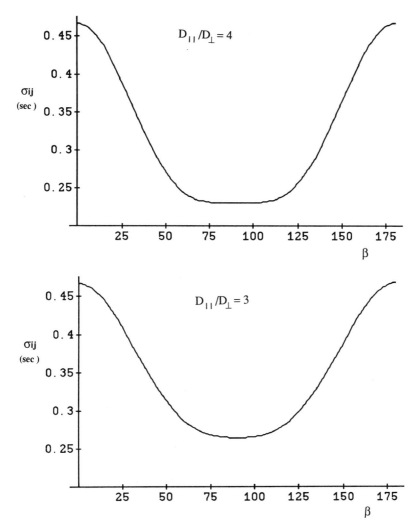

Figure 2. The curves show the dependence of σ_{ij} for a proton-proton vector on the angle β to the z axis of a cylinder for the indicated ratios of parallel/perpendicular diffusion rates. The σ_{ij} were calculated using the equation in the text for the case of a proton-proton distance of 0.3 nm and $\omega\tau_\perp = 19$ corresponding to a field strength of 500 MHz and $\tau_\perp = 6$ nsec. $\tau_\perp$ is equal to $1/6D_\perp$.

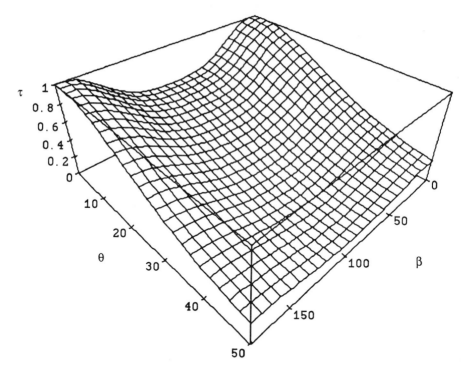

Figure 3. The three-dimensional plot shows the dependence of the effective correlation time on the angle, β, of a vector to the z axis of a cylinder, θ is the square root of the root mean square polar angle of motion for the case of $D_{\parallel}/D_{\perp} = 3.5$.

depends not only on the interproton distance but on the orientation of the proton-proton vector relative to the symmetry axis of the DNA. There are additional effects due to the motion of the internuclear vector relative to the symmetry axis.

A DNA double helix can be modeled to a reasonable level of accuracy as a cylinder with a diameter of 2.05 nM and a length of 0.34 nM per base pair. Thus, a dodecamer will have a length to diameter ratio of about 2. The diffusion rate of a vector along the long symmetry axis will be different than the diffusion rate of a factor along an axis perpendicular to the symmetry axis.

Consider the case of one proton-proton vector pointing along the symmetry, z, axis and one perpendicular to the z axis along the x axis. Since there will be more rapid motion about the z axis this will cause a proton-proton vector aligned along the x axis to have a shorter effective correlation time than a proton-proton vector aligned along the z axis.

There have been at least three variations of derivations of the effective correlation time of a vector at an angle β to the z axis when the only motion is diffusion of the cylinder and there is not motion of the vector relative to the cylinder. The treatments of Woessner[4], Szabo[5] and Spiess[6] are all quite similar. Woessner's model for the effective correlation time for a vector at an angle β to the z axis can be given by

$$
\begin{aligned}
\tau_e \; = \; & 3\left\{\cos[\beta]^2\left(1 - \cos[\beta]^2\right) / \left(D_{\|} + 5D_{\perp}\right)\right\} \\
+ \; & 3\left\{\left(1 - \cos[\beta]^2\right)^2 / \left(4\left(4D_{\|} + 2D_{\perp}\right)\right)\right\} \\
+ \; & \left\{\left(-1 + 3\cos[\beta]^2\right)^2 / \left(24D_{\perp}\right)\right\}
\end{aligned}
\tag{2}
$$

with $D_{\|}$ the diffusion constant about the z axis, $D_{\perp}$ the diffusion constant perpendicular to the z axis and β the angle between the internuclear vector and the z axis and τ_e the effective correlation time.

This expression has been used to calculate the dependence of the effective correlation time on orientation relative to the z axis which is shown in Fig. 1 for the range of $D_{\|}/D_{\perp}$ of 1 to 5 which corresponds roughly to double stranded DNA helixes of length 6 to about 20^{16}. The results indicate that the effective correlation time, and hence rate of NOE buildup, can vary by about 2.5 fold depending on the orientation of the proton-proton vector to the helical z axis and the ratio $D_{\|}/D_{\perp}$. A 2.5 fold variation in NOE buildup rate corresponds to about a 17% error in distance which means a 0.4 nM distance might be in error by more than 0.06 nM.

It is noted that the H5-H6 vector of a cytosine is more or less perpendicular to the z axis, 86° in canonical B-form, whereas a H2'-H2'' vector is at 55° to the z axis in canonical B-Form DNA. Therefore, in the absence of internal motion the NOE buildup rate of these two pairs of protons will not scale with internuclear distance. A recent report by Reid et al.[7] claims that there is no internal motion in duplex DNA since the NOE buildup rates for these two pairs of protons, in a dodecamer sample, scale to the sixth power of their respective interproton distances. This claim is not consistent with the cylindrical shape of duplex DNA. It is also noted that if the claim of Reid and co-workers is correct than the H2'-H2'' cross peaks of all residues should have the same build up rates. Similarly, all of the H5-H6 cross peaks should have the same build up rates. Careful examination of the published data and that of W. Massefski, Jr. and A.G. Redfield (personal communication) shows that this is not the case.

The actual value of $D_{\|}/D_{\perp}$ for a double stranded DNA such as a dodecamer is not known. Simple hydrodynamics would suggest that $D_{\|}/D_{\perp}$ goes as L^2/D^2. More sophisticated hydrodynamic theories would give corrections to this straightforward expression[8-10] but do not dramatically change the value of $D_{\|}/D_{\perp}$ from the simple model.

Fig. 2 contains plots of σ_{ij} versus β for the case of $\omega\tau_{\perp} = 19$ which corresponds to a dodecamer investigated at 500 MHz with $\tau_{\perp}$ being 6 nsec which is the value experimentally determined by dynamic depolarized light scattering[10] and $\tau_\perp = 1/6D_\perp$. It is seen that σ_{ij} versus β is about the same for both $D_\parallel/D_\perp = 3$ and $D_\parallel/D_\perp = 3$ and $D_\parallel/D_\perp = 4$.

Internal motion in an asymmetric molecule like double stranded DNA can have pronounced effects on NOE buildup rates as well as relative NOE buildup rates. Using an approach similar to that recently proposed by Eimer et al.[10] an expression for the effective correlation time of an internuclear vector can be obtained which is

$$
\begin{aligned}
\tau_e = & \left\{ 1/4 + 9/\left(8\exp[4\theta^2]\right) + (3\mathrm{Cos}[2\beta])/\left(2\exp[2\theta^2]\right) \right. \\
& + \left. (9\mathrm{Cos}[4\beta])/\left(8\exp[4\theta^2]\right) \right\}/(24D_\perp) \\
& + \left\{ 3\left(1 + 1/\left(2\exp[4\theta^2]\right)\right) - (2\mathrm{Cos}[2\beta])/\exp[2\theta^2] \right) \\
& + \left(\exp[-4\theta^2 - 4\zeta^2]\left(\mathrm{Cos}[4\beta]\right)/2\right) \right\}/\left(16(2D_\perp + 4D_\parallel)\right) \\
& + \left\{ 3\left(\exp[-4\theta^2] - \exp[-4\theta^2 - \zeta^2]\mathrm{Cos}[4\beta]\right)\right\}/\left(8\left(5D_\perp + D_\parallel\right)\right) \quad (3)
\end{aligned}
$$

with $D_\parallel$ the diffusion constant about the z axis, $D_\perp$ the diffusion constant perpendicular to the z axis, β the angle between the internuclear vector and the z axis, θ is the square root of the root mean square polar angle of motion and ζ the square root of the root mean square azimuthal angle of motion. When there is no internal motion the above equation reduces to Woessner's equation, given above, for the effective correlation time.

The above expression has been used to calculate the effective correlation time for a cylinder having $D_\parallel/D_\perp = 3.5$ as a function of β and θ with the results shown in Fig. 3. These results show that in the presence of internal motion the orientational dependence of the effective correlation time diminishes. When the extent of internal motion approaches 25° or so the orientational dependence of the correlation time is somewhat diminished and when it approaches 45° it is negligible. Thus, in the presence of sufficient internal motion proton-proton pairs with different β could have NOE buildup rates which scale as a function of the internuclear distance alone. It is also noted that the effective correlation times for all vectors regardless of β decrease as the internal motion increases.

At the present time there is not sufficient data to reliably estimate the extent of internal motion in DNA duplexes and in particular the extent of internal motion for different proton-proton vectors. The existing data, from a number of experiments, many of which are cited in (10), is consistent with internal motion of most proton-proton and carbon-proton vectors on the order of 10-20°. More reliable experimental information could be obtained from detailed study of ^{13}C and ^{15}N single- and multiple-quantum relaxation parameters of both the sugars and bases of DNA duplexes. Another source of

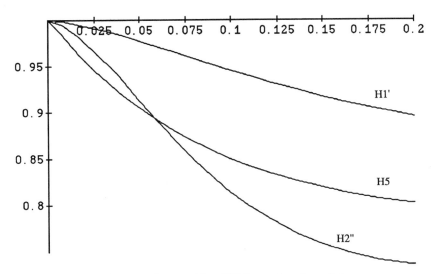

Figure 4. Intraresidue NOE transfers from H6.

information about the extent of internal motion could be molecular dynamics trajectories of DNA in solution.

AN EXAMPLE

To illustrate the magnitude and direction of the effects which can occur we have carried out calculations on a model system. To pick a system which is a fair case for illustration we have chosen the "B-72" form of DNA[14] also used in other demonstrations[15]. The test case is a single cytidine residue with internuclear distances and orientations of "B-72". It was assumed that $D_{\parallel}/D_{\perp}$ is 3.5 and that $\tau_{\perp}$ is 5 nsec. The time course of the NOE buildups were obtained by solving the simultaneous linear differential equations[12,13] using the Runge-Kutta method. This model case is intended to illustrate the sort of effects which can occur rather than be illustrative of any real situation in an actual DNA duplex. The inclusion of only the spins of one nucleotide, for example, does not allow representation of an adequate sink for the magnetization. Also, magnetization transfer due to strong to moderate coupling is neglected. Inclusion of such effects would tend to make the buildup curves for the 2' and 2" protons more similar since the 2' and 2" magnetization would become faster as would the 5' to 5" transfer.

Fig. 4 shows the rate of buildups of some cross peaks due to magnetization transfer from H6. It is noted that obtaining the time course via solution of the simultaneous linear equations apparently gives more curvature to the

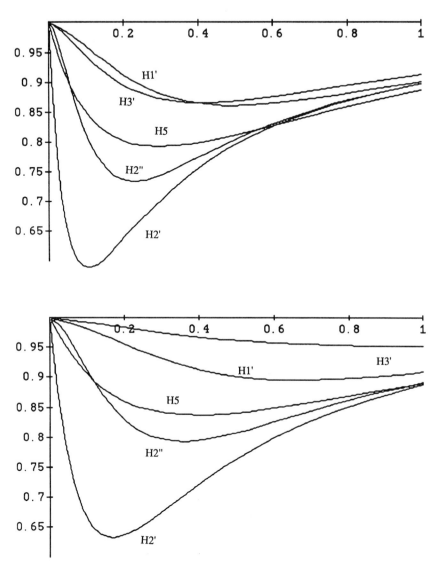

Figure 5. Intraresidue NOE transfers from H6: top — no orientation effects; bottom — with orientation effects.

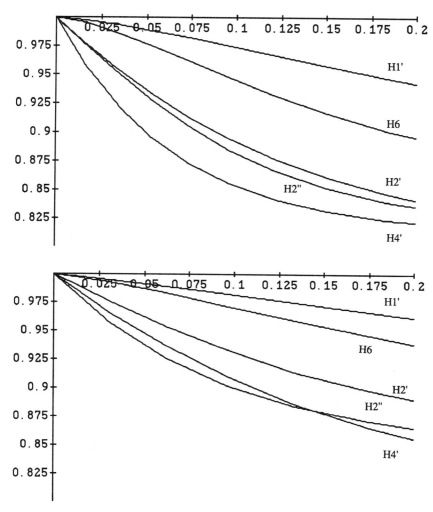

Figure 6. Intraresidue NOE transfers from H3′: top — no orientation effects; bottom — with orientation effects.

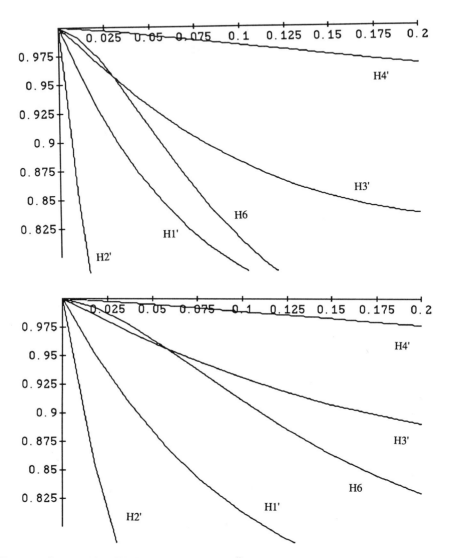

Figure 7. Intraresidue NOE transfers from H2″: top — no orientation effects; bottom — with orientation effects.

buildup curves than the use of single exponentials for the rate constants in a relaxation matrix approach.

A comparison of the NOE buildup rates of several protons due to magnetization transfer from H6 is shown in Fig. 5 calculated with and without orientation effects. It is seen that several of the protons have distinctly different buildup rates when the orientation effects are taken into consideration. Of note is the time course of the H3'. Essentially all of the H6 to H3' transfer is via diffusion by the H6 to H2' to H3' pathway. Since the H2'-H3' vector is at an angle of about 80° the diffusion pathway is slowed due to the orientation effect decreasing the effective correlation time for the H2'-H3' vector.

Fig. 6 shows the NOEs to several protons due to magnetization transfer from H3'. In this case the orientation effect is such that the buildup of the H2″ NOE is faster than that of the H4' in the presence but not in the absence of orientation effects. Changes in some of the other NOE buildup rates are also apparent.

Fig. 6 shows the analogous results due to magnetization transfer from the H2″ proton. The largest change is in the buildup of the H6 NOE which is due to a slowing of the H2″-H2'-H6 diffusion pathway. Both of these vectors are at about 55° to the symmetry axis.

The above calculations show that orientation effects can significantly alter both direct and diffusion NOE buildup rates. The inclusion of internal motion will tend to moderate the orientation effects as presented above. We are now extending this treatment to actual DNA heteroduplex dodecamers. This next generation treatment will use the angles and internal motion determined by molecular dynamics simulations and then calculate the NOE buildup rates for all protons by solving the approximately 200 simultaneous linear differential equations. The NOEs from the simulation data, with and without distance constraints, will be compared with the actual experimental data. We are also preparing carbon-13 labeled DNA to obtain experimental data on the rates of internal motion of DNA in solution.

Nevertheless, the orientation of internuclear vectors in DNA and other assymetric molecules can have a pronounced effect on NOE buildup rates such that the buildup rates depend on the internuclear vector and not only the internuclear distance. Neglect of the orientation effect will lead to systematic errors, such as underwinding of the DNA helix, that can be accounted for in a straightforward manner when the vector nature of the problem is considered.

ACKNOWLEDGEMENTS

This research was supported, in part, by grants from the State of Connecticut, Department of Higher Education, Bristol-Meyers-Squibb, and NIH 1T32 GM-08271.

REFERENCES

1. G. M. Clore and A. M. Gronenborn, *CRC Crit. Rev. Biochem. and Mol. Biol.* **24**, 479 (1989).
2. K. Wüthrich, "NMR of Proteins and Nucleic Acids", Wiley, New York (1986).
3. S. B. Landy and B. D. Nageswara Rao, *J. Magn. Reson.* **83**, 29 (1983).
4. D. E. Woessner, *J. Chem. Phys.* **37**, 647 (1962).
5. G. Lipardi and A. Szabo, *J. Am. Chem. Soc.* **104**, 4546 (1982).
6. H. W. Spiess, *Basic Princ. NMR* **15**, 54 (1978).
7. B. R. Reid, K. Banks, P. Flynn and W. Nerdal, *Biochemistry* **15**, 10001 (1989).
8. M. M. Tirado, M. C. Lopez Martinez, and J. Garcia de la Torre, *Biopolymers* **23**, 611 (1984).
9. T. Yoshizaki and H. Ymamkawa, *J. Chem.Phys.* **72**, 57 (1980).
10. W. Eimer, J. R. Williamson, S. G. Boxer, and R. Pecora, *Biochemistry* **29**, 799 (1990).
11. Mathematica is a trademark of Wolfram Research Inc., Champaign, Illinois.
12. B. A. Borgias, M. Gochin, D. J. Kerwood and T. L. James, *Progress in NMR Spectroscopy* **22**, 83 (1989).
13. F. J. M. Van De Ven and C. W. Hilbers, *Eur. J. Biochem.* **178**, 1 (1988).
14. S. Arnott and D. W. L. Hukins, *Biochem. Biophys. Res. Commun.* **47**, 1504 (1972).
15. G. M. Clore and A. M. Gronenborn, *J. Magn. Reson.* **84**, 398 (1989).
16. J. M. Withka, S. Swaminathan, and P. H. Bolton, *J. Magn. Reson.* **00**, 000 (1990).

IMPROVEMENT IN RESOLUTION WITH NONLINEAR METHODS APPLIED TO NMR SIGNALS FROM MACROMOLECULES

A. Polichetti[a], P. Barone[b], V. Viti[a], and L. Fiume[c]

[a]Laboratorio di Fisica
Istituto Superiore di Sanità
Roma, Italy
[b]Istituto Applicazioni del Calcolo
CNR
Roma, Italy
[c]Dipartimento di Patologia Sperimentale
Bologna, Italy

ABSTRACT

Spectral estimators alternative to the Fast Fourier Transform have elicited a great deal of interest for performing the spectral anaylsis of NMR signals in time domain. In our approach, we used Prony's method, in connection with the covariance algorithm, to calculate the autoregressive coefficients. The main problem in the use of this method is the determination of the model order p. We have examined the behavior of the residuals as a function of p and then we studied the stability of the result around relative minima. We have applied this method to the 31P NMR signals from the antiviral drug ara-AMP conjugated to human lactosaminated albulmin as hepatotropic carrier, obtaining very high resolution spectra.

INTRODUCTION

Spectral estimators alternative to the Fast Fourier Transform and particularly those methods that fit the free induction decay (FID) with a limited number of exponentially damped sinusoids have elicited a great deal of interest for performing the spectral analysis of NMR signals in time domain[1,2]. In the present work, we have applied Prony's method in connection with the covariance algorithm to obtain a very high resolution spectrum from ^{31}P NMR signals.

Computational Aspects of the Study of Biological Macromolecules by Nuclear Magnetic Resonance Spectroscopy, Edited by J.C. Hoch *et al.*, Plenum Press, New York, 1991

421

METHOD

In Prony's method, the observed signal is fitted by means of a sum of complex exponentials, the true signal $\hat{x}_n$ plus the observation noise w_n

$$
\begin{aligned}
x_n &= \sum_{k=1}^{\tilde{p}} A_k e^{i\theta_k} e^{(\alpha_k + i2\pi\nu_k)n\Delta t} + w_n \\
&= \hat{x}_n + w_n
\end{aligned}
\tag{1}
$$

with A_k, θ_k, α_k and ν_k, the k-th amplitude, phase, decay and frequency, respectively.

It is possible to show that the true signal satisfies the following difference equation

$$
\hat{x}_n = -\sum_{k=1}^{\tilde{p}} a_k \hat{x}_{n-k}
\tag{2}
$$

whose characteristic equation is

$$
\sum_{k=0}^{\tilde{p}} a_k z^{\tilde{p}-k} = 0 \qquad \text{with } a_0 = 1
\tag{3}
$$

with distinct roots $z_1, \ldots, z_{\tilde{p}}$; as a consequence, for the observed signal, we have

$$
x_n = -\sum_{k=1}^{\tilde{p}} a_k x_{n-k} + \sum_{k=1}^{\tilde{p}} a_k w_{n-k} + w_n
\tag{4}
$$

that represents an autoregressive moving average $(\text{ARMA}(\tilde{p},\tilde{p}))$ model with equal autoregressive (AR) and moving average (MA) coefficients.

The following equations are found for frequencies and decays

$$
\alpha_k = \frac{1}{\Delta t} log|z_k|
\tag{5}
$$

$$
\nu_k = \frac{1}{2\pi\Delta t} tg^{-1}\left(\frac{Im\ z_k}{Re\ z_k}\right).
\tag{6}
$$

The method consists in the following steps: i) approximation of the $\text{ARMA}(\tilde{p},\tilde{p})$ model with an AR(p) one, with $p>\tilde{p}$; ii) estimation of the AR coefficients; iii) calculation of the roots of the characteristic polynomial to determine frequencies and decays; iiii) calculation of amplitudes and phases through linear least square routines. For computing the AR coefficients the covariance algorithm has been used because in principle it seems very useful in application to nonstationary signals; it consists on the minimization of the prediction error energy

$$
\mathcal{E} = \sum_{n=p}^{N-1} |e_n|^2 = \sum_{n=p}^{N-1} \left| \sum_{k=0}^{p} a_k x_{n-k} \right|^2
\tag{7}
$$

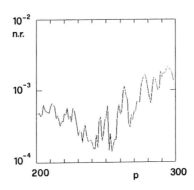

Figure 1. Norm of residuals as a function of p.

where

$$e_n = \sum_{k=0}^{p} a_k x_{n-k} \tag{8}$$

is the forward prediction error.

The least square minimum for $\mathcal{E}$ is found by setting to zero the derivatives of $\mathcal{E}$ with respect to the AR coefficients, obtaining the following linear equations

$$\sum_{k=0}^{p} a_k \left(\sum_{n=p}^{N1} x_{nk} x_{ni}^* \right) = 0$$

$$\text{for } 1 \leq i \leq p. \tag{9}$$

RESULTS AND DISCUSSION

The analysis of simulated signals, similar to those considered in the following, performed by using 4 different algorithms (singular value decomposition, Burg, Marple's LS, and covariance) have shown that the best results both in terms of computer time saving and stability of the solution are obtained when applying the covariance algorithm, thus confirming the goodness of this latter algorithm.

We have applied this method to the ^{31}P NMR signals from the antiviral drug arabinofuranosyladenine 5'-monophosphate (ara-AMP) conjugated to human lactosaminated albumin as hepatotropic carrier[3]. Ara-AMP is conjugated to the lysines and histidines of the albumin molecule through a carboxymethylation reaction pathway[4].

The main problem in the use of this method is the determination of the model order p. We have examined the behavior of the residuals as a function of p and of the number of observations and then we have studied the stability of the result around relative minima. Furthermore, we have studied how the size of the signal affects the resolution.

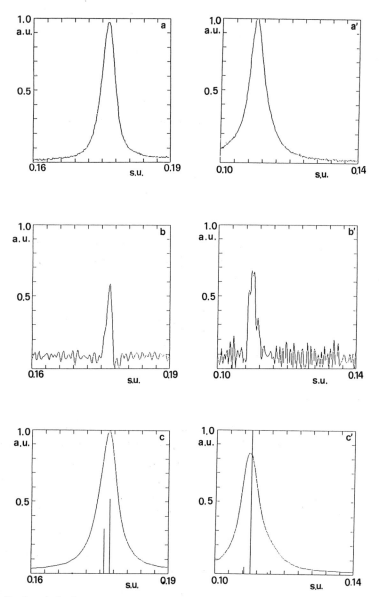

Figure 2. Lysine (a,b,c) and histidine (a',b',c') regions: a,a') FT; b,b') FT + gaussian multiplication; c,c') Prony-covariance; (a.u. = arbitrary units; s.u. = sampling units, 0.0001 s.u. = 1Hz).

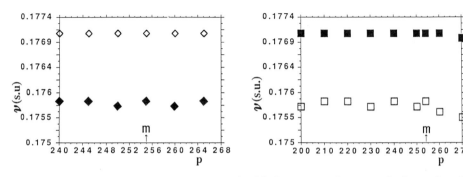

Figure 3. Frequencies ν of the lines obtained with Prony-covariance method as a function of the model order p for two different sizes (1024 and 800 data points); m indicates the minimum of the norm of the residuals.

Fig. 1 shows the norm of residuals as a function of p for the cited compound. It is possible to observe a minimum at $p = 254$; for this value the norm of residuals is $1.4 \cdot 10^{-4}$.

Fig. 2 shows the lysine and histidine regions as obtained by means of a) FT; b) FT + gaussian multiplication; and c) Prony-covariance method. In this latter figures, the position of the lines obtained are also reported; their heights are proportional to the signal energies. It is possible to observe that while with the first and second methods the region of the lysines is not resolved, with the Prony covariance method two lines are clearly observable.

This solution is stable, as it is possible to observe in Fig. 3, both with respect to the number of data points and with respect to the model order p.

CONCLUSIONS

Spectra with higher resolution can be obtained with the Prony covariance method relative to the ones obtained with other more traditional methods. Work is now in progress to obtain very high resolution spectra, for the histidine and lysine bound ara-AMP, in order to assess the differences between two kinds of conjugates obtained through two different reaction pathways.

REFERENCES

1. H. Barkhuijsen, R. de Beer, W. M. N, H, Bovée, D. van Ormondt, *J. Magn. Reson.* **61**, 465 (1985).
2. P. Barone, L. Guidoni, E. Massaro, V. Viti, *J. Magn. Reson.* **73**, 23 (1987).
3. L. Fiume, C. Busi, A. Mattioli, G. Spinosa, *CRC Critical Reviews in Therapeutic Drug Carrier Systems* **4**, 265 (1988).
4. L. Fiume, B. Bassi, A. Bongini, *Pharm. Acta Helv.* **63**, 137 (1988).

STELLA AND CLAIRE: A SERAGLIO OF PROGRAMS FOR HUMAN-AIDED ASSIGNMENT OF 2D 1H NMR SPECTRA OF PROTEINS

Gerard J. Kleywegt, Rolf Boelens, and Robert Kaptein

Department of Organic Chemistry
University of Utrecht
Padualaan 8
3584 CH Utrecht
The Netherlands

ABSTRACT

Algorithms for the automation of the assignment of two-dimensional proton NMR spectra of proteins are presented. STELLA is a program suite for automatic, semi-automatic or manual peak-picking. CLAIRE is a software package for human-aided assignment. The programs are demonstrated using data sets pertaining to phoratoxin B, a protein consisting of 46 amino-acid residues.

INTRODUCTION

Automation of the assignment of complex multidimensional NMR spectra, in particular of those pertaining to biomacromolecules, has developed into a holy grail pursued by scientists in many laboratories the world over. For the case of 2D 1H NMR spectra of proteins, most approaches tend to implement Wüthrich's sequential assignment strategy[1]. Automating this paradigm yields four distinct tasks:

1. data abstraction, e.g., spectrum matching or alignment, peak-picking, possibly combined with noise-reduction or symmetry-detection operations;

2. pattern detection, i.e., finding possible spin systems;

3. detection of inter-pattern correlations, i.e., finding inter-residue cross peaks in a 2D NOE spectrum;

Computational Aspects of the Study of Biological Macromolecules by Nuclear Magnetic Resonance Spectroscopy, Edited by J.C. Hoch *et al.*, Plenum Press, New York, 1991

427

4. generation of assignments, namely those that are consistent with the amino-acid sequence of the protein under study and with the information obtained in the previous two steps.

The first two of these steps have received most attention; as yet there has been only a handful of reports on program suites that tackle the problem in toto[2-7]. Here, we present two software packages which serve as "clever" tools in the process of assigning protein spectra. Although they comprise all four steps of this process, the user is still in control at all times.

STELLA — an acronym for "Synergistic approach Toward the evaluation of Local maxima in Low-symmetry spectrA" — is our peak-picking package[8] consisting of three programs (Fig. 1). SMART2 is the actual peak-picker, which may be used as is for entirely automatic peak-picking.

PPMAN2 is a program for manual peak-picking as well as peak-editing, which can be used for entirely manual peak-picking. These two approaches, however, each have a serious drawback: the former is error-prone and the latter is time-consuming and tedious. More interesting, therefore, is an intermediate approach in which user and software cooperate synergistically. Obviously, some hints from the human NMR expert would be of tremendous help in alleviating the software's verdancy and in extending it the wherewithal to discriminate between genuine cross peaks on the one hand and noise peaks and artifacts on the other. It is precisely this kind of input that can be provided through the third program, LEARN2. Basically, the user indicates some "typical examples" of real and spurious cross peaks and labels them accordingly. The intensity matrices of these example peaks are stored and used by SMART2 to decide whether a local maximum that it encounters is a real or a spurious peak (using a K-Nearest-neighbor approach). Fig. 2 gives an impression of the results that can be obtained if user and software cooperate in this fashion.

CLAIRE — an acronym for "CLuster of programs for the Assignment of Individual REsonances" — is our software package (see Table 1) for computer-assisted assignment of 2D ^{1}H NMR spectra of proteins[4,9]. The *modus operandi* when using Claire is roughly as follows:

1. first, one has to record a 2D HOHAHA and a 2D NOE spectrum (both in water) of the protein under investigation. This should be done in such a manner that both spectra match within approximately one channel for a 1024 * 1024 data set[9].
2. second, one should peak-pick both spectra, for example using Stella. Peak-picking is a critical operation and should be carried out carefully.
3. third, SPIN2D can be used to detect patterns (sets of channel numbers which are correlated through cross peaks in both spectra) and FILPAT to filter out patterns which are virtually similar to one or more of the other patterns.

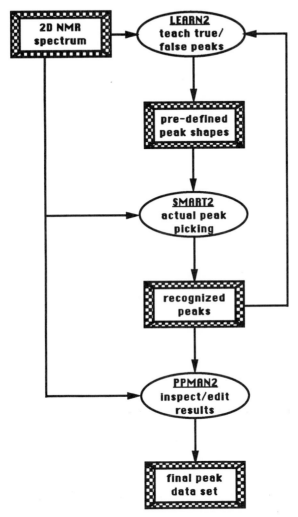

Figure 1. General outline of the peak-picking process using Stella's programs.

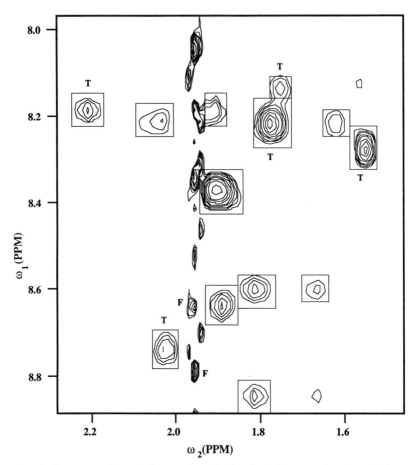

Figure 2. A demonstration of the results of a peak-picking session in which user and software work together. The peaks marked "T" were provided as examples of real peaks, those labelled "F" as examples of spurious peaks. The boxes ("drawn" automatically by the peak-picking program SMART2) indicate which peaks were recognized.

Table 1. Overview of programs in CLAIRE

Program	Purpose	Typical Runtime[a]
AUTOPI	interactive parameter input for all other programs	n/a
CONPAT	interactive pattern editing	n/a
SPIN2D	pattern detection	4:00
FILPAT	pattern filtering	0:28
XREF2D	detection of inter-pattern NOEs and pattern reshuffling	0:22
ASSI2D	generation of possible assignments	0:13
CHECKA	generation of assignments for individual protons, etc.	0:32
XPAT2D	extension of patterns, search for neighboring amides	0:14
AROM2D	search for aromatics, Asn and Gln spin systems	0:25

[a]Runtimes (where applicable) are in cpu minutes:seconds on a VAXstation 3100 running in batch mode.

4. fourth, the user has to perform another critical operation, namely sorting out the patterns. This may entail deleting channels if they are not deemed to belong to the pattern (spin system), merging patterns, extending patterns that might belong to long-side-chain residues (e.g., with program XPAT2D), etc. Additionally, the user should try to locate aromatic residues as well as any Asn- and Gln-type residues (program AROM2D may be of help here) and input as much knowledge as is safely possible (for instance, identification of glycines). This step would usually be executed through program CONPAT, our interactive pattern editor, which has a number of useful options.

At this stage, the first two steps have been completed and one may enter the actual assignment process, which will typically be an iterative procedure in which the user employs XREF2D, ASSI2D, CHECKA and CONPAT.

ASSIGNMENTS

In the actual assignment process, the user will typically use XREF2D (to compute the inter-pattern correlation matrix), ASSI2D (to generate possible assignments) and CONPAT (to inspect the suggestions of ASSI2D and to implement them if they look convincing):

1. XREF2D computes a matrix that contains the number of cross peaks in the 2D NOE peak set which involve the NH channel of one pattern and the channels of another pattern. Subsequently, the order of the patterns is altered, so as to juxtapose patterns which are obviously

highly correlated (through cross peaks). The underlying heuristic is that if there are "many" NOEs involving the NH channel of a pattern i and the channels of another pattern j, then it is likely that pattern i is the C-terminal neighbour of pattern j in the protein.

2. ASSI2D integrates all information available (patterns, inter-pattern correlations, amino-acid sequence, possible identities for patterns) to generate all possible assignments which are (a) consistent with this information and (b) satisfy a number of criteria imposed by the user. Assignments are not actually implemented; it is left to the user to assess them and to effectuate those that are judged to be correct.

3. CONPAT is a versatile, interactive pattern editor (which might also be of use in traditional "paper-and-pencil" assignment procedures). It provides for the essential visual feedback to the original data, i.e., the 2D NMR spectra, even spectra that are not used by the other CLAIRE programs (such as COSY and D_2O spectra). Apart from offering a cascade of pattern-manipulation options, CONPAT contains a number a very useful display options, most notably the pattern-contour option (Fig. 3-5) and the scroll-contour option (Fig. 6). The idea of the pattern-contour option is that the user selects a set of N channel numbers (for instance, a pattern, or a pattern plus one or two potential $NH(i+1)$ channels) and produces N^2 small contour plots of small environments near all the possible diagonal and cross-peak positions. The scroll-contour option can be employed to extend patterns, to find candidate $NH(i+1)$ channels, to locate the aliphatic brethren of an aromatic spin system, etc. With this option it is possible to only look at a selected set of channels (plus a few to the left and right) and to "walk" through the spectrum in small steps.

4. CHECKA is used for a variety of tasks, such as generating assignments for individual resonances (channel numbers) of assigned patterns, assigning HOHAHA and 2D NOE peaks and searching for proline spin systems.

APPLICATION TO PHORATOXIN B

We have applied Stella and Claire in the assignment of real spectra of phoratoxin B[9]. This protein consists of 46 amino acids and displays homology to crambin, a protein we used previously to test the basic algorithms[4]. Peaks were picked with Stella, yielding 667 peaks for the HOHAHA spectrum and 1730 for the 2D NOE spectrum (τ_m =200 ms). SPIN2D detected 508 patterns, but only 92 of them survived FILPAT. The patterns were sorted out with CONPAT, three out of four aromatic residues were detected as well as

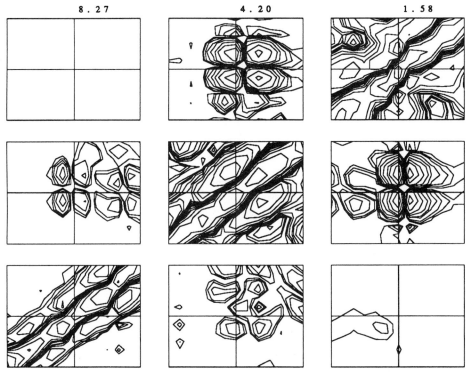

Figure 3. The pattern that corresponds to the spin system of residue Ala9 of phoratoxin B in the COSY spectrum displayed using the pattern-contour option of CONPAT. The individual boxes are thirteen channels wide in both directions and they are centered about the diagonal and cross-peak positions for this pattern. Hence, for example, the top-left box extends from channel 220 to 232 in the horizontal direction (ω_2) and from channel 767 to 779 in the vertical direction (ω_1). It thus corresponds to the NH-C$^\beta$H cross-peak position (since this is a COSY spectrum, obviously no cross peak is observed). Since the channel numbers of the pattern refer to the 2D NOE spectrum, the actual cross peaks in other spectra may be slightly displaced.

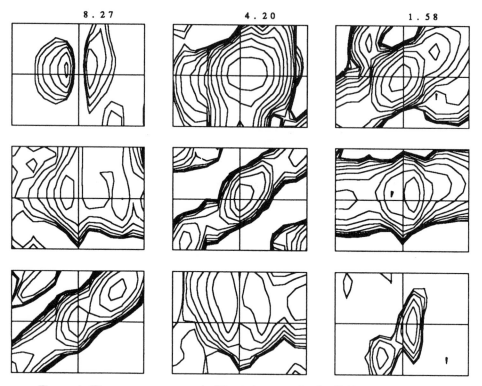

Figure 4. The same pattern as in Fig. 3, but now in the HOHAHA spectrum.

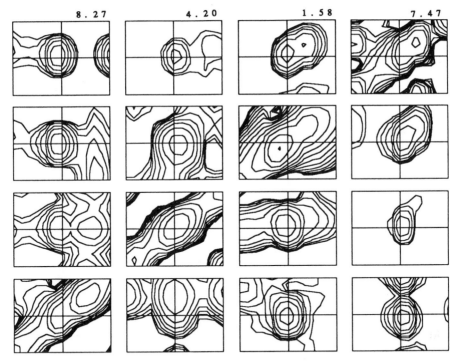

Figure 5. The pattern of Ala9 is shown again (as in Fig. 3), but now in the 2D NOE spectrum and with the addition of the NH channel of Arg10 (top row and right column). One clearly observes an intense NH(i)-NH($i+1$) peak, a slightly weaker C^αH(i)-NH($i+1$) peak and a strong C^βH(i)-NH($i+1$) peak.

both asparagine residues and all six glycines. For most other patters, the possible identities could be narrowed down to only a few (about four, on average).

The order in which the assignment proceeded (using combinations of XREF2D, ASSI2D, CONPAT, XPAT2D and CHECKA) is shown in Table 2. Residues 25 to 30 were assigned manually. This is the most difficult stretch, since it is here that the chemical-shift heterogeneities between phoratoxin A (also present in the sample) and B occur.

When comparing the final assignments with those from the literature[10], there turn out to be eleven residues for which we found fewer assignments (from one up to six) and five for which we found one extra assignment. It is worth noting that our results were basically obtained by a worker without much previous exposure to protein NMR, without the aid of any paper plots and without resorting to "old tricks" such as recording spectra at slightly different conditions. Therefore we expect that Claire will be a useful tool in the hands of more experienced NMR spectroscopists. CONPAT is a versatile program that can also be of help to workers who carry out "paper-and-pencil" assignment and to those who study biopolymers other than

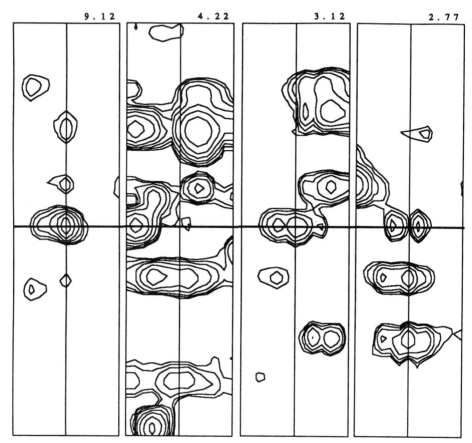

Figure 6. A demonstration of the scroll-contour option of program CONPAT. We have selected the NH, C^αH and both C^βH channels of the pattern corresponding to Asn14. For each of these channels (plus six channels to the left and right of them) we have plotted small strips in the amide region, in this case from channel 170 to 250. By a mere keypress it is possible to "scroll" up or down (in steps of, for instance, seventy channels), thus facilitating very close scrutiny of the amide regions of the selected channels. In this case, this results in detection of the NH channel of Thr15 (bold line) since we observe strong NH(i)-NH($i+1$) and C^βH(i)-NH($i+1$) contacts.

Table 2. Overview of the assignment of phoratoxin B spectra using CLAIRE

Cycle	Assigned residues			
1	Thr8→Phe18			
2	Gly18→Gly21	Ser42→Asp45	Gly31→Lys33	Gly37
3	Ile34→Ile35	Thr38→Lys39		
4	Cys40→Asp41			
5	His46	Ser2→Cys4[a]		
6	Ser22→Arg23[b]			
7	Ile25→Ser30[c]			

[a,b]Pro5, Pro24 found through combined use of CHECKA and CONPAT.
[c]Assigned manually.

proteins. At present, efforts are underway to extend the approach presented here to the analysis of 3D NMR spectra.

REFERENCES

1. K. Wüthrich, "NMR of Proteins and Nucleic Acids", Wiley, New York (1986).
2. C. Cieslar, G. M. Clore, and A. M. Gronenborn, *J. Magn. Reson.* **80**, 119 (1988).
3. C. D. Eads, and I. D. Kuntz, *J. Magn. Reson.* **82**, 467 (1989).
4. G. J. Kleywegt, R. M. J. N. Lamerichs, R. Boelens, and R. Kaptein, *J. Magn. Reson.* **85**, 186 (1989).
5. F. J. M. Van de Ven, *J. Magn. Reson.* **86**, 633 (1990).
6. F. J. M. Van de Ven, P. O. Lycksell, A. Van Kammen, and C. W. Hilbers, *Eur. J. Biochem.* **190**, 583 (1990).
7. P. Catasti, E. Carrara, and C. Nicolini, *J. Comput. Chem.* **11**, 805 (1990).
8. G. J. Kleywegt, R. Boelens, and R. Kaptein, *J. Magn. Reson.* **88**, 601 (1990).
9. G. J. Kleywegt, R. Boelens, M. Cox, M. Llinás, and R. Kaptein, "Computer-Assisted Assignment of 2D ^{1}H NMR Spectra of Proteins. Basic Algorithms and Application to Phoratoxin B", *J. Biomol. NMR* **1**(1), 23 (1991).
10. G. M. Clore, D. K. Sukumaran, M. Nilges, and A. M. Gronenborn, *Biochemistry* **26**, 1732 (1987).

MolSkop: TOWARDS NMR MOLECULAR SCOPE

S. Kumazawa, S. Endo, T. Yamazaki, K. Fujita, and
K. Nagayama
JEOL Ltd.
Musashino, Akishima
Tokyo 196, Japan

The market is now supplying balanced computer platforms for chemical research which integrate both computation power and graphic ability. This new type of computer well fits NMR applications which require (1) graphics functionality for interactive spectral analysis, (2) rapid calculation of molecular structures based on NMR parameters and (3) the real time quick view of the spectra in two or three dimensions or the obtained molecular structures. Computers have always been one of the major motive forces driving the development of modern NMR instruments by which the performance of data throughput is dramatically improved. It is time for us to have an integrated system which is composed of a NMR spectrometer and a versatile computer. The system is expected to take care of (protein) chemists in tasks from the data acquisition to the display of molecular structures through rapid numerical computation. If it were fully supported to run in an automatic manner, it would be deserved to be called a 'NMR molecular scope'. This seems to be a bold extension of the microscope concept in optics.

Midway to the final goal, we have made a developing product, MolSkop, which helps researchers to do tasks necessary for determining protein structures with the use of various kinds of computational tools. The major principles for the fundamental design are; (1) menu-driven interfaces friendly to users, (2) interactive and/or semiautomatic handling of NMR data, computation and graphical representation, (3) easy access to various solver software in the field of computer chemistry, (4) real time monitor of conformation calculations, (5) balance between labors demanding for spectral analysis, conformation calculation and evaluation of obtained results with graphics, and (6) implementation of methods established as the standard of basic NMR research. To fulfill the design principle, requirements for the computer system are as follows.

Computational Aspects of the Study of Biological Macromolecules by Nuclear Magnetic Resonance Spectroscopy, Edited by J.C. Hoch *et al.,* Plenum Press, New York, 1991

Hardware architecture

high speed computation	> 20 MIPS, > 2 MFLOPS (double precision)
high speed graphics	> 5×10^5 3D vectors/s, > 5×10^4 3D polygons/s
high speed storage	> 400 MB disk
large mass storage	> 1 GB (auxiliary device)
local area network	10 Mbit/s (Ethernet)

Software requirements

general software platform	UNIX, C, FORTRAN
graphic interface	X-Window, 3D-graphics library (Doré , PEX, etc.)
network software	TCP/IP, NFS, FTP

With the considerations mentioned above, the MolSkop architecture has been determined. It consists of three main sectors, NMR2/Distance, DADAS90 and VIEWER. A schematics of data flow connecting these three sectors is shown in Fig. 1 together with various kinds of input and output files. NMR2/Distance is a program extended from NMR2 of NMRi to include automatic assignment of individual and sequential proton resonances. We adopted the commercial NMR data system, NMR2, for MolSkop to be easily interfaced to most vendors of NMR spectrometers. The characteristics, for example, assignment procedures with interactive graphics based on a peak table as the central data-base for knowledge, are described in detail by F. Delaglio in an article of this issue. NMR2/Distance can well assist the analysis of protein NMR spectra and lead to the input to structure calculations managed by DADAS90. DADAS90 is a program subject to determination of protein conformations in the dihedral angle space (DAS) with NMR information such as NOE distance constraints and dihedral angle constraints. By reducing the independent variables by one eighth of that of Cartesian coordinates and with the use of analytical minimization with first and second derivatives, the DADAS90 guarantees very rapid efficient convergence of calculated structures. The full description of DADAS90 appears elsewhere in this volume (by S. Endo, H. Wako, K. Nagayama, and N. Gō). A super computer version of a DADAS program (DIANA) is also presented here (by W. Braun).

DADAS90 is now running on the TITAN series of Stardent and on VAX of DEC and will be ported to various powerful engineering workstations (EWS) soon. The performance of DADAS90 on a Titan 3000 with 2 CPUs is 2 hours for BPTI (Mw=6,500), 3 hours for Tendamistat (Mw=8,500) and 5 hours for RNase H (Mw=18,000).

VIEWER includes various kinds of menu-driven modules. Functions of VIEWER are shown in Fig. 2 together with its operation flow. First the I/O module reads a molecular structure described in the dihedral angle

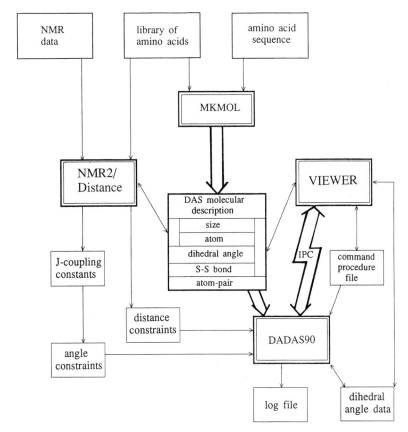

Figure 1. Schematics and data flow of the MolSkop.
IPC: inter process communication, DAS: dihedral angle space.

space (DAS molecular description) which is made from an amino acid sequence and a library of amino acids as shown in Fig. 1; second the NMR constraints data after the completion of the spectral analysis and assignment is transferred to the distance geometry program, DADAS90; third the distance geometry calculation starts while the structures generated in the intermediate stages of calculation are monitored in real time; fourth when the structure is converged, their best fit superposition are examined and evaluated in comparison with input data; and finally the DISPLAY module shows obtained structures with various kinds of molecular graphic models, and stereo views. If necessary for automatic operation and so on, a series of the procedure can be easily repeated by macro commands. The real time monitor to display conformations during DADAS90 calculation is executed with the interprocess communication (Fig. 1). This is an important feature of MolSkop which can interactively check the intermediate conformations to avoid the trap of the local minima. The places where violation of distance

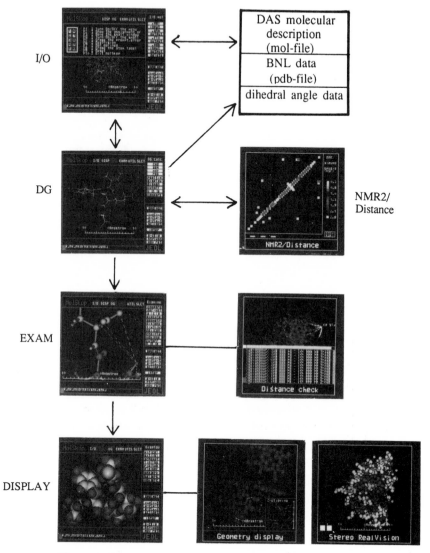

I/O

DG

EXAM

DISPLAY

DAS molecular
description
(mol-file)

BNL data
(pdb-file)

dihedral angle data

NMR2/
Distance

Figure 2. Images of main menus and displays in the MolSkop.

constraints occurs are easily viewed on the screen. Like conventional molecular graphics, obtained conformations or structures registered in the Protein Data Bank (PDB) of Brookhaven National Laboratory are displayed with various kinds of models such as wire, lines, ribbon, tube, cylinder and plate, ball and stick, dot surface and CPK models. For the convenience of preparing figures in reports, colored letters with arbitrary choice of fonts can be written on the graphics screen.

RIBONUCLEASE H: FULL ASSIGNMENT OF BACKBONE PROTON RESONANCES WITH HETERONUCLEAR 3D NMR AND SOLUTION STRUCTURE

Kuniaki Nagayama, Toshio Yamazaki, Mayumi Yoshida[†],
Shigenori Kanaya[‡], and Haruki Nakamura[‡]

Biometrology Lab
JEOL Ltd.
Musashino, Akishima
Tokyo 196, Japan
[†]Tokyo Research Laboratories
Kyowa Hakko, Machida
Tokyo 194, Japan
[‡]Protein Engineering Research Institute
Furuedai, Suita
Osaka 565, Japan

The atomic scale investigation of proteins by NMR primarily depends on the assignment of the individual resonance peaks. The assignment strategy now utilized for smaller proteins (<10 kDa) consists of two steps[1]: 1) The identification of the spin system characterizing each of the amino acids by COSY and 2) The sequential connection of the identified spin systems along the amino acid sequence by NOESY. To extend the strategy to larger proteins, the overlapping of chemical shifts of the densely scattered resonance should be eliminated. A new strategy recently developed in our group utilizes heteronuclear 3D NMR[2,3] and isotope enrichment to a large extent to solve the problem.

Several kinds of isotopically enriched samples were prepared to perform the complete sequential assignment: i) uniformly labeled with ^{15}N, ii) uniformly labeled with ^{13}C, iii) selectively labeled to a particular amino acid with ^{15}N, and iv) positive labeling of a particular amino acid with ^{1}H in a perdeuterated protein. With these samples, resonance peaks of the backbone nuclei were first assigned. The step of spin identification with COSY was replaced by the [1]H-^{15}N HMQC[4] experiment with the ^{15}N-selectively

Computational Aspects of the Study of Biological Macromolecules by Nuclear Magnetic Resonance Spectroscopy, Edited by J.C. Hoch *et al.*, Plenum Press, New York, 1991

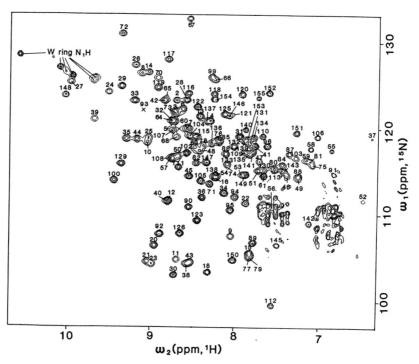

Figure 1. Two dimensional ^{1}H-^{15}N HMQC spectrum of a sample solution (2 mM, pH 5.5) of ribonuclease H enriched in amide position with ^{15}N. The figures assigned to each of the cross peaks arising from amides are the residue numbers.

labeled protein (iii). The sequential assignment of the backbone nuclei with NOESY was replaced by a combination of four types of 3D NMR; ^{1}H-^{15}N NOESY-HMQC, ^{1}H-^{15}N TOCSY-HMQC, ^{1}H-^{13}C HMQC-NOESY, and ^{1}H-^{13}C HMQC-TOCSY. The uniformly labeled proteins (i and ii) provided ideal samples to enhance spectral sensitivity in these experiments. The combination of four kinds of 3D NMRs could provide general information to correlate the connectivities arising from NH$_i$-C$^\alpha$H$_{i-1}$, NH$_i$-C$^\alpha$H$_i$, and NH$_{i+1}$-C$^\alpha$H$_i$ in the finger-print region, leading to completion of sequential assignment of the backbone with an aid of the knowledge of step 1. The assignment of side-chain protons can be performed with the simplified COSY spectra obtained for the ^{1}H-^{15}N TOCSY-HMQC and ^{1}H-^{13}C HMQC-TOCSY. The standard pulse sequences for 3D NMR were utilized[5] with slight modifications in the suppression of the water peak and its long tail and recovery of the α-CH protons otherwise hidden under water.

The new method was applied to a protein having a molecular weight of 17.6 kDa (155 amino acid residues), ribonuclease H from *Escherichia coli*. Except for a cysteine residue, Cys63, which did not show the ^{15}N^{1}H con-

Table 1. Chemical Shifts of Backbone Protons (NH, CH)
of Ribonuclease H (*E. coli*)

Res	#	NH	C$^\alpha$H	Res	#	NH	C$^\alpha$H
1	M	–	4.17	41	R	7.77	3.88
2	L	8.71	4.55	42	T	8.85	4.08
3	K	8.73	4.31	43	T	8.53	5.27
4	Q	8.31	5.12	44	N	9.19	3.92
5	V	8.79	4.83	45	N	8.56	4.03
6	E	8.64	4.80	46	R	7.38	3.58
7	I	8.60	5.33	47	M	7.71	4.44
8	F	9.13	5.67	48	E	8.44	3.86
9	T	8.05	5.59	49	L	7.24	3.96
10	D	9.06	5.15	50	M	8.69	3.93
11	G	8.70	3.39/5.04	51	A	7.72	3.31
12	S	8.80	4.78	52	A	6.47	3.99
13	C	8.44	4.95	53	I	7.94	3.13
14	L	9.05	4.47	54	V	8.25	3.30
15	G	7.82	3.83/4.11	55	A	6.85	3.34
16	N	8.24	5.05	56	L	7.57	3.76
17	P	–	4.92	57	E	8.76	3.94
18	G	8.32	3.90/4.30	58	A	7.10	4.11
19	P	–	4.92	59	L	7.15	4.26
20	G	8.96	3.76/5.09	60	K	8.59	4.38
21	G	9.04	4.22/5.36	61	E	7.66	4.48
22	Y	7.87	5.71	62	H	8.46	4.78
23	G	8.99	3.90/5.17	63	C	–	5.12
24	A	9.52	5.87	64	E	8.75	5.02
25	I	9.04	4.87	65	V	8.82	4.70
26	L	9.21	5.28	66	I	8.25	4.75
27	R	9.98	5.47	67	L	8.57	4.70
28	Y	8.59	5.06	68	S	8.72	5.73
29	R	9.37	3.77	69	T	8.50	5.11
30	G	8.72	3.61/4.14	70	D	8.94	5.45
31	R	7.95	4.68	71	S	8.31	4.52
32	E	8.76	5.42	72	Q	9.37	3.98
33	K	9.21	4.66	73	Y	8.72	4.48
34	T	8.14	5.31	74	V	8.06	3.60
35	F	9.31	4.99	75	R	7.08	2.52
36	S	8.40	4.58	76	Q	8.22	3.15
37	A	6.37	4.23	77	G	7.82	3.14/3.82
38	G	8.58	3.49/5.36	78	I	8.40	3.74
39	Y	9.69	5.48	79	T	7.80	4.11
40	T	8.80	3.72	80	Q	7.53	4.47

Table 1. Chemical Shifts of Backbone Protons (NH, CH)
of Ribonuclease H (*E. coli*) (continued)

Res	#	NH	CᵅH	Res	#	NH	CᵅH
81	W	7.10	4.28	121	V	8.03	4.20
82	I	8.48	3.49	122	K	8.46	4.54
83	H	7.93	4.13	123	G	8.46	3.91/4.07
84	N	7.44	4.38	124	H	8.57	4.64
85	W	8.20	4.38	125	A	8.14	4.45
86	K	8.12	2.70	126	G	8.66	3.82
87	K	7.36	4.09	127	H	8.40	5.23
88	R	7.25	4.48	128	P	–	4.39
89	G	7.78	3.93	129	E	9.37	3.63
90	W	8.56	3.42	130	N	7.66	4.45
91	K	6.82	5.05	131	E	7.83	3.92
92	T	8.90	4.38	132	R	8.02	4.09
93	A	9.10	4.07	133	C	8.10	4.01
94	D	8.00	4.64	134	D	7.78	4.44
95	K	8.06	3.67	135	E	7.83	3.92
96	K	7.62	4.72	136	L	8.30	4.02
97	P	–	4.61	137	A	8.35	3.62
98	V	7.57	3.44	138	R	8.25	3.80
99	K	8.29	4.05	139	A	7.89	4.04
100	N	9.45	4.28	140	A	7.88	3.98
101	V	7.65	3.60	141	A	7.79	3.75
102	D	8.59	4.09	142	M	7.11	4.29
103	L	7.28	4.16	143	N	7.42	5.04
104	W	8.55	4.58	144	P	–	4.10
105	Q	8.42	3.66	145	T	7.49	4.56
106	R	7.02	4.18	146	L	8.08	4.75
107	L	8.76	4.12	147	E	8.35	4.82
108	D	8.76	4.56	148	D	10.04	4.88
109	A	7.88	4.20	149	T	7.88	4.14
110	A	7.67	4.30	150	G	8.02	3.73/4.21
111	L	8.68	4.10	151	Y	7.27	4.24
112	G	7.54	3.95/4.13	152	Q	7.60	4.11
113	Q	7.52	4.27	153	V	7.86	3.84
114	H	7.37	5.04	154	E	8.26	4.36
115	Q	8.48	4.74	155	V	7.73	4.04
116	I	8.57	4.54				
117	K	8.81	4.48				
118	W	8.26	4.50				
119	E	8.94	4.32				
120	W	7.93	5.62				

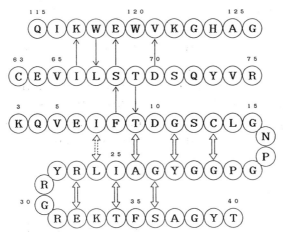

Figure 2. Schematic view of a β-sheet core structure found in the global fold of RNase H.

nectivity in the HMQC spectrum of selectively labeled RNase H, the full assignment of the backbone nuclei, ^{1}H, ^{13}C, and ^{15}N was completed. The assignment of residue members for amide (^{15}N^{1}H) connectivities in HMQC spectrum is shown in Fig. 1. In the HMQC spectrum cross peaks arising from histidines are missing because an auxotrophic strain[6] was utilized in our study. The chemical shifts thus obtained for backbone protons are summarized in Table 1.

In our strategy the laborious task for assignment of backbone nuclei has become a simple one because the amides, which usually show sensitive and sharp resonances in HMQC spectra, play roles of keystones in this amide-directed assignment method. In contrast with the backbone sequential assignment, the side-chain assignment is complicated and difficult because ^{1}H and ^{13}C nuclei on the sidechain manifest broader and weaker resonances due to the strong dipole coupling. Our auxotrophic strain required His, Ile, and Val. Therefore the three amino acids were easily protonated during the incubation in the perdueterated medium leading to specific observation of strong proton resonances and hence to assignment in backbone and side chain protons with conventional COSY. The aliphatic protons of the other amino acid residues were assigned with 3D NMR. About 80% of the β-CH protons were already assigned. The protons on the side chain terminals such as methyls, amides and aromatics were also early recognized and assigned owing to their high sensitivity.

Based on the strengths and patterns of cross peaks between backbone atoms in 3D NOESY and TOCSY spectra, the secondary structure of the protein was determined. Five β-strands and five α-helices were recognized. The location of the secondary structure completely agreed with the results recently obtained by X-ray crystallography[7]. NOE cross peaks between backbone protons revealed that the β-strands formed a β-sheet, where the first three made an anti-parallel β-sheet and the other two strands ran parallel to the first β-strand as shown in Fig. 2. The global fold of the protein was obtained from the calculation with DADAS90 using distance constraints among methyl and aromatic protons deduced from the ^{1}H-^{13}C 3D NOESY experiments. The conformation calculation has been carried out with a computer platform, MolSkop, of which interesting features are briefly described in another article.

REFERENCES

1. K. Wüthrich, "NMR of Proteins and Nucleic Acids", John-Wiley, New York (1986).
2. K. Nagayama, T. Yamazaki, M. Yoshida, S. Kanaya, and H. Nakamura, *J. Biochem.* **108**, 149-152 (1990).
3. T. Yamazaki, M. Yoshida, S. Kanaya, H. Nakamura, and K. Nagayama, *Biochemistry*, **30**, 6036-6047 (1991).
4. L. Müller, *J. Am. Chem. Soc.* **101**, 4481-4482 (1977).
5. D. Marion, et al., *J. Am. Chem. Soc.* **111**, 1515-1517 (1989).
6. S. Kanaya, A. Kohara, M. Miyagawa, T. Matsuzaki, K. Morikawa, and M. Ikehara, *J. Biol. Chem.* **264**, 11546-11549 (1989).
7. K. Katayanagi, et al., *Nature*, **347**, 306-309 (1990).

SAMPLING PROPERTIES OF SIMULATED ANNEALING AND DISTANCE GEOMETRY

Michael Nilges, John Kuszewski, and Axel T. Brünger

The Howard Hughes Medical Institute and
Department of Molecular Biophysics and Biochemistry
Yale University
New Haven, CT 06511

ABSTRACT

Properties of two different methods for calculating three-dimensional structures of macromolecules from nuclear magnetic resonance data, distance geometry, and simulated annealing, are studied. It is shown that a simulated annealing refinement of structures generated with distance geometry is sufficient to remove bias introduced in the distance geometry stage. A new efficient simulated annealing protocol which does not require initial structures from distance geometry is presented. The influence of distance selection on quality and distribution of structures generated with distance geometry is studied.

INTRODUCTION

Several recent studies indicate that sampling properties of metric matrix distance geometry (DG) algorithms are insufficient in a more fundamental way than previously thought[1-3]. In particular, it has been suggested that the distance geometry approach may not be able to find folds of the polypeptide chain which are compatible with the NOE derived distance data[3], and that a simulated annealing (SA) stage following the distance geometry stage may not suffice to insure appropriate sampling of the conformational space[2].

We investigate in the present study if the SA stage as implemented in the hybrid DG-SA method[4] is sufficient to remove bias introduced in the distance geometry stage. Further, a new SA protocol is presented which does not require starting structures obtained with a DG program and thus avoids any DG-related sampling problems, and is simpler and more efficient than earlier protocols[5,6]. We also study the effects of metrization on the distribution and quality of structures produced by DG.

Computational Aspects of the Study of Biological Macromolecules by Nuclear Magnetic Resonance Spectroscopy, Edited by J.C. Hoch *et al.,* Plenum Press, New York, 1991

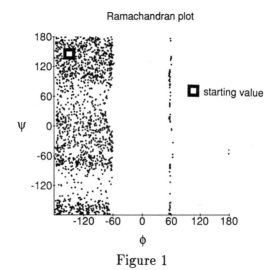

Ramachandran plot

Figure 1

SAMPLING OF SA REFINEMENT

A particularly biased starting structure is used to study the amount of randomization by the SA stage of the DG-SA hybrid method, namely, a polyalanine 20mer in an extended conformation ($\phi = -139°$, $\psi = 135°$). 100 structures were calculated in the absence of any tertiary distance information. The only difference between calculations was the random number seed used to assign initial velocities. While it is intuitively clear that molecular dynamics at high temperature would randomize the conformation it has not been shown in detail for the particular conditions used in the SA refinement in (4) (very low van der Waals interaction initially, short dynamics run at 1000° K).

As a reference set we generated 100 structures by randomly assigning values for ϕ and ψ, where sterically forbidden regions in the $\phi - \psi$ map were excluded. Only structures which had little van der Waals (vdW) overlap were accepted. The mean end-to-end distance of the structures generated in this way (≈ 28 Å) agrees with values obtained by others on the same system (J. Thomason, personal communication).

Fig. 1 shows a Ramachandran plot for all 100 calculated SA structures. The square in the top left corner indicates the starting conformation. As can easily be seen, all of the allowed region is visited. In Fig. 2, we compare how randomly the structures are distributed in space globally. For each C^α atom in the sequence, the number of C^α atoms within a distance of 7.6 and 11.4 Å (2 and 3 C_i^α - C_{i+1}^α distances) is counted. This number is averaged over 100 structures. The two curves are almost identical, the random structures are only slightly more densely packed than the SA structures. This also indicates

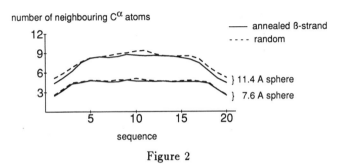

Figure 2

Table 1. average properties of [Ala]$_{20}$ conformations

calculation	r.m.s. bonds[*] [Å]	r.m.s. angles[*] [°]	D$_{1-20}$[†] [Å]	r.m.s.d.[#] [Å]
DG, no metrization	12.8	68.1	48.2	11.7
DG, ordered metr.	0.7	48.2	7.0	3.4
DG, random metr.	0.6	47.8	5.5	3.0
SA	0.0	0.0	28.5	6.7
random ϕ, ψ	0.0	0.0	27.5	6.4

[*]root mean square deviation from ideal value
[†]average distance from first to last C$^\alpha$
[#]average pairwise root mean square difference

that the missing attractive component in the vdW interaction does *not* lead to overly extended structures.

Further properties of the calculated structures are compiled in Table 1, together with those of the DG structures. All investigated properties of structures generated with SA, starting from a single starting conformation (an extended β-strand), are nearly indistinguishable from those of random structures. The short dynamics run at high temperature together with the employed force field and the variation of the vdW force constant thus ensures very good sampling and seems easily sufficient to remove bias in DG structures. Our findings are in contrast to (2), where a different SA algorithm[7] was used. One reason for the sampling properties of our SA protocol[4] is probably the way the vdW weight is varied during the refinement. Due to its very low initial value, the structures can rotate almost freely around rotatable bonds.

A NEW SA PROTOCOL FOR STRUCTURE DETERMINATION

One requirement for employing a calculational strategy like the DG-SA hybrid method is that the general fold of the molecule is correctly determined by the DG phase. Re-folding on a very large scale cannot be expected during the short dynamics run, and alternate folds as described in (3) will indeed be missed if they are not picked up in the DG phase. For this reason, a new SA protocol was developed with the aim to ensure good global sampling of the conformational space. The protocol is outlined briefly below.

1. generate starting conformation: ϕ and ψ random, $\chi_i = 180°$

2. 15 ps dynamics at 1000° K
   NOE pseudo potential with "soft" asymptotic behaviour[5]
   switch between harmonic and asymptotic region at constant value of 0.5 Å
   slope of the asymptote set to 0.1
   weight on the quartic vdW term[4] set to a very low value (0.001 — 0.01 kcal mol^{-1} Å^{-4}) to allow atoms to pass through each other

3. 10 ps of dynamics at 1000° K
   tilt asymptote of the NOE potential from 0.1 to 1
   increase weight on the vdW interaction to 0.1 kcal mol^{-1} Å^{-4}.

4. cooling and minimization as described before[4].

The protocol relies mainly on the fact that the overall fold of a protein is rather well determined by the NOE distance restraints even if the vdW interaction is very low, and does not attempt to fold the polypeptide chain from an extended strand. The protocol is distributed with the refinement program X-PLOR[8]. One calculation for NP-5 (478 atoms) with data from (9) takes 22 minutes on a Convex C2 computer.

INFLUENCE OF METRIZATION ON DG STRUCTURE

The effects of different ways to generate the distance matrix on the distribution and quality of structures produced by DG were studied on a polyalanine 20mer without any tertiary distance restraints. Three sets of calculations were performed. 10 structures were calculated without metrization[10], which takes the effect of a randomly chosen distance on the remaining distance bounds into account. The results of metrization depends on the order of choosing distances. In the second set of calculations, the distances were simply chosen in sequence, in the third set, in random order.

The results, which are essentially in agreement with those of others[2,11], are compiled in Table 1. Structures calculated without metrization show large deviations from input bounds, as evidenced by the large r.m.s. deviation from bond lengths. Metrization improves the quality of the structures in this respect drastically. The pairwise r.m.s. differences for the DG structures with metrization is too small. This is a consequence of the too tight packing of the DG structures,and indicates that scaling of the coordinates after embedding may be necessary.

Inspection of the structures on a graphics screen revealed that for structures calculated with ordered metrization the N-terminus is packed considerably more tightly than the C-terminus, a consequence of the ordered way in which distances are chosen. This artefact is removed by randomizing the order of choosing distances.

REFERENCES

1. A. T. Brünger, G. M. Clore, A. Gronenborn, and M. Karplus, (1987), *Protein Eng.* **1**, 399-406 (1987).
2. W. J. Metzler, D. R. Hare, and A. Pardi, *Biochemistry* **28**, 7045-052 (1989).
3. R. M. Levi, D. A. Bassolino, D. B. Kitchen, and A. Pardi, *Biochemistry* **28**, 9361-9372 (1989).
4. M. Nilges, G. M. Clore, and A. M. Gronenborn, *FEBS Lett.* **229**, 317-324 (1988).
5. M. Nilges., A. M. Gronenborn, A. T. Brünger, and G. M. Clore, *Protein Eng.* **2**, 27-38 (1988).
6. M. Nilges, G. M. Clore, and A. M. Gronenborn, *FEBS Lett.* **239**, 129-136 (1988).
7. W. Nerdal, D. R. Hare, and B. R. Reed, *J. Mol. Biol.* **201**, 717-739 (1988).
8. A. T. Brünger X-Plor version 2.1, User Manual, Yale University (1990).
9. A. Pardi, D. R. Hare, M. E. Selsted, R. D. Morrison, D. A. Bassolino, and A. C. Bach, *J. Mol. Biol.* **201**, 625-636 (1988).
10. T. F. Havel and K. Wüthrich, *Bull. Math. Biol.* **46**, 673-698 (1984).
11. T. F. Havel, *Biopolymers* **29**, 1565-1585 (1990).

LIST OF WORKSHOP PARTICIPANTS

Martin Billeter
Institut für Molekularbiologie
   und Biophysik
ETH Hönggerberg
8093 Zürich
SWITZERLAND

Rolf Boelens
Bijvoet Center for Biomolecular
   Research
NMR Spectroscopy
Padualaan 8
3584 CH Utrecht
THE NETHERLANDS

Philip Bolton
Chemistry Department
Wesleyan University
Middletown, CT 06457
USA

Benoit Boulait
Section de Chimie
Université de Lausanne
Rue de la Barre 2
CH-1005 Lausanne
SWITZERLAND

Werner Braun
Institut für Molekularbiologie
   und Biophysik
ETH Honggerberg
CH-8093 Zürich
SWITZERLAND

David Case
Department of Molecular Biology
Research Institute of Scripps Clinic
MB1 Rm. 102
10666 N. Torrey Pines Road
La Jolla, CA 92037
USA

G. Marius Clore
National Institutes of Health
Laboratory of Chemical Physics
Building 2, Room 123, NIDDK
Bethesda, Maryland 20892
USA

David Cowburn
The Rockefeller University
1230 York Avenue
New York, New York 10021
USA

Frank Delaglio
National Institutes of Health
Laboratory of Chemical Physics
Building 2, Room 123, NIDDK
Bethesda, Maryland 20892
USA

Marc A. Delsuc
Laboratoire de R.M.N.
ICSN-CNRS
Gif-Sur-Yvette 91190
FRANCE

Lydon Emsley
Université de Lausanne
Rue de la Barre 2
CH-1005 Lausanne
SWITZERLAND

S. Endo
JEOL Ltd.
Biometrology Laboratory
Akishima, Tokyo 196
JAPAN

Richard R. Ernst
Laboratorium für Physikalische
    Chemie
ETH Zentrum
CH-8092 Zürich
SWITZERLAND

Mark J. Forster
National Institute for Biological
    Standards and Control
Blanche Lane, South Mimms
Hertsfordshire, EN6 3QG
UNITED KINGDOM

Henrik Gesmar
University of Copenhagen
The H.C. Ørsted Institute
5 Universitetsparken
DK-2100, Copenhagen Ø
DENMARK

Nobuhiro Gō
Department of Chemistry
Faculty of Science
Kyoto University
Kitashirakawa, Sakyo-Ku
Kyoto 606
JAPAN

Hans Grahn
Department of Organic Chemistry
Umeå University
S-90187 Umeå
SWEDEN

Christian Griesinger
Institute for Organic Chemistry
Johann Wolfgang Goethe Univesity
Niederurseler Hang
6000 Frankfurt-am-Main 50
FEDERAL REPUBLIC OF
    GERMANY

Angela Gronenborn
National Institutes of Health
Laboratory of Chemical Physics
Bldg. 2, Rm. 123, NIDDK
Bethesda, Maryland 20892
USA

Dennis Hare
Hare Research, Inc.
14810 216th Avenue
Woodinville, WA 98072
USA

Timothy Harvey
Department of Biochemistry
University of Oxford
South Parks Road
Oxford OX1 3QU
ENGLAND

Jeffrey C. Hoch
Rowland Institute for Science
100 Cambridge Parkway
Cambridge, MA 02142
USA

Tadeusz Holak
Max-Planck-Institut für Biochemie
AM Klopferspitz 7
D-8033 Martensried bei München
FEDERAL REPUBLIC OF
GERMANY

Peter Huber
Section de Chimie
Université de Lausanne
Rue de la Barre 2
CH-1005 Lausanne
SWITZERLAND

Thomas L. James
Department of Pharmaceutical
Chemistry
University of California
San Francisco, CA 94143
USA

Oleg Jardetzky
Stanford Magnetic Resonance
Laboratory
Stanford University
Stanford, CA 94305-5055
USA

Hans Kalbitzer
Max-Planck Institute für
Medizinische Forshung
Abt. Biophysik
Jahnstrasse-29
D-6900 Heidelberg
FEDERAL REPUBLIC OF
GERMANY

Robert Kaptein
Bijvoet Center for Biomolecular
Research
NMR Spectroscopy
Padualaan 8
3584 CH Utrecht
THE NETHERLANDS

Lewis Kay
National Institutes of Health
Laboratory of Chemical Physics
Building 2, Room 123, NIDDK
Bethesda, MD 20892
USA

Mogens Kjær
Department of Chemistry
Carlsberg Laboratory
Gamle Carlsbergvej 10
DK-2500 Valby, Copenhagen
DENMARK

Gerard Kleywegt
Department of Organic Chemistry
University of Utrecht
Padualaan 8
3584 CH Utrecht
THE NETHERLANDS

Per J. Kraulis
Department of Molecular Biology
Biomedical Center
Uppsala University
Box 590, S-751 24 Uppsala
SWEDEN

Ernest D. Laue
Department of Biochemistry
University of Cambridge
Tennis Court Road
Cambridge CB2 1QW
UNITED KINGDOM

Robin J. Leatherbarrow
Department of Chemistry
Imperial College
South Kensington
London SW7 2AY
ENGLAND

George C. Levy
NMR and Data Processing
    Laboratory
Syracuse University
Syracuse, NY 13244-4100
USA

Dominique Marion
Centre de Biophysique Moleculaire
CRNS
Avenue de la Scientifique
    Recherche
F-45071 Orléans, Cedex 2
FRANCE

John Markley
Biochemistry Department
College of Agricultural and
    Life Sciences
420 Henry Mall
University of Wisconsin
Madison, WI 53706
USA

Kuniaki Nagayama
JEOL Ltd.
Biometrology Laboratory
Akishima, Tokyo 196
JAPAN

Michael Nilges
Howard Hughes Medical Institute
Department of Molecular Biophysics
    and Biochemistry
Yale University
260 Whitney Avenue JWG 603
P.O. Box 6666
New Haven, Connecticut 06511
USA

David Norman
University of Oxford
Department of Biochemisty
South Parks Road
Oxford OX1 3QU
ENGLAND

Edward T. Olejniczak
Pharmaceutical Discovery
Abbott Laboratories
Abbott Park, IL 60064
USA

Hartmut Oschkinat
Max-Planck-Institit für Biochemie
AM Klopferspitz 7
D-8033 Martensried bei München
FEDERAL REPUBLIC OF
    GERMANY

Flemming M. Poulsen
Department of Chemistry
Carlsberg Laboratory
Gamle Carlsbergvej 10
DK-2500 Valby, Copenhagen
DENMARK

James Prestegard
Yale University
Chemistry Department
P.O. Box 6666
New Haven, CT 06511
USA

Alan S. Stern
Rowland Institute for Science
100 Cambridge Parkway
Cambridge, MA 02412
USA

Christina Redfield
Inorganic Chemistry Laboratory
University of Oxford
South Parks Road
Oxford OX1 3QR
ENGLAND

Michael Sutcliffe
P.O. Box 138
Medical Sciences Building
University Road
Leicester, LE1 9HN
ENGLAND

James Peter Robertson
University of Oxford
Department of Biochemistry
South Parks Road
Oxford OX1 3QU
ENGLAND

Andrew Torda
Physical Chemistry Department
ETH Zentrum
CH-8092 Zürich
SWITZERLAND

Ole W. Sørensen
Laboratorium für Physikalische
    Chemie
Eidgenössische Technische
    Hochschule Zentrum
8092 Zürich
SWITZERLAND

Vincenza Viti
Laboratorio di Fisica
Instituto Superiore di Sanità
Viale Regina Elena 00161
Roma
ITALY

Ruud Scheek
University of Groningen
Department of Physical Chemistry
Nijenborgh 16
9747 AG Groningen
THE NETHERLANDS

Erik Zuiderweg
Biophysics Research Division
University of Michigan
2200 Bonisteel Blvd.
Ann Arbor, MI 48109
USA

INDEX

one-dimensional NMR, 27, 39, 163
order parameters, 317

PROTEAN, 363
parallel computers, 87
parallel processing, 105
parametric estimation, 27
pattern recognition, 175
Prony's method, 421
protein structure, 209, 219, 253,
    269, 279, 375, 395
  error analysis, 253, 375

relaxation matrix, 317, 331, 349,
    391, 395
relayed magnetization transfer,
    selective, 151
resolution enhancement, 1, 105
restrained molecular dynamics, 209,
    219, 253, 269, 331
ribonuclease H, 445

secondary structure, 445
selective deuteration, 363
signal averaging, 1
signal-to-noise enhancement, 1

simulated annealing, 233, 395, 451
solution structure, 57, 331, 349, 363
    error analysis 331
structure refinement, 87, 209, 219,
    317, 331, 349, 439
symmetry enhancement, 175
symmetry recognition, 291
systolic loops, 87

TOCSY, 446
three-dimensional NMR, 27, 39,
    127, 361, 445
  data processing, 175
  heteronuclear, 127, 175
  homonuclear, 127, 163
transputer, 87
triple resonance, 127
trp repressor, 363
two-dimensional NMR, 27, 39, 51,
    163, 175, 361, 427, 445
  prehistoric, 1
  selective, 151

variable target function, 199, 233

x-ray crystallography, 57